João B. P. Soares and
Timothy F. L. McKenna

Polyolefin Reaction Engineering

Related Titles

Isayev, A. I. (ed.)

Encyclopedia of Polymer Blends

Volume 2: Processing

Series: Encyclopedia of Polymer Blends
2011
ISBN: 978-3-527-31930-5

Isayev, A. I. (ed.)

Encyclopedia of Polymer Blends

Volume 1: Fundamentals

Series: Encyclopedia of Polymer Blends
2010
ISBN: 978-3-527-31929-9

Xanthos, M. (ed.)

Functional Fillers for Plastics

Second, Updated and Enlarged Edition

2010
ISBN: 978-3-527-32361-6

Elias, H.-G.

Macromolecules

Series: Macromolecules (Volumes 1–4)

2009
ISBN: 978-3-527-31171-2

Matyjaszewski, K., Gnanou, Y., Leibler, L. (eds.)

Macromolecular Engineering

Precise Synthesis, Materials Properties, Applications

2007
ISBN: 978-3-527-31446-1

Meyer, T., Keurentjes, J. (eds.)

Handbook of Polymer Reaction Engineering

2005
ISBN: 978-3-527-31014-2

Severn, J. R., Chadwick, J. C. (eds.)

Tailor-Made Polymers

Via Immobilization of Alpha-Olefin Polymerization Catalysts

2008
ISBN: 978-527-31782-0

Asua, J. (ed.)

Polymer Reaction Engineering

2007
ISBN: 978-4051-4442-1

João B. P. Soares and Timothy F. L. McKenna

Polyolefin Reaction Engineering

WILEY-VCH Verlag GmbH & Co. KGaA

The Authors

Prof. Dr. João B. P. Soares
University of Waterloo
Department of Chemical Engineering
University Avenue West 200
Waterloo, ON N2L 3G1
Canada

Prof. Dr. Timothy F. L. McKenna
C2P2 UMR 5265
ESCPE Lyon, Bat 308F
43 Blvd du 11 Novembre 1918
69616 Villeurbanne Cedex
France

All books published by **Wiley-VCH** are carefully produced. Nevertheless, authors, editors, and publisher do not warrant the information contained in these books, including this book, to be free of errors. Readers are advised to keep in mind that statements, data, illustrations, procedural details or other items may inadvertently be inaccurate.

Library of Congress Card No.: applied for

British Library Cataloguing-in-Publication Data
A catalogue record for this book is available from the British Library.

Bibliographic information published by the Deutsche Nationalbibliothek
The Deutsche Nationalbibliothek lists this publication in the Deutsche Nationalbibliografie; detailed bibliographic data are available on the Internet at <http://dnb.d-nb.de>.

© 2012 Wiley-VCH Verlag & Co. KGaA, Boschstr. 12, 69469 Weinheim, Germany

All rights reserved (including those of translation into other languages). No part of this book may be reproduced in any form – by photoprinting, microfilm, or any other means – nor transmitted or translated into a machine language without written permission from the publishers. Registered names, trademarks, etc. used in this book, even when not specifically marked as such, are not to be considered unprotected by law.

Print ISBN: 978-3-527-31710-3
ePDF ISBN: 978-3-527-64697-5
ePub ISBN: 978-3-527-64696-8
Mobi ISBN: 978-3-527-64695-1
oBook ISBN: 978-3-527-64694-4

Cover Design Adam-Design, Weinheim
Typesetting Laserwords Private Limited, Chennai, India
Printing and Binding Markono Print Media Pte Ltd, Singapore

To our wives, Maria Soares and Salima Boutti-McKenna, for their love, dedication, and patience while we wrote this book, not to mention the interminably long hours we spent discussing polyolefins in their presence. This book belongs to both of you, but you don't need to read it – you have heard all about it already.

João Soares and Timothy McKenna

Contents

Acknowledgments *XI*

Preface *XIII*

Nomenclature *XVII*

1   **Introduction to Polyolefins** *1*
1.1   Introduction *1*
1.2   Polyethylene Resins *4*
1.3   Polypropylene Resins *10*
         Further Reading *13*

2   **Polyolefin Microstructural Characterization** *15*
2.1   Introduction *15*
2.2   Molecular Weight Distribution *17*
2.2.1   Size Exclusion Chromatography *17*
2.2.2   Field Flow Fractionation *27*
2.3   Chemical Composition Distribution *29*
2.3.1   Crystallizability-Based Techniques *29*
2.3.2   High-Performance Liquid Chromatography *40*
2.4   Cross-Fractionation Techniques *43*
2.5   Long-Chain Branching *46*
         Further Reading *51*

3   **Polymerization Catalysis and Mechanism** *53*
3.1   Introduction *53*
3.2   Catalyst Types *56*
3.2.1   Ziegler–Natta Catalysts *56*
3.2.2   Phillips Catalysts *61*
3.2.3   Metallocenes *62*
3.2.4   Late Transition Metal Catalysts *67*
3.3   Supporting Single-Site Catalysts *70*

3.4	Polymerization Mechanism with Coordination Catalysts	76
	Further Reading	86

4 Polyolefin Reactors and Processes 87
4.1 Introduction 87
4.2 Reactor Configurations and Design 89
4.2.1 Gas-Phase Reactors 90
4.2.1.1 Fluidized Bed Gas-Phase Reactors 91
4.2.1.2 Vertical Stirred Bed Reactor 97
4.2.1.3 Horizontal Stirred Gas-Phase Reactor 99
4.2.1.4 Multizone Circulating Reactor 102
4.2.2 Slurry-Phase Reactors 104
4.2.2.1 Autoclaves 105
4.2.2.2 Slurry Loop Reactors 106
4.2.3 Solution Reactors 107
4.2.4 Summary of Reactor Types for Olefin Polymerization 108
4.3 Olefin Polymerization Processes 109
4.3.1 Polyethylene Manufacturing Processes 112
4.3.1.1 Slurry (Inert Diluent) Processes 112
4.3.1.2 Gas-Phase Processes 115
4.3.1.3 Mixed-Phase Processes 118
4.3.1.4 Solution Processes 119
4.3.2 Polypropylene Manufacturing Processes 121
4.3.2.1 Slurry (Inert Diluent) Processes 122
4.3.2.2 Gas-Phase Processes 122
4.3.2.3 Mixed-Phase Processes 125
4.4 Conclusion 128
References 128
Further Reading 129

5 Polymerization Kinetics 131
5.1 Introduction 131
5.2 Fundamental Model for Polymerization Kinetics 134
5.2.1 Single-Site Catalysts 134
5.2.1.1 Homopolymerization 134
5.2.1.2 Copolymerization 145
5.2.2 Multiple-Site Catalysts 149
5.2.3 Temperature Dependence of Kinetic Constants 152
5.2.4 Number of Moles of Active Sites 154
5.3 Nonstandard Polymerization Kinetics Models 156
5.3.1 Polymerization Orders Greater than One 156
5.3.2 Hydrogen Effect on the Polymerization Rate 161
5.3.3 Comonomer Effect on the Polymerization Rate 173
5.3.4 Negative Polymerization Orders with Late Transition Metal Catalysts 179

5.4	Vapor-Liquid-Solid Equilibrium Considerations	*181*
	Further Reading *184*	
6	**Polyolefin Microstructural Modeling** *187*	
6.1	Introduction *187*	
6.2	Instantaneous Distributions *188*	
6.2.1	Molecular Weight Distribution *188*	
6.2.1.1	Single-Site Catalysts *188*	
6.2.1.2	Multiple-Site Catalysts *199*	
6.2.2	Chemical Composition Distribution *212*	
6.2.2.1	Single-Site Catalysts *212*	
6.2.2.2	Multiple-Site Catalysts *222*	
6.2.3	Comonomer Sequence Length Distribution *232*	
6.2.4	Long-Chain Branching Distribution *237*	
6.2.5	Polypropylene: Regio- and Stereoregularity *250*	
6.3	Monte Carlo Simulation *251*	
6.3.1	Steady-State Monte Carlo Models *252*	
6.3.2	Dynamic Monte Carlo Models *262*	
	Further Reading *268*	
7	**Particle Growth and Single Particle Modeling** *271*	
7.1	Introduction *271*	
7.2	Particle Fragmentation and Growth *274*	
7.2.1	The Fragmentation Step *275*	
7.2.2	Particle Growth *284*	
7.3	Single Particle Models *286*	
7.3.1	Particle Mass and Energy Balances: the Multigrain Model (MGM) *287*	
7.3.2	The Polymer Flow Model (PFM) *292*	
7.3.3	An Analysis of Particle Growth with the MGM/PFM Approach *295*	
7.3.4	Convection in the Particles – High Mass Transfer Rates at Short Times *301*	
7.4	Limitations of the PFM/MGM Approach: Particle Morphology *304*	
	References *307*	
	Further Reading *307*	
8	**Developing Models for Industrial Reactors** *311*	
8.1	Introduction *311*	
	References *321*	
	Further Reading *322*	
	Index *325*	

Acknowledgments

Personally I'm always ready to learn, although I do not always like being taught.

Sir Winston Churchill (1874–1965)

Several of the concepts covered in this book arose from our daily interactions with students and colleagues in academia and industry. They are too many to be named individually here, but we would like to express our sincere gratitude to their outstanding contributions that are summarized in this work. We did like being thought by all of you.

First, we would like to thank our former mentors, who trusted and guided us when we were starting our careers, and kept encouraging us throughout these years. Their mentoring, support, and friendship are greatly appreciated.

This book could not have been written without the dedication of our graduate students, post-doctoral fellows, and research assistants, who toiled day after day in our laboratories to propose and test hypotheses, challenge us with unexpected new results, and in the process advance our understanding of polyolefin reaction engineering. Several of their results are interspersed throughout this book and constitute main contributions to the field of olefin polymerization science and engineering. We are very thankful to their hard work, perseverance, and confidence in us as their supervisors.

We would also like to thank our academic and industrial collaborators who over the years helped us better understand olefin polymerization and polyolefin characterization, often kindly allowing us to use their laboratory facilities (for free!) to complement the work done in our institutions. We are indeed indebted to these extraordinary colleagues and look forward to continue working with them in the future.

Finally, we would like to thank the polyolefin companies all over the world that have hired us as consultants and instructors of our industrial short course on *Polyolefin Reaction Engineering*. This book is a result, in large part, from the stimulating discussions we had with the scientists and engineers who took these courses. If it is true, as said by Scott Adams, the creator of the comic strip *Dilbert*, that *"Give a man a fish, and you'll feed him for a day. Teach a man to fish, and he'll buy a funny hat. Talk to a hungry man about fish, and you're a consultant"*, then we hope that talking to the course participants over these years has at least stimulated them to look deeper into the vast sea of polyolefin reaction engineering.

Preface

> It is the mark of an instructed mind to rest satisfied with the degree of precision which the nature of the subject permits and not to seek exactness where only an approximation of the truth is possible.
> *Aristotle (384–322 BC)*
>
> The art of being wise is the art of knowing what to overlook.
> *William James (1842–1910)*

The manufacture of polyolefins with coordination catalysts has been a leading force in the synthetic plastic industry since the early 1960s. Owing to the constant developments in catalysis, polymerization processes, and polyolefin characterization instruments, it continues to be a vibrant area of research and development today.

We have been working in this area for over 15 years, always feeling that there was a need for a book that summarized the most important aspects of polyolefin reaction engineering. This book reflects our views on this important industry. It grew out of interactions with the polyolefin industry through consulting activities and short courses, where we first detected a clear need to summarize, in one single source, the most generally accepted theories in olefin polymerization kinetics, catalysis, particle growth, and polyolefin characterization.

As quoted from Aristotle above, we will *rest satisfied with the degree of precision which the nature of the subject permits* and hope that our readers agree with us that this is indeed *the mark of an instructed mind*. It was not our intention to perform an extensive scholarly review of the literature for each of the topics covered in this book. We felt that this approach would lead to a long and tedious text that would become quickly outdated; several excellent reviews summarizing the most recent findings on polyolefin manufacturing and characterization are published regularly and are more adequate for this purpose. Instead, we present our interpretation of the field of polyolefin reaction engineering. Since any selection process is always subjective, we may have left out some approaches considered to be relevant by others, but we tried to be as encompassing as possible, considering the limitations of a book of this type. We have also sparsely used references in the main body of the chapters but added reference sections at their end where we discussed some alternative theories, presented exceptions to the general approach followed in the chapters, and suggested additional readings. The reference sections are not meant to be exhaustive compilations of the literature but sources of supplemental readings and a door to the vast literature in the area. We hope this approach will make this book a pleasant reading and also provide the reader with additional sources of reference.

Chapter 1 introduces the field of polyolefins, with an overview on polyolefin types, catalyst systems, and reactor configurations. We also introduce our general philosophy of using mathematical models to link polymerization kinetics, mass and heat transfer processes at several length scales, and polymer microstructure characterization for a complete understanding of olefin polymerization processes.

We discuss polyolefin microstructure, as defined by their distributions of molecular weight, chemical composition, stereo- and regioregularity, and long-chain branching, in Chapter 2. It is not an overstatement to say that among all synthetic polymers, polyolefins are the ones where microstructure control is the most important concern. Polyolefin microstructure is a constant theme in all chapters of this book and is our best guide to understanding catalysis, kinetics, mass and heat transfer resistances, and reactor behavior.

Chapter 3 is dedicated to polymerization catalysis and mechanisms. The field of coordination catalysis is huge and, undoubtedly, the main driving force behind innovation in the polyolefin manufacturing industry; to give it proper treatment, a separate book would be necessary. Rather, we decided to focus on the most salient aspects of the several classes of olefin catalysts, their general behavior patterns and mechanisms, and how they can be related to polymerization kinetics and polyolefin microstructural properties.

The subject of Chapter 4, polymerization reactors, is particularly dear to us, polymer reactor engineers. In fact, polyolefin manufacturing is a "dream come true" for polymer reactor engineers because practically all possible configurations of chemical reactors can be encountered. A great deal of creativity went into reactor design, heat removal strategies, series and parallel reactor arrangements, and cost reduction schemes of polyolefin reactors. We start the chapter by discussing reactor configurations used in olefin polymerization and then continue with a description of the leading processes for polyethylene and polypropylene production.

Chapter 5 is the first chapter dedicated to the mathematical modeling of olefin polymerization. We start our derivations with what we like to call the fundamental model for olefin polymerization kinetics and develop, from basic principles, its most general expressions for the rates of catalyst activation, polymerization, and catalyst deactivation. The fundamental model, albeit widely used, does not account for several phenomena encountered in olefin polymerization; therefore, some alternative polymerization kinetic schemes are discussed at the end of this chapter.

In Chapter 6, we develop mathematical models to describe the microstructure of polyolefins. This is one of the core chapters of the book and helps connect polymerization kinetics, catalysis, and mass and heat transfer resistances to final polymer performance. We opted to keep the mathematical treatment as simple as possible, without compromising the most relevant aspects of this important subject.

Particle fragmentation and growth are covered in Chapter 7. These models are collectively called *single particle models* and can be subdivided into polymer growth models and morphology development models. The two most well-established particle growth models are the polymeric flow model and the multigrain model. These models are used to describe heat and mass transfer in the polymeric particle after fragmentation takes place. The fragmentation of the catalyst particles themselves (described with morphology development models) is much harder to model, and there is still no well-accepted quantitative model to tackle this important subject. We review the main modeling alternatives in this field.

Finally, Chapter 8 is dedicated to macroscopic reactor modeling. This chapter is, in a way, the most conventional chapter from the chemical engineering point of view, since it involves well-known concepts of reactor residence time distribution, micromixing and macromixing, and reactor heat removal issues. The combination of macroscopic reactor models, single particle models, detailed polymerization kinetics, and polymer microstructural distributions, however, is very challenging and represents the ultimate goal of polyolefin reactor engineers.

Nomenclature

> What's in a name? William Shakespeare (1564–1616)

Acronyms

CCD	chemical composition distribution
CEF	crystallization elution fractionation
CFC	cross-fractionation
CGC	constrained geometry catalyst
CLD	chain length distribution
CRYSTAF	crystallization analysis fractionation
CSLD	comonomer sequence length distribution
CSTR	continuous stirred tank reactor
CXRT	computed X-ray tomography
DEAC	diethyl aluminum chloride
DIBP	di-iso-butylphthalate
DSC	differential scanning calorimetry
EAO	ethylaluminoxane
EB	ethyl benzoate
EDX	energy dispersive X-ray spectroscopy
EGMBE	ethylene glycol monobutylether
ELSD	evaporative light scattering detector
EPDM	ethylene-propylene-diene monomer rubber
EPR	ethylene–propylene rubber
FBR	fluidized bed reactor
FFF	field flow fractionation
FTIR	Fourier-transform infrared
GPC	gel permeation chromatography
HDPE	high-density polyethylene
HMDS	hexamethyldisilazine
HPLC	high-performance liquid chromatography
HSBR	horizontal stirred bed reactor

IR	infrared
LALLS	low-angle laser light scattering
LCB	long-chain branch
LDPE	low-density polyethylene
LLDPE	linear low-density polyethylene
LS	light scattering
MALLS	multiangle laser light scattering
MAO	methylaluminoxane
MDPE	medium-density polyethylene
MFI	melt flow index
MFR	melt flow rate
MGM	multigrain model
MI	melt index
MWD	molecular weight distribution
MZCR	multizone circulating reactor
NMR	nuclear magnetic resonance
NPTMS	n-propyltrimethoxysilane
ODCB	orthodichlorobenzene
PDI	polydispersity index
PFM	polymer flow model
PFR	plug flow reactor
PP	polypropylene
PSD	particle size distribution
RND	random number generated in the interval [0,1]
RTD	residence time distribution
SCB	short-chain branch
SEC	size exclusion chromatography
SEM	scanning electron microscopy
SLD	sequence length distribution
SPM	single particle model
tBAO	t-butylaluminoxane
TCB	tricholorobenzene
TEA	triethyl aluminum
TEM	transmission electron microscopy
TGIC	temperature gradient interaction chromatography
TMA	trimethyl aluminum
TOF	turnover frequency
TREF	temperature rising elution fractionation
UHMWPE	ultrahigh-molecular weight polyethylene
ULDPE	ultralow-density polyethylene
VLDPE	very low-density polyethylene
VISC	viscometer
VSBR	vertical stirred bed reactor

Symbols

a	Mark–Houwink equation constant, Eq. (2.7)
a_s	specific surface area of the support
A	monomer type A
A	total reactor heat transfer area
A_i	Arrhenius law preexponential factor for reaction of type i
A_S	support specific surface area
Al	cocatalysts
$[AS^*]$	concentration of active sites per unit surface area in the microparticle
B	monomer type B
B_n	average number of long-chain branches per polymer chain
C	catalyst precursor or active site
C^*	active site
$[C_0]$	initial concentration of active sites
C_d	deactivated catalytic site
C_p	heat capacity
D_b	bulk diffusivity
D_{eff}	effective diffusivity in the macroparticle
d_p	polymer (or catalyst) particle diameter
D_p	diffusivity in the primary particle
D_r	dead polymer chain
$D_{r,i}$	dead polymer chain of length r having i long-chain branches
$D_{r,i}^=$	dead polymer chain of length r having i long-chain branches and a terminal unsaturation (macromonomer)
$E(t)$	reactor residence time distribution
E_i	Arrhenius law activation energy for reaction of type i
$f^=$	molar fraction of macromonomers in the reactor
f_i	molar fraction of monomer type i in the polymerization medium
f_r	frequency Flory chain length distribution, Eq. (6.13)
$\overline{f_r}$	overall frequency chain length distribution for chain having long-chain branches, Eq. (6.101)
f_{rk}	frequency chain length distribution for chains with k long-chain branches per chain, Eq. (6.86)
$f_{\log r\,k}$	frequency chain length distribution for chains with k long-chain branches per chain, log scale, Eq. (6.88)
F	monomer molar flow rate to the reactor
F_A	comonomer molar fraction in the copolymer
$\overline{F_A}$	average comonomer molar fraction in the copolymer
F_{Br}	molar fraction of comonomer B as a function of chain length
$F_{M,\text{in}}$	molar flow rate of the monomer feed to the reactor
$F_{M,\text{out}}$	molar flow rate of the monomer exiting the reactor
g	branching index, Eq. (2.18)
g'	viscosity branching index, Eq. (2.17)
ΔG	Gibbs free energy change

h	average convective heat transfer coefficient between the macroparticle and surroundings
ΔH	enthalpy change
ΔH_p	average enthalpy of polymerization
ΔH_r	enthalpy of reaction
ΔH_u	enthalpy of melting for a crystallizable repeating unit, Eq. (2.26)
ΔH_{vap}	enthalpy of vaporization
I_1	Bessel function of the first kind and order 1
k_a	site activation rate constant
k_c, k_c^-	forward and reverse rate constants, respectively, for the formation of dormant site with Ni-diimine catalysts, Table 5.8
k_d	first-order deactivation rate constant
k_d^*	second-order deactivation rate constant
k_f	forward rate constant for reversible monomer coordination or β-agostic interaction; thermal conductivity
k_{fL}	effective thermal diffusivity in the macroparticle
k_{fp}	thermal conductivity of the polymer layer around the catalyst fragment in the microparticle
k_{iH}	rate constant for initiation of metal hydride active sites
k_p	propagation rate constant
k_p'	apparent propagation rate constant, Eq. (5.115)
$\hat{k}_p$	pseudo-propagation rate constant
$\tilde{k}_p$	apparent propagation rate constant
k_{pi}	propagation rate constant for monomer type i (Bernoullian model)
k_{pij}	propagation rate constant for chain terminated in monomer type i coordinating with monomer type j (terminal model)
k_{pijk}	propagation rate constant for chain terminated in monomer types i and j coordinating with monomer type k (penultimate model)
k_{pm}	propagation rate constant for meso insertion (propylene)
k_{pr}	propagation rate constant for racemic insertion (propylene)
k_r	reverse rate constant for reversible monomer coordination or β-agostic interaction
k_{tAl}	rate constant for transfer to cocatalyst
$k_{t\beta}$	rate constant for β-hydride elimination
k_{tH}	rate constant for transfer to hydrogen
k_{tM}	rate constant for transfer to monomer
K	Mark–Houwink equation constant, Eq. (2.7)
K_a	initiation frequency, $k_a[Al]$
K_{eq}	equilibrium constant for dormant sites, Eq. (5.67)
K_{g-1}, K_{g-1}^*	gas–liquid partition coefficients, Eq. (5.113)

K_{g-s}, K^*_{g-s}, K'_{g-s}	gas–solid partition coefficients, Eqs (5.111) and (5.114)
K_{l-s}	liquid–solid partition coefficient, Eq. (5.112)
K_H	Henry law constant
K_T	lumped chain-transfer constant, Eq. (5.74)
K_T^1	lumped chain-transfer constant, Eq. (5.70)
K_T^H	lumped chain-transfer constant, Eq. (5.71)
m_i	mass fraction of polymer made on site type i
m_k	mass fraction of chains with k long-chain branches
m_p	mass of polymer, polymer yield
$\dot{m}_{vap}$	vaporization rate
mw	molecular weight of repeating unit; in the case of copolymers, the average molecular weight of the repeating units
M	molecular weight
M	monomer
M_C	molar mass of catalyst
M_n	number average molecular weight
M_v	viscosity average molecular weight
M_w	weight average molecular weight
MW	polymer molecular weight
n	number of long-chain branches per chain; number of active site types
$n_c(v)$	polymer particle size distribution
$n_{\tilde{C}_0}$, n_{C_0}	number of moles of catalyst
n_{LCB}	average number of long-chain branches in a polymer sample
n_M	number of moles of monomer
n_w	weight average number of long-chain branches per chain
N_A	Avogadro number
N_i	flux of species i
N_s	number of macroparticles per unit volume of the reactor
Nu	Nusselt number
P_A, P_B	probability of propagation of monomers A and B, respectively
P_H^*	metal hydride active site
P_M	partial pressure of monomer
P_p	propagation probability
P_r	living chain with length r
P_r^i	living polymer chain with length r terminated in monomer type i(A or B for binary copolymers) or 1-2 or 2-1 insertions for polypropylene
$P^*_{r,i}$	living polymer chain of length r having i long-chain branches
$\tilde{P}_1$	dormant site due to β-agostic interaction, Table 5.5
$\tilde{P}_r$	dormant site for Ni-diimine catalysts, Table 5.8
P_t	termination probability
PDI	polydispersity index

Symbol	Description
$\overline{PDI}$	polydispersity index for chains containing long-chain branches
PDI_{F_A}	polydispersity index as a function of copolymer composition
PDI_k	polydispersity index for chain with k long-chain branches, Eq. (6.106)
Pr	Prandtl number
$\dot{Q}$	heat generation rate
r	polymer chain length
r_i	comonomer reactivity ratio
r_L	radial position in macroparticle (multigrain model)
r_n	number average chain length
$\overline{r}_n$	number average chain length for chains containing long-chain branches, Eq. (6.107)
$\tilde{r}_n$	number average molecular weight that would result in the absence of long-chain branch formation reactions, see footnote 13 in Chapter 6
r_{nF_A}	number average chain length as a function of copolymer composition
r_{nk}	number average chain length for chains with k long-chain branches, Eq. (6.103)
r_s	radial position in the microparticle (multigrain model)
r_w	weight average chain length
$\overline{r}_w$	weight average chain length for chains containing long-chain branches, Eq. (6.108)
r_{wF_A}	weight average chain length as a function of copolymer composition
r_{wk}	weight average chain length for chains with k long-chain branches, Eq. (6.104)
r_{z_k}	z-average chain length for chains with k long-chain branches, Eq. (6.105)
$\overline{r_0^2}$	root-mean-square end-to-end distance of a polymer chain
R	gas constant
R_c	catalyst fragment radius (multigrain model)
R_i	reaction rate of species i
$\langle R_g^2 \rangle_b$	squared radius of gyration of branched chains
$\langle R_g^2 \rangle_l$	squared radius of gyration of linear chains
R_L	macroparticle radius (multigrain model)
R_p	polymerization rate
$\overline{R_p}$	average polymerization rate per unit volume of the reactor
$\overline{R_p}'$	average polymerization rate per polymer particle
R_S	microparticle radius (multigrain model)
R_t	chain-transfer rate
Re	Reynolds number
ΔS	entropy change
Sc	Schmidt number
Sh	Sherwood number
t	time

$\bar{t}$	average reactor residence time
t_R	reactor residence time
$t_{\frac{1}{2}}$	catalyst half-time
T	temperature
T_c	crystallization temperature
T_{i0}	reactor inlet temperature
T_m	melting temperature
T_m^0	melting temperature of an infinitely long polyethylene chain
T_S	temperature in the microparticle
T_w	reactor coolant temperature
U	global heat-transfer coefficient
V	Monte Carlo control volume; reactor volume
V_e	elution volume
V_i	interstitial volume
V_p	pore volume
V_R	reactor volume
$w_{\log r}$	weight Flory chain length distribution, log scale, Eq. (6.24)
w_r	weight Flory chain length distribution, Eq. (6.17)
$\overline{w_r}$	cumulative weight Flory chain length distribution
$\overline{w_r}$	overall weight chain length distribution for chain having long-chain branches, Eq. (6.102)
w_{r,F_A}	Stockmayer bivariate distribution, Eq. (6.60)
w_{rk}	weight chain length distribution for chains with k long-chain branches per chain, Eq. (6.87)
$w_{\log rk}$	weight chain length distribution for chains with k long-chain branches per chain, log scale, Eq. (6.89)
$w_{\log MW, F_A}$	Stockmayer bivariate distribution, log scale, Eq. (6.61)
$w_{r,F_A k}$	trivatiate distribution of chain length, chemical composition, and long-chain branching, Eq. (6.117)
w_{ry}	Stockmayer bivariate distribution, Eq. (6.56)
$w_{\log MW}$	weight Flory molecular weight distribution, log scale, Eq. (6.32)
w_{MW}	weight Flory molecular weight distribution, Eq. (6.30)
W_C	mass of catalyst
x_c	mass fraction of catalyst in a supported catalyst
x_i	molar fraction of comonomer i in the copolymer
y	deviation from average comonomer molar fraction in the copolymer, Eq. (6.57)
y_k	molar fraction of chains with k long-chain branches
Y_i	ith moment of living polymer
$[Y_0]$	total concentration of active sites or living polymer chains
Z	compressibility factor

Greek Letters

α	polymer chain hydrodynamic volume constant, Eq. (2.4); long-chain branching parameter, Eq. (6.93)
$\beta, \hat{\beta}$	Stockmayer bivariate distribution parameters, Eqs. (6.58) and (6.62), respectively
ε	exponent relating g' to g, Eq. (2.19); macroparticle void fraction
η	catalyst site efficiency, Eq. (5.45)
$[\eta]$	intrinsic viscosity
ϕ	fraction of actives sites with growing polymer chains, Eq. (5.55)
ϕ_i	fraction of living chains terminated in monomer type i
ϕ_k	Catalan numbers, Eq. (6.95)
Φ	polymer chain hydrodynamic volume constant defined in Eq. (2.4)
κ	size exclusion partition coefficient
λ, λ_n	number of long-chain branches per 1000 C atoms
μ	long-chain branching parameter, Eq. (6.94); viscosity
ρ_C	support (catalyst) density
$\overline{\rho C_p}$	average value of the heat capacity per unit volume of the macroparticle
τ	Flory most probable chain length distribution parameter; ratio of all chain-transfer rates to the propagation rate, Eq. (6.29); g macroparticle tortuosity
$\hat{\tau}$	Flory most probable molecular weight distribution parameter, Eq. (6.31)
τ_B	chain length distribution parameter for polymers containing long-chain branches, Eq. (6.90)
τ_d	characteristic diffusion time in the macroparticle

Superscripts and Subscripts

^	pseudokinetic constant
−	average
12, 21	1-2 or 2-1 propylene insertions
A, B	monomer types
bulk	bulk conditions
C	catalyst, monomer type C in the case of terpolymerization
l	liquid phase
M	monomer
MC	Monte Carlo simulation rates and constants
P	polymer
s	solid polymer phase

1
Introduction to Polyolefins

> It is a near perfect molecule [...].
>
> *Jim Pritchard, Phillips Petroleum Company*

1.1
Introduction

Polyolefins are used in a wide variety of applications, including grocery bags, containers, toys, adhesives, home appliances, engineering plastics, automotive parts, medical applications, and prosthetic implants. They can be either amorphous or highly crystalline, and they behave as thermoplastics, thermoplastic elastomers, or thermosets.

Despite their usefulness, polyolefins are made of monomers composed of only carbon and hydrogen atoms. We are so used to these remarkable polymers that we do not stop and ask how materials made out of such simple units achieve this extraordinary range of properties and applications. The answer to this question lies in how the monomer molecules are connected in the polymer chain to define the molecular architecture of polyolefins. By simply manipulating how ethylene, propylene, and higher α-olefins are bound in the polymer chain, polyolefins with entirely new properties can be produced.

Polyolefins can be divided into two main types, polyethylene and polypropylene, which are subdivided into several grades for different applications, as discussed later in this chapter.[1] Taking a somewhat simplistic view, three components are needed to make a polyolefin: monomer/comonomer, catalyst/initiator system, and polymerization reactor. We will start our discussion by taking a brief look at each of these three components.

Commercial polyethylene resins, despite their name, are most often copolymers of ethylene, with varying fractions of an α-olefin comonomer. The most

---

1) Ethylene-propylene-diene(EPDM) terpolymers, another important polyolefin type, are elastomers with a wide range of applications. They are made using catalysts and processes similar to those used to produce polyethylene and polypropylene. Even though they are not discussed explicitly in this book, several of the methods explained herein can be easily adopted to EPDM processes.

commonly used α-olefins are 1-butene, 1-hexene, and 1-octene. They are used to decrease the density and crystallinity of the polyolefin, changing its physical properties and applications. Industrial polypropylene resins are mostly isotactic materials, but a few syndiotactic grades are also available. There are two main types of propylene copolymers: random propylene/ethylene copolymers[2] and impact propylene/ethylene copolymers.

At the heart of all polyolefin manufacturing processes is the system used to promote polymer chain growth. For industrial applications, polyethylene is made with either free radical initiators or coordination catalysts, while polypropylene is produced only with coordination catalysts. Low-density polyethylene (LDPE) is made using free radical processes and contains short chain branches (SCB) and long chain branches (LCB). Its microstructure is very different from that of polyethylenes made with coordination catalysts. Coordination catalysts can control polymer microstructure much more efficiently than free radical initiators and are used to make polyolefins with a range of properties unimaginable before their discovery. The quotation at the beginning of this chapter was motivated by the revolutionary synthesis of unbranched, linear polyethylene, the "near perfect molecule," using a Phillips catalyst.

Catalyst design is behind the success of modern industrial olefin polymerization processes because the catalyst determines how the monomers will be linked in the polymer chain, effectively defining the polymer microstructure and properties. Industrial and academic research on olefin polymerization catalysis have been very dynamic since the original discoveries of Ziegler and Natta (Ziegler–Natta catalysts) and Hogan and Banks (Phillips catalysts), with many catalyst families being developed and optimized at a rapid pace. There are basically four main types of olefin polymerization catalysts: (i) Ziegler–Natta catalysts, (ii) Phillips catalysts, (iii) metallocene catalysts, and (iv) late transition metal catalysts. Ziegler–Natta and Phillips catalyst were discovered in the early 1950s, initiating a paradigm shift in olefin polymerization processes, while metallocene and late transition metal catalysts (sometimes called *post-metallocenes*) were developed in the 1980s and 1990s, respectively. In Chapter 3, olefin polymerization catalysts and mechanisms are discussed in detail, while polymerization kinetic models are developed in Chapter 5. For our purposes in this chapter, it suffices to say that polymerization with all coordination catalyst types involves monomer coordination to the transition metal active site before insertion in the metal–carbon bond at one end of the polymer chain. The coordination step is responsible for the versatility of these catalysts: since the incoming monomer needs to coordinate to the active site before propagation may occur, the electronic and steric environment around it can be changed to alter polymerization parameters that control the chain microstructure, such as propagation and chain transfer rates, comonomer reactivity ratios, stereoselectivity, and regioselectivity. This concept is illustrated in Figure 1.1.

2) Propylene/1-butene copolymers and propylene/ethylene/α-olefin terpolymers are also manufactured, but in a smaller scale.

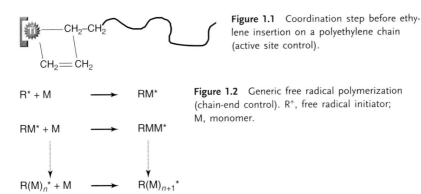

Figure 1.1 Coordination step before ethylene insertion on a polyethylene chain (active site control).

Figure 1.2 Generic free radical polymerization (chain-end control). R*, free radical initiator; M, monomer.

This mechanism controls the polymer microstructure better than free radical polymerization, where the chemical nature of the free radical initiator stops being important after a few propagation steps since the polymerization locus moves away from the initiator molecule, as illustrated in Figure 1.2. Free radical processes for LDPE production are not the main topic of this book and are only discussed briefly as comparative examples.

Even though a few processes for ethylene polymerization use homogeneous catalysts in solution reactors, most olefin polymerization processes operate with heterogeneous catalysts in two-phase or three-phase reactors. This adds an additional level of complexity to these systems since inter- and intraparticle mass and heat transfer resistances during polymerization may affect the polymerization rate and polymer microstructure. If significant, mass and energy transport limitations create nonuniform polymerization conditions within the catalyst particles that lead to nonuniform polymer microstructures. Many other challenging problems are associated with the use of heterogeneous catalysts for olefin polymerization, such as catalyst particle breakup, agglomeration, growth, and morphological development, all of which are discussed in Chapter 7.

All the phenomena mentioned above take place in the polymerization reactor. The variety of polyolefin reactors can be surprising for someone new to the field: polyolefins are made in autoclave reactors, single- and double-loop reactors, tubular reactors, and fluidized-bed reactors; these processes may be run in solution, slurry, bulk, or gas phase. Each reactor configuration brings with it certain advantages, but it also has some disadvantages; the ability to select the proper process for a given application is an important requirement for a polyolefin reaction engineer. Chapter 4 discusses these different reactor configurations and highlights some important polyolefin manufacturing processes. Reactor models that take into account micromixing and macromixing effects, residence time distributions, and mass and heat transfer phenomena at the reactor level are needed to simulate these processes, as explained in Chapter 8.

Catalyst type, polymerization mechanism and kinetics, inter- and intraparticle mass and heat transfer phenomena, and macroscale reactor modeling are essential for the design, operation, optimization, and control of polyolefin reactors. Figure 1.3

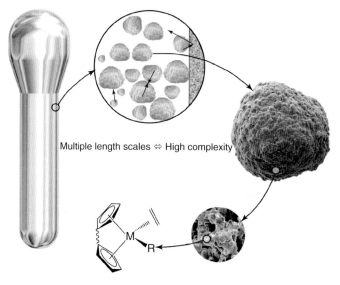

Figure 1.3 Modeling scales in polyolefin reaction engineering.

shows how these different length scales are needed to help us understand olefin polymerization processes.

In summary, much science and technology is hidden behind the apparent simplicity of everyday polyethylene and polypropylene consumer goods. No other synthetic polymer is made with such a variety of catalyst types, reactor configurations, and microstructural complexity. In this book, we will explain how, from such simple monomers, polyolefins have become the dominant commodity plastic in the twenty-first century.

1.2
Polyethylene Resins

Polyethylene resins are classified into three main types: LDPE, linear low-density polyethylene (LLDPE), and high-density polyethylene (HDPE). This traditional classification distinguishes each polyethylene type according to its density range: 0.915–0.940 g cm^{-3} for LDPE, 0.915–0.94 g cm^{-3} for LLDPE,[3] and 0.945–0.97 g cm^{-3} for HDPE, although these limits may vary slightly among different sources. Lower-density polyethylene resins (<0.915 g cm^{-3}) are sometimes called ultra low-density polyethylene (ULDPE) or very low-density polyethylene (VLDPE). HDPE with molecular weight averages of several millions is called ultrahigh molecular weight polyethylene (UHMWPE). To reduce the

---

3) Sometimes polyethylene resins in the range 0.926–0.940 g cm^{-3} are called medium-density polyethylene (MDPE).

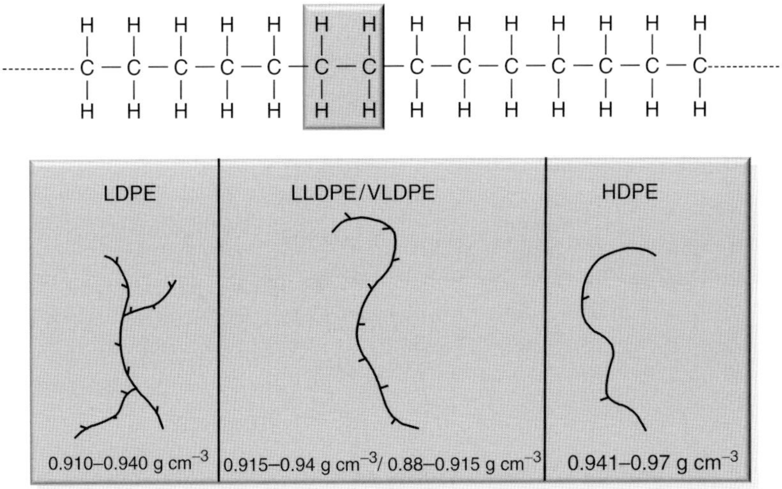

Figure 1.4 Classification of polyethylene types according to branching structure and density.

number of acronyms in this book, we have often grouped MDPE, LLDPE, ULDPE, and VLDPE under the generic term LLDPE and have made no distinction between HDPE and UHMWPE; these resins are very similar from a structural point of view, as explained below.

This division according to polyethylene density or molecular weight, although standard for commercial resins, tells us little about their microstructures. A more descriptive classification, based on their microstructural characteristics, is presented in Figure 1.4.

HDPE and LLDPE are made with coordination catalysts, while LDPE is made with free radical initiators. LDPE has both SCB and LCB, while polyethylenes made by coordination polymerization generally have only SCBs. Some polyethylene resins made with specific metallocenes or Phillips catalysts may also have some LCBs, but their LCB topology is distinct from that of LDPE resins. Most commercial HDPE and LLDPE grades are made with Ziegler–Natta or Phillips catalyst. Phillips catalysts are very important for the production of HDPE but are not used for LLDPE manufacture. Metallocenes can be used in making both HDPE and LLDPE, but metallocene resins are very different from the ones made with either Ziegler–Natta or Phillips catalyst, as explained below. The market share of metallocene resins is still relatively small but has been increasing steadily since the 1990s. Resins made with late transition metal catalysts have had no significant commercial applications to date.

The mechanism of SCB and LCB formation in LDPE is different from that in coordination polymerization; in LDPE, LCBs are formed by transfer-to-polymer reactions, while SCBs result from backbiting reactions. Contrarily, SCBs in HDPE and LLDPE are produced by the copolymerization of α-olefins added to the reactor as comonomers. LCBs, when present, are also formed by copolymerization reactions

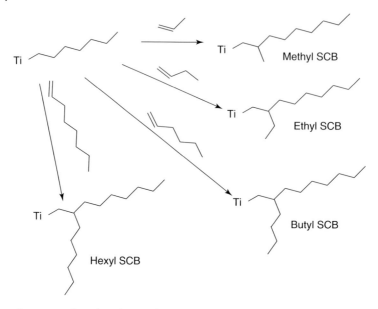

Figure 1.5 Short-chain branch formation mechanism in coordination polymerization. The chains are shown growing on a titanium active site.

with a polymer chain having a terminal group containing a reactive double bond (macromonomer).

Figure 1.5 illustrates the SCB formation mechanism during the copolymerization of ethylene and α-olefins. The SCB behaves as a defect in the polymer chain, decreasing polymer density, crystallite size, and melting temperature. Therefore, the higher the molar fraction of α-olefin in the polymer chain, the lower its crystallinity. HDPE resins have very low α-olefin comonomer fractions (typically below a few mole percentage), while the comonomer content increases from LLDPE to ULDPE to VLDPE.

Density is, therefore, a reflection of the α-olefin molar fraction in the polyolefin chain, and it also depends, to a lower degree, on its molecular weight; all other factors being the same, polyolefins with higher molecular weight averages tend to have a slightly lower density than those with lower molecular weight averages.

Density has been used for decades to classify polyethylene resins, but it is a poor descriptor for these materials because the microstructure of commercial polyethylenes is too complex to be captured with a single density value. Let us first focus on the chemical composition distribution (CCD) of LLDPEs, that is, the distribution of α-olefin fraction in the polymer chains. Most commercial LLDPEs are made with heterogeneous Ziegler–Natta catalysts. These catalysts have more than one type of active site, each one producing polymer chains with different average comonomer fractions and molecular weights. In addition, active sites that favor α-olefin incorporation also result in polymers with lower average molecular weights. As a consequence, the CCDs of Ziegler–Natta LLDPE resins are very

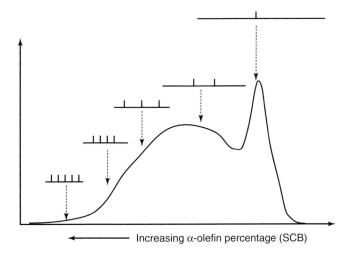

Figure 1.6 Generic chemical composition distribution of an LLDPE resin made with a heterogeneous Ziegler–Natta catalyst.

broad, generally bimodal, and the average α-olefin content is correlated to the polymer molecular weight, as illustrated in Figure 1.6.

Two distinct regions can be identified in Figure 1.6: a sharp high-crystallinity peak (low α-olefin fraction) and a broad low-crystallinity peak (high α-olefin fraction). These two regions are associated with at least two types of active sites, one with much lower reactivity ratio toward α-olefin incorporation than the other. As the relative amounts of polymer made under these two modes vary, polyethylene resins vary from HDPE, with a unimodal, high-crystallinity peak and sometimes a small, lower-crystallinity tail, to MDPE, LLDPE, ULDPE, and VLDPE, with increasingly pronounced lower-crystallinity peak. The shape of the CCD is a strong function of the catalyst type, but it also depends on ethylene/α-olefin ratio, α-olefin type, and polymerization temperature.

From the discussion of Figure 1.6, it is apparent that classifying these complex microstructures according to a single density value is inadequate. The picture becomes even more complex when we take a look at the joint distribution of molecular weight and chemical composition (MWD × CCD) for Ziegler–Natta LLDPE, such as the one depicted in Figure 1.7. This tridimensional plot summarizes the complexity inherent to most commercial polyolefin resins. It also demonstrates that microstructural characterization techniques are indispensable tools to understand these polymers, as discussed in Chapter 2.

The MWD × CCD correlation exemplified in Figures 1.6 and 1.7 is not desirable for certain polyolefin applications. A notable example are bimodal pipe resins, where better mechanical properties are achieved if the higher molecular weight chains also have a higher α-olefin fraction than the lower molecular weight component. The reason for this improved performance has been linked to the presence of tie molecules, a subject that is, unfortunately, beyond the scope of this book. The reader is directed to the references at the end of the chapter for more information on

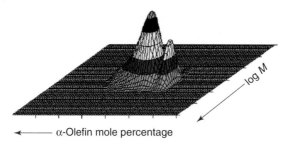

Figure 1.7 Joint distribution of molecular weight and chemical composition (MWD × CCD) for an LLDPE made with heterogeneous Ziegler–Natta catalysts.

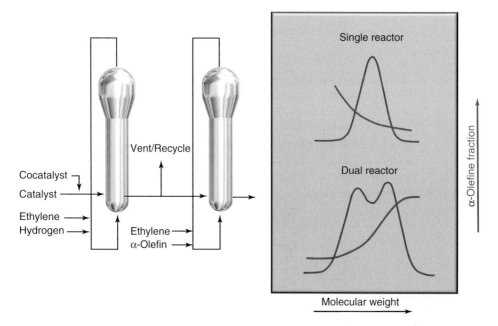

Figure 1.8 Polyethylene with regular and reverse comonomer incorporation made in single- and dual-reactor systems, respectively.

this subject. The usual MWD × CCD correlation observed in polyethylene resins made with heterogeneous Ziegler–Natta catalysts can be *partially* reversed using at least two reactors in series, as depicted in Figure 1.8. The polymers are called *bimodal polyethylenes* because they have broad, and sometimes bimodal, molecular weight distribution (MWD). The first reactor makes low molecular weight HDPE in the absence, or under very low concentration, of α-olefin. Hydrogen, the standard chain transfer agent in olefin polymerization, is used in the first reactor to lower the polymer molecular weight. The polymer made in the first reactor is then transferred continuously to the second reactor, which is operated under higher α-olefin concentration in the absence, or under a much lower concentration, of hydrogen, thus producing an LLDPE component with higher average molecular

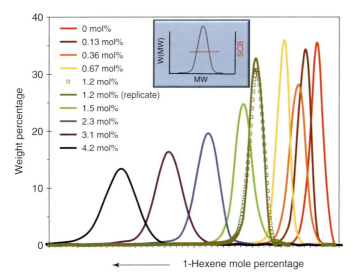

Figure 1.9 Chemical composition distributions of ethylene/1-hexene copolymers made with a metallocene catalyst.

weight than the HDPE component made in the first reactor. Many polyethylene industrial processes include two reactors in series to broaden the range of product properties, as described in Chapter 4.

The advent of metallocene catalysts added a new dimension to commercial polyolefin resins. Metallocenes are single-site catalysts that are used to make polyethylenes with completely different microstructures from those made with Ziegler–Natta and Phillips catalysts, but are still classified loosely as HDPE and LLDPE. Polyethylenes made with metallocene catalysts have uniform microstructures, with narrow MWDs and CCDs. Figure 1.9 shows the CCDs for a series of ethylene/1-hexene copolymers made with a metallocene catalyst. All distributions are narrow and unimodal in sharp contrast to the behavior observed for Ziegler–Natta LLDPEs. Notice the uniform incorporation of 1-hexene and the absence of the high-crystallinity peak. In addition, their average copolymer composition is independent of their MWD.

From a polymerization reaction engineering point of view, polyolefins made with metallocene catalysts provide an excellent opportunity for model development because they have "well-behaved" microstructures. Chapter 6 shows that models developed for single-site catalysts can also be extended to describe the more complex microstructures of polyolefins made with multiple-site complexes such as Ziegler–Natta and Phillips catalysts.

Substantially linear polyethylenes are an important new class of polyolefins. These polymers are also made with single-site catalysts, but their main characteristic is the presence of LCBs formed by terminal branching. They are called *substantially linear* because their LCB frequencies are typically below a few LCBs per 1000 carbon atoms. Terminal branching is illustrated in Figure 1.10, showing a polymer

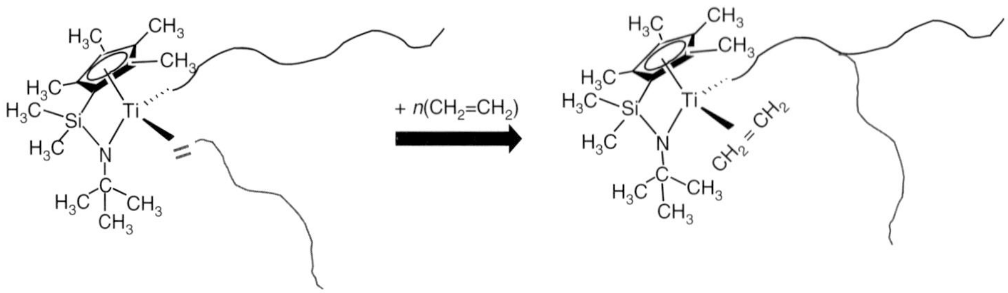

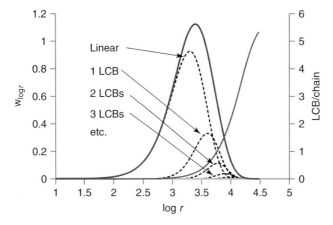

Figure 1.10 Long-chain branch formation through terminal branching promoted by a single-site catalyst.

Figure 1.11 Model predictions for the long-chain branch and chain length (r) distributions of substantially linear polyethylene.

chain with a reactive terminal bond being copolymerized with ethylene to form an LCB. The branching topology resulting from this mechanism is very different from that of LDPE resins. Figure 1.11 illustrates a model prediction for the chain length distribution and LCB frequency for a polyolefin formed via this mechanism. More details on this LCB formation mechanism are provided in Chapter 6.

1.3
Polypropylene Resins

Because propylene is an asymmetrical monomer, polypropylene can be produced with different stereochemical configurations. The most common types of polypropylene, shown in Figure 1.12, are isotactic, syndiotactic, and atactic. In isotactic polypropylene, the methyl groups are placed on the same side of the backbone; in syndiotactic polypropylene, on alternating sides; and in atactic polypropylene, the methyl groups are arranged randomly along the chain. Atactic polypropylene

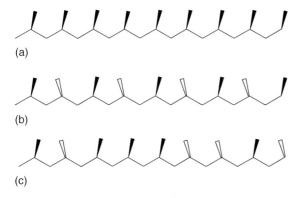

Figure 1.12 Main polypropylene types: (a) isotactic, (b) syndiotactic, and (c) atactic.

is amorphous and has little commercial value. Both isotactic and syndiotactic polypropylene are semicrystalline polymers with high melting temperatures. Isotactic polypropylene dominates the market, likely because it is easily produced with heterogeneous Ziegler–Natta and metallocene catalysts; syndiotactic polypropylene can be produced only with some metallocene catalysts and has much less widespread commercial use.

Modern Ziegler–Natta catalysts used for propylene polymerization make isotactic polypropylene with a very small fraction of atactic polypropylene. Non-specific sites are responsible for the formation of atactic chains in Ziegler–Natta catalysts. Many years of catalyst development were required to minimize the fraction of these catalyst sites, as discussed in Chapter 3.

Ziegler–Natta catalysts also are used to make chains with very high regioregularity, favoring 1-2 insertions and head-to-tail enchainment (Figure 1.13). Defects such as a 2-1 insertion following a 1-2 insertion create irregularities along the polymer chain, decreasing its crystallinity and melting temperature. Several metallocene catalysts produce polypropylene with very high isotacticity but lower regioregularity, which causes their melting temperatures to be lower than of those made with Ziegler–Natta catalysts. Metallocene catalysts can also make polypropylenes with other stereostructures such as atactic–isotactic block chains, but these products have not found commercial applications yet.

Impact propylene/ethylene copolymers[4] are produced using at least two reactors in series with heterogeneous Ziegler–Natta catalysts or supported metallocenes. The first reactor makes isotactic polypropylene, and the second produces propylene–ethylene random copolymer. The copolymer component is amorphous or has very low crystallinity, and is intimately dispersed in the homopolymer phase, even though the two phases are immiscible. The copolymer phase dissipates

---

4) These materials are sometimes (erroneously) called *block copolymers*. In fact, they are heterophasic materials composed of isotactic polypropylene and random propylene/ethylene copolymer chains.

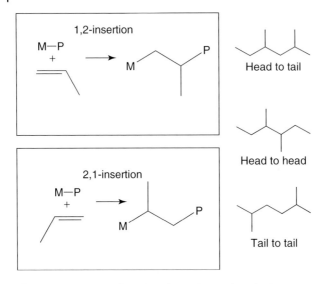

Figure 1.13 Regioregularity in polypropylene polymerization.

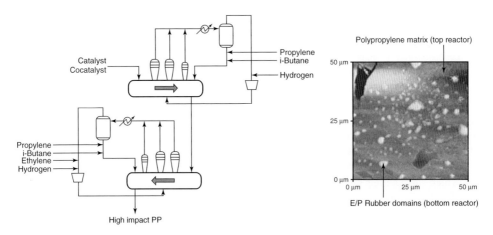

Figure 1.14 Process for the production of impact polypropylene.

energy during impact, greatly increasing the impact resistance of these resins. Several processes have been designed to produce impact polypropylene of high quality, as discussed in Chapter 4. Figure 1.14 schematically illustrates a process for the production of impact polypropylene.

Finally, the same comments made for Ziegler–Natta versus metallocene polyethylene apply to polypropylene resins. Metallocene catalysts make polypropylene with narrower MWD and, in the case of copolymers, narrower CCD.

Further Reading

There are several excellent books on polyethylene and polypropylene properties and applications. The Handbook of Polyethylene is a good source of information on HDPE, LLDPE, and LDPE.

Peacock, A.J. (2000) *Handbook of Polyethylene*, Marcel Dekker, New York.

Two multiauthored books on polypropylene are also recommended.

Karian, H.G. (ed.) (2003) *Handbook of Polypropylene and Polypropylene Composites*, Marcel Dekker, New York.Karger-Kocsis, J. (ed.) (1999) *Polypropylene. An A-Z Reference*, Kluwer Academic Publishers, Dordrecht.

The Encyclopedia of Polymer Science and Technology has excellent entries on HDPE (Benham, E. and McDaniel, M. (2002) Ethylene polymers, HDPE, in *Encyclopedia of Polymer Science and Technology* (ed. H.F. Mark), John Wiley & Sons, New York. doi: 10.1002/0471440264.pst408.pub2), LDPE (Maraschin, N. (2002) Ethylene polymers, LDPE, in *Encyclopedia of Polymer Science and Technology* (ed. H.F. Mark), John Wiley & Sons, New York. doi: 10.1002/0471440264.pst121), LLDPE (Simpson, D.M. and Vaughan, G.A. (2002) Ethylene polymers, LLDPE, in *Encyclopedia of Polymer Science and Technology* (ed. H.F. Mark), John Wiley & Sons, New York. doi: 10.1002/0471440264.pst122), ethylene/propylene elastomers (Noordemeer, J.W.M. (2002) Ethylene-propylene elastomers, in *Encyclopedia of Polymer Science and Technology* (ed. H.F. Mark), John Wiley & Sons, New York. doi: 10.1002/0471440264.pst125), and polypropylene.

Steward, C. (2002) Propylene polymers, in *Encyclopedia of Polymer Science and Technology* (ed. H.F. Mark) John Wiley & Sons, New York. doi: 10.1002/0471440264.pst301

These references are constantly updated by the authors and can be accessed on-line, making them a valuable source of information on these polymers.

We have also published general overviews on olefin polymerization reaction engineering, covering polymerization mechanisms and catalysis, properties, characterization, and mathematical modeling.

Soares, J.B.P. and Simon, L.C. (2005) Coordination polymerization, in *Handbook of Polymer Reaction Engineering* (eds T. Meyer and J. Keurentjes), Wiley-VCH, Weinheim, pp. 365–430.Soares, J.B.P., McKenna, T.F., and Cheng, C.P. (2007) Coordination polymerization, in *Polymer Reaction Engineering* (ed. J.M. Asua), Blackwell Publishing, pp. 29–117.

2
Polyolefin Microstructural Characterization

> It is the simplest of polymers, it is the most complex of polymers.
> Bill Knight and Willem deGroot, Dow Chemical (with some help from Charles Dickens)

2.1
Introduction

The simplicity of polyolefins is only apparent. No other synthetic polymer has such a wide variety of microstructures and can be used in so diverse applications as polyolefins, despite being built of very simple monomeric units. Not surprisingly, the ability to analyze polyolefin microstructures is essential for mastering polyolefin manufacturing technology.

The main microstructural distributions for polyethylene are the molecular weight distribution (MWD), the chemical composition distribution (CCD), the comonomer sequence length distribution (SLD), and long-chain branching (LCB). Because propylene molecules are asymmetric, regiochemical and stereochemical distributions are also very important for polypropylene.

Table 2.1 lists several analytical techniques applied routinely to analyze polyolefins. They can be divided into methods used to measure whole distributions or only distribution averages. The MWD of polyolefins is most commonly determined with high-temperature size exclusion chromatography (SEC), but field flow fractionation (FFF) may be used occasionally for UHMWPE resins. Both techniques rely on the size of the polymer molecules in solution as their basic fractionation mechanism. There are several analytical methods that use differences on chain crystallizability in solution to separate polyolefins according to their chemical composition or tacticity. Such techniques include temperature rising elution fractionation (TREF), crystallization analysis fractionation (CRYSTAF), and crystallization elution fractionation (CEF). Differential scanning calorimetry (DSC) may also be used to estimate these distributions. More recently, high-performance liquid chromatography (HPLC) techniques have been developed to analyze the CCD and tacticity distribution of polyolefins; these methods are based on

Polyolefin Reaction Engineering, First Edition. João B. P. Soares and Timothy F. L. McKenna.
© 2012 Wiley-VCH Verlag GmbH & Co. KGaA. Published 2012 by Wiley-VCH Verlag GmbH & Co. KGaA.

2 Polyolefin Microstructural Characterization

Table 2.1 Main polyolefin characterization techniques.

	MWD	M_v	M_w	CCD, tacticity	$\overline{F_A}$	SLD	LCB
SEC	X						
+IR	X				X		
+VISC	X	X					X
+LS	X		X				X
FFF	X						
TREF, CRYSTAF, or CEF				X			
+IR				X	X		
+VISC		X		X			
+LS			X	X			
HPLC				X			
DSC				X		X	
NMR					X	X	X

interactions between polymer chains and a stationary phase, not on chain crystallizability, and are promising alternatives to the conventional crystallizability-based methods.

When combined with microstructure-sensitive detectors, the techniques described above permit the determination of the distribution for the main variable being analyzed (MWD in the case of SEC; CCD for TREF and related techniques), while also measuring another average property across the distribution: for instance, when a size exclusion chromatographer is equipped with an infrared (IR) detector, the average copolymer composition ($\overline{F_A}$) may be obtained as a function of molecular weight; when it is combined with a viscometer (VISC) and a light scattering (LS) detector, the viscosity average molecular weight (M_v), weight average molecular weight (M_w), and LCB frequency can also be inferred in a single analysis. Similar *hyphenated techniques* using detectors that are sensitive to different microstructural characteristics may be used with TREF, CRYSTAF, CEF, and HPLC. The adoption of multiple detectors is becoming increasingly common because of the additional information they provide in a single analysis.

Carbon-13 nuclear magnetic resonance (^{13}C NMR) is the standard approach to quantify SLDs of all types in polyolefins: comonomer, 1-2 versus 2-1 insertions, and meso versus racemic insertions. It can also be used for chain-end analysis and LCB determination. NMR techniques for the determination of SLDs of polyolefins have been the subject of several excellent textbooks and are not discussed in this chapter.

Several of the analytical methods described in this chapter are also available in preparative mode. While analytical methods require only a few milligrams of sample, preparative fractionation uses several grams of polymer to collect fractions for further analysis by other techniques. For instance, a polyolefin may be fractionated either by molecular weight or chemical composition to collect narrow-MWD or narrow-CCD cuts, respectively, that can then be investigated with other analytical methods or tested for its mechanical or rheological properties. Since the principles governing preparative and analytical fractionations are the same, preparative methods are not covered herein, but the reader can find a few key references at the end of the chapter.

The extensive description of all characterization techniques used with polyolefins would require another (large) book. For the sake of brevity, only the most important aspects of these techniques are explained in this chapter, but enough details are given to allow the reader to appreciate their importance in the context of polyolefin reaction engineering.

2.2
Molecular Weight Distribution

2.2.1
Size Exclusion Chromatography

The MWD is the most fundamental microstructural distribution of any polymer because it has a large influence on its mechanical and rheological properties. High-temperature SEC, also known as *high-temperature gel permeation chromatography* (GPC), is the most widely used technique for MWD determination of polyolefins. High analysis temperatures are required because most commercial polyolefins are only soluble at temperatures above 120 °C in chlorinated solvents such as trichlorobenzene (TCB) or orthodichlorobenzene (ODCB).

SEC fractionates polymer chains using a series of columns that are generally packed with cross-linked poly(styrene-co-divinylbenzene) gels with varying pore diameters. Figure 2.1 shows an SEC diagram. SEC fractionation can be succinctly described as follows: the polymer sample, present as a dilute solution in a vial, is injected into the mobile phase as a narrow pulse and flows through a series of packed columns. The elution time of a polymer chain depends on its volume in solution (*hydrodynamic volume*); chains with higher volumes penetrate into fewer pores and elute first, while chains with smaller volumes penetrate deeper and into more pores, and elute later. A mass detector monitors the polymer concentration exiting from the last column in the series, and a calibration curve[1] is applied to relate elution volume to molecular weight. *Retention time* is the time taken by a polymer fraction to elute from the SEC columns. Likewise, *elution volume* is the volume of mobile phase required to elute a given polymer fraction from the columns.

---

1) If an on-line laser light scattering (LLS) detector is available, no calibration curve is required.

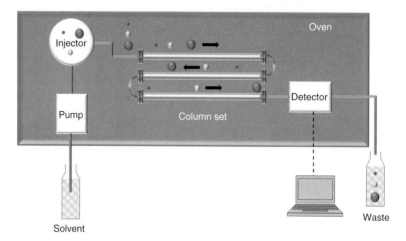

Figure 2.1 Size exclusion chromatography (SEC) schematic.

The elution volume (V_e) of polymer chains is related to the interstitial volume (V_i) between the packing particles in the SEC column and to their pore volume (V_p) by the expression,

$$V_e = V_i + \kappa V_p \qquad (2.1)$$

where κ is the partition coefficient for the polymer chains between the stationary and mobile phases; κ depends on the pore size of the packing particles and on the polymer hydrodynamic volume. When $\kappa = 0$, the chains are too large to penetrate into any of the pores (*total exclusion limit*); when $\kappa = 1$, they are small enough to diffuse into all pores (*total permeation limit*). These two limits set the resolution bounds in SEC, as shown in the picture at the upper right corner of Figure 2.2. Chains whose volumes are below the total permeation limit or above the total exclusion limit cannot be fractionated by SEC. The resolution range can be adjusted by the proper selection of the SEC column set.

The partition coefficient κ is related to the thermodynamics of the polymer chains in solution by the equation,

$$\kappa = \exp\left(\frac{\Delta G}{RT}\right) = \exp\left(\frac{\Delta H}{RT} + \frac{\Delta S}{R}\right) \qquad (2.2)$$

Ideally, only entropic effects govern SEC fractionation; no interaction should take place between polymer chains and the stationary phase ($\Delta H = 0$). Therefore, Eq. (2.2) is reduced to

$$\kappa \cong \exp\left(\frac{\Delta S}{R}\right) \qquad (2.3)$$

and the value of κ becomes proportional to the entropy decrease experienced by the polymer chains as they diffuse into the pores of the packing particles; consequently, $0 \leq \kappa \leq 1$.

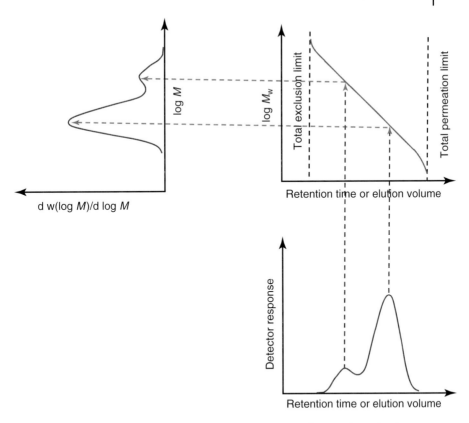

Figure 2.2 MWDs are obtained from the SEC retention time or elution volume distribution using a calibration curve.

The polymer hydrodynamic volume is a function of its chain length, branching topology, solvent type, and temperature. Polymer solution concentration may also influence the hydrodynamic volume, but SEC analyses are run under conditions that approximate infinite dilution to eliminate this factor.

The volume of linear chains in solution is directly proportional to their chain lengths or molecular weights. Therefore, a calibration curve relating molecular weight to elution volume (or retention time) it is relatively easy to generate for linear polymers. In principle, SEC may be calibrated with narrow-MWD standards of the polymer one wishes to analyze, to obtain the calibration curve shown in Figure 2.2, which can then be used to convert the raw SEC profile shown at the bottom of Figure 2.2 into the MWD presented in the upper left corner of the same figure.

This approach, however, suffers from the limitation that the calibration curve is only applicable to one specific polymer type. In the case of polyolefins, for instance, it would be necessary to determine calibration curves for polypropylene and polyethylene, and among the polyethylenes, calibration curves for those with different types, molar fractions, and sequence distributions of α-olefins, since these variables may affect the polymer hydrodynamic volume. As an additional difficulty, narrow-MWD standards are not commonly available for a variety of polyolefins.

One of the great advantages of SEC as an analytical technique for determining polymer MWD is that it can be calibrated using a *universal calibration* curve that is independent of polymer type, provided that the chains are linear and behave like random coils in solution.[2] The universal calibration curve was proposed by Grubisic et al. in the 1960s and relies on the principle that polymer chains with the same hydrodynamic volume also have the same retention time and elution volume.

The product of the polymer molecular weight, M, and intrinsic viscosity, $[\eta]$, is proportional to the polymer hydrodynamic volume,

$$[\eta]M = \Phi(\overline{r_0^2})^{3/2}\alpha^3 \qquad (2.4)$$

where $\overline{r_0^2}$ is the root-mean-square end-to-end distance of the polymer chain, and α and Φ are constants that depend on the type of solvent and polymer. When the logarithm of the product $[\eta]M$ for linear polymers is plotted versus elution volume, an elegant relationship arises that is independent of polymer type, as illustrated by the universal calibration curve depicted in Figure 2.3.

Narrow-MWD polystyrene samples[3] spanning a wide range of molecular weights can be produced via living anionic polymerization and are ideal calibration standards for SEC. For instance, the hydrodynamic volume of polyethylene (PE) can be compared to that of a polystyrene standard (PS) with the equation,

$$M_{PE}[\eta]_{PE} = M_{PS}[\eta]_{PS} \qquad (2.5)$$

If the SEC is equipped with a VISC detector, Eq. (2.5) can be used directly to find out the molecular weight of polyethylene corresponding to that of a polystyrene chain of the same hydrodynamic volume,

$$M_{PE} = M_{PS}\frac{[\eta]_{PS}}{[\eta]_{PE}} \qquad (2.6)$$

If an on-line VISC is not available, the Mark–Houwink equation is the appropriate choice for estimating the value of $[\eta]$,

$$[\eta] = KM^a \qquad (2.7)$$

---

2) The universal calibration curve, however, is only "universal" in the sense of being shared by several linear polymers but needs to be determined for each SEC setup. Different SEC instruments and column sets will have distinct universal calibration curves, which also need to be recalculated as the column set ages.

3) The choice of polystyrene as standards for SEC calibration is convenient, but any other narrow-MWD standard may be used to generate the universal calibration curve. SEC calibration using broad-MWD standards is also possible but is less commonly used.

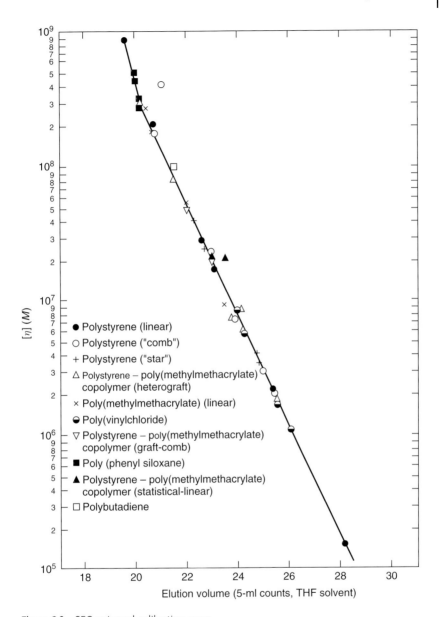

Figure 2.3 SEC universal calibration curve.

where a and K are constants that depend on temperature, polymer, and solvent types, as described in introductory polymer science textbooks. Consequently, the molecular weight of polyethylene (or any other linear polymer) can be calculated from the molecular weight of polystyrene by combining Eqs. (2.6) and (2.7),

$$\log M_{PE} = \frac{1+a_{PS}}{1+a_{PE}} \log M_{PS} + \left(\frac{1}{1+a_{PE}}\right) \log \frac{K_{PS}}{K_{PE}} \qquad (2.8)$$

The molecular weight of the polystyrene standards, M_{PS}, is generally related to GPC elution volume using a polynomial calibration curve,

$$\log M_{PS} = \sum_{i=0}^{n} A_i \times V_e^{i-1} \qquad (2.9)$$

Therefore, the molar mass of polyethylene can be related to its elution volume with a simple algebraic expression,

$$\log M_{PE} = \frac{1+a_{PS}}{1+a_{PE}} \sum_{i=0}^{n} A_i \times V_e^{i-1} + \left(\frac{1}{1+a_{PE}}\right) \log \frac{K_{PS}}{K_{PE}} \qquad (2.10)$$

Equation (2.8) can sometimes be further simplified for systems that have similar polymer–solvent interactions and, therefore, similar values for the parameter a, such as polyethylene and polystyrene. In this case, Eq. (2.8) is reduced to

$$\log M_{PE} = \log M_{PS} + \left(\frac{1}{1+a_{PE}}\right) \log \frac{K_{PS}}{K_{PE}}$$

$$\Rightarrow M_{PE} = \left[\left(\frac{1}{1+a_{PE}}\right) \log \frac{K_{PS}}{K_{PE}}\right] \times M_{PS} = A' \times M_{PS} \qquad (2.11)$$

and Eq. (2.10) becomes

$$\log M_{PE} = \left[A_0 + \left(\frac{1}{1+a_{PE}}\right) \log \frac{K_{PS}}{K_{PE}}\right] + \sum_{i=1}^{n} A_i \times V_e^{i-1}$$

$$= A_0' + \sum_{i=1}^{n} A_i \times V_e^{i-1} \qquad (2.12)$$

Consequently, the calibration curve for polyethylene is obtained through a simple linear shift of the calibration curve for polystyrene, if we can assume that $a_{PE} \approx a_{PS}$.

Once the MWD has been determined, the molecular weight averages can be calculated using their usual definitions,

$$M_n = \frac{\sum M_i N_i}{\sum N_i} \qquad (2.13)$$

$$M_w = \frac{\sum M_i^2 N_i}{\sum M_i N_i} \qquad (2.14)$$

where M_n and M_w are the number and weight average molecular weights, respectively, M_i is the molecular weight of a GPC cut, and N_i is the number of polymer molecules in the same GPC cut.

Since MWDs measured by GPC are usually reported as normalized weight distributions in logarithm scale, it is more convenient to use the following expressions to calculate their averages,

$$M_n = \frac{1}{\int \frac{w(\log M)}{M} \cdot d(\log M)} \tag{2.15}[4]$$

$$M_w = \int M \cdot w(\log M) \cdot d(\log M) \tag{2.16}[5]$$

where $w(\log M)$ is the normalized MWD on a weight basis in logarithmic scale.

A refractive index detector (RI) is commonly used to measure the mass concentration of polymers eluting from the GPC columns. More recently, single-frequency IR detectors have also been used as mass detectors for GPC. Their main advantages over the traditional RI detectors are better baseline stability and lower sensitivity to temperature fluctuations in the detector cell. VISC detectors are also a common choice because they permit the direct determination of the polymer intrinsic viscosity required for the use of the universal calibration curve, as already discussed above. Finally, low-angle laser light scattering (LALLS) and multiangle laser light scattering (MALLS) detectors allow the direct measurement of the M_w for the polymer fraction in the detector cell, therefore eliminating the need of a calibration curve. When the chains reach the detector cell, they have already been fractionated by hydrodynamic volume and can be considered to be approximately monodisperse, except in the case of highly branched polymers with very complex microstructures; consequently, $M_w \approx M_n \approx M$, and a calibration curve is not required.

The analysis of polyolefins containing LCBs, such as LDPE and substantially linear polyolefins, becomes more involved because the hydrodynamic volume for these polymers is a function not only of molecular weight but also of branching density and type. The *viscosity branching index*, g', is defined as

$$g' = \frac{[\eta]_b}{[\eta]_l} < 1 \tag{2.17}$$

where $[\eta]_l$ and $[\eta]_b$ are the intrinsic viscosities of linear and branched polymers with the same molecular weight, respectively.

Similarly, the *branching index*, g, is expressed in terms of the dimensions of branched and linear polymer of the same molecular weight,

[4]
$$M_n = \frac{\sum M_i N_i}{\sum N_i} = \frac{\sum W_i}{\sum N_i}$$
$$= \frac{\sum W_i}{\sum \frac{W_i}{M_i}} = \frac{\sum \frac{W_i}{\sum W_i}}{\sum \frac{W_i}{M_i \sum W_i}}$$
$$= \frac{1}{\sum \frac{w_i}{M_i}} = \frac{1}{\int \frac{w(M)}{M} dM}$$
$$= \frac{1}{\int \frac{w(\log M)}{M} d(\log M)}$$

[5]
$$M_w = \frac{\sum M_i^2 N_i}{\sum M_i N_i} = \frac{\sum W_i M_i}{\sum W_i}$$
$$= \frac{\sum \frac{W_i M_i}{\sum W_i}}{\sum \frac{W_i}{\sum W_i}} = \frac{\sum w_i M_i}{1}$$
$$= \sum w_i M_i = \int M \cdot w(M) \cdot dM$$
$$= \int M \cdot w(\log M) \cdot d(\log M)$$

$$g = \frac{\langle R_g^2 \rangle_b}{\langle R_g^2 \rangle_l} < 1 \tag{2.18}$$

where $\langle R_g^2 \rangle_l$ and $\langle R_g^2 \rangle_b$ are the squared radius of gyration of linear and branched chains of the same molecular weight, respectively. Both parameters quantify the polymer coil volume contraction due to LCB and are related by the expression,

$$g' = g^\varepsilon \tag{2.19}$$

where the exponent ε has been reported to vary from 0.5 to 1.5 and depends on LCB topology, solvent, and polymer type. Unfortunately, ε is also a function of molecular weight, which further complicates the use of Eq. (2.19).

Zimm and Stockmayer have proposed relations for the branching index, g, and the number of LCBs per chain, n. For instance, for trifunctional, randomly branched, monodisperse polymers the following relationship applies:

$$g = \left[\left(1 + \frac{n}{7}\right)^{1/2} + \frac{4n}{9\pi} \right]^{-1/2} \tag{2.20}$$

Similarly, for trifunctional, randomly branched, and *polydisperse* polymers,

$$g = \frac{6}{n_w} \left\{ 1/2 \left(\frac{2 + n_w}{n_w} \right)^{1/2} \ln \left[\frac{(2 + n_w)^{1/2} + n_w^{1/2}}{(2 + n_w)^{1/2} - n_w^{1/2}} \right] - 1 \right\} \tag{2.21}$$

where n_w is weight average number of LCBs per chain. Relations for other LCB topologies are also available but are not shown here for brevity's sake.

Figure 2.4 shows how g varies as the chains become more branched for mono- and polydisperse polymers. Assuming a linear polymer with molecular weight $M_0 = 10\,000$ and $\varepsilon = 0.7$ (a typical value for polyethylene), the plot also predicts the molecular weight of the branched chains that would elute at the same time from the GPC columns using the identity $M_b[\eta]_b = M_l[\eta]_l$,

$$M_b = M_l \frac{[\eta]_l}{[\eta]_b} = \frac{M_l}{g'} = \frac{M_l}{g^\varepsilon} \tag{2.22}$$

Therefore, a mixture of chains with different molecular weights and branching frequencies may exist at the detector cell at a given retention time. In this case, the universal calibration curve cannot be used accurately anymore, and the Mark–Houwink equation does not apply. Chapter 6 shows how the equations discussed above can be combined with a microstructural model for substantially linear polyolefins to predict the polymer chain mixture eluting from the GPC at a given retention time.

The GPC "signature" for the presence of LCBs in a polyolefin is the deviation from the linear relationship expected for the $\log[\eta] \times \log M$. According to the Mark–Houwink equation, a straight line should describe this relationship for linear polymers

$$\log[\eta] = \log K + a \log M \tag{2.23}$$

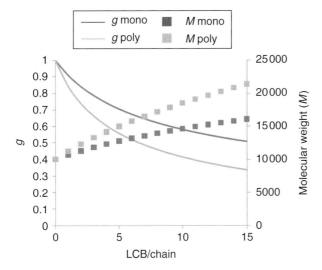

Figure 2.4 LCB effect on polymer hydrodynamic volume.

Deviations from this behavior generally indicate the presence of LCBs.[6] Figure 2.5 shows the MWD and $\log[\eta] \times \log M$ plot for an LCB polyethylene, as clearly attested by the deviation from the straight line relationship shown for NBS 1475, a linear polyethylene GPC calibration standard (the MWD for NBS is not shown in Figure 2.5 for simplicity).

A relatively simple approach to estimate the average branching frequency of polyolefins by GPC/VISC analysis involves the following steps:

1) From the $\log[\eta] \times \log M$ data (for instance, Figure 2.5), calculate the viscosity branching index g' at each molecular weight using Eq. (2.17).
2) Calculate the branching index g at each molecular weight using Eq. (2.19).
3) Use Eq. (2.20) or (2.21) to estimate the number of average LCBs per molecule at each molecular weight.
4) The average number of LCBs per chain (B_n) and number of LCBs per 1000 carbon atoms (λ_n) for the whole polymer are calculated with Eqs. (2.24) and (2.25).

$$B_n = M_n \sum_r \frac{w_r}{M_r} n_r \tag{2.24}$$

$$\lambda_n = \frac{14000}{M_n} \times B_n \tag{2.25}$$

6) Deviations from the Mark–Houwink equation can also be expected for very low and very high molecular weights but are not very significant for most commercial polyolefins.

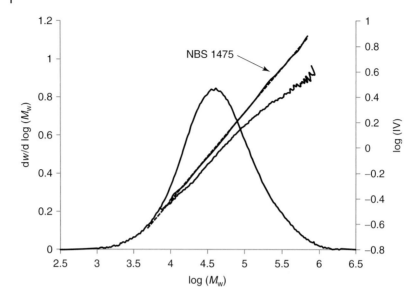

Figure 2.5 MWD and intrinsic viscosity (IV) plot for a polyethylene containing LCBs.

where w_r is the weight fraction of chains having chain length r, n_r is the number of LCBs per molecule of the chains having chain length r, M_r is the molecular weight of chains having chain length r, and M_n is the number average molecular weight of the polymer sample.[7]

The GPC/VISC method described above suffers from the disadvantage that the value of the parameter ε in Eq. (2.19) must be known accurately, which most often is not possible due to its dependency on LCB topology and molecular weight. The addition of an MALLS detector to the GPC system eliminates this limitation because the radius of gyration of the chains eluting from the columns can be determined during the analysis, allowing the direct use of the Zimm–Stockmayer relations shown in Eqs. (2.20) and (2.21). The use of GPC triple-detector systems is becoming increasingly more common in face of the increasing microstructural complexity of modern polyolefins.

We saw in Chapter 1 that for resins such as LLDPE made with heterogeneous Ziegler–Natta catalysts, the average α-olefin fraction varies inversely with molecular weight. Dual reactor systems may be used to change (to a certain extent) this dependency or, as done more recently, blends of single-site catalysts may be employed to obtain a copolymer with reverse comonomer incorporation where the α-olefin content increases with increasing polymer molecular weight. Because this relation has such an important impact on the mechanical properties of

---

7) Equation (2.24) can be used with any polymer, but Eq. (2.25) applies only to polyethylene, since it assumes a molar mass of 28 g·mol^{-1} for the repeating unit.

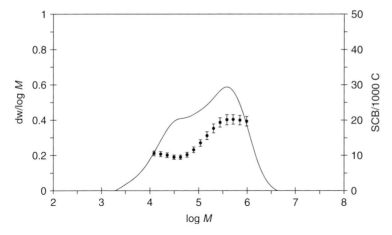

Figure 2.6 MWD and average SCB frequency for an ethylene/1-hexene copolymer measured by GPC/IR.

polyolefins, IR or Fourier-transform infrared (FTIR) detectors may be added to GPC to measure the comonomer fraction as a function of molecular weight. Figure 2.6 shows the GPC/IR plot for an ethylene/1-hexene copolymer made with two single-site catalysts supported on the same carrier. The IR plot measures how the SCB frequency (or comonomer incorporation) varies as a function of molecular weight in the copolymer. This information is very important when designing new resins, such as this one that shows reverse comonomer incorporation.

In GPC/IR or GPC/FTIR methods that are currently prevalent in industry and academia, the SCB frequency is measured in a heated flow cell attached to the exit of the GPC columns, using the C-H stretching bands found between 3000 and 2800 cm^{-1}. The ratio between the methyl and methylene groups at 2965 and 2928 cm^{-1} can be used to quantify the SCB frequency of the polymer being analyzed via a calibration curve ($A^{2965}(CH_3)/A^{2928}(CH_2) \times SCB/1000\ C$) constructed using polymers with known SCB frequency, although more sophisticated approaches using chemometrics have also been published in the literature.

2.2.2
Field Flow Fractionation

High-temperature FFF may be occasionally used to determine the MWD of polyolefins, especially for resins with very high molecular weights, such as UHMWPE, which would otherwise be very difficult to analyze by GPC. Polyolefin chains with molecular weights much higher than 10^6 generally pose significant experimental difficulties during GPC analysis because of the column plugging and overpressure caused by the elevated solution viscosity resulting from their high molecular weights. In addition, the high shear stress in the GPC columns may promote chain scission of these very long chains and lead to invalid MWD results. Contrary to

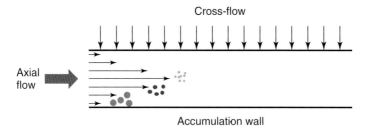

Figure 2.7 Schematic for field flow fractionation.

GPC, FFF channels contain no packing, broadening the range of hydrodynamic radii that can be studied with this technique.

In FFF, a pulse of a polymer solution is injected into a carrier liquid and flows through an empty rectangular thin channel. An external physical field is applied perpendicular to the axial flow, acting as a driving force for the fractionation of the polymer molecules according to molecular weight and/or other microstructural properties. The axial flow is laminar, establishing a parabolic velocity profile in the rectangular channel. The nature of the external field determines the type of fractionation achieved in FFF; the most common are thermal, flow, and gravitational/centrifugal. Flow fields are more commonly used for polyolefins.

In flow-FFF (Figure 2.7), the carrier liquid flows perpendicularly through the porous walls of the channel, establishing an external flow field that is normal to the axial flow (cross-flow). A narrow pulse of polymer solution is injected at one end of the column, and the axial flow is stopped for a short period. The cross-flow forces the polymer molecules to accumulate near the opposite wall (accumulation wall), generating a transversal polymer concentration gradient. According to Fick's law, this concentration gradient will act as the driving force for molecular diffusion in the opposite direction of the external field. After a short time, a steady state is reached, in a process called *relaxation*, with the establishment of a polymer layer of a given thickness near the accumulation wall. Chains with smaller hydrodynamic radii (smaller molecular weights) have higher diffusion coefficients and will move farther away from the accumulation wall than chains with higher hydrodynamic radii. When the axial flow is restarted, generating a parabolic velocity profile inside the channel, smaller chains will, on average, be transported with higher axial velocities than larger chains and will elute from the channel at shorter retention times. This mode of operation, where the intensity of the driving force is constant throughout the channel, is called the *classic or normal mode* of operation, but there are several alternatives to this procedure. Commercial systems use a slightly simplified setup called *asymmetric flow-FFF*, with only one porous wall, but the fractionation principles are the same as the ones shown in Figure 2.7. Since FFF is used only rarely with polyolefins, it is not discussed any further in this chapter. A few references on this topic are listed at the end of the chapter for the interested reader.

2.3 Chemical Composition Distribution

2.3.1 Crystallizability-Based Techniques

The CCD is the second most important microstructural distribution of polyolefins. The most commonly used methods for CCD determination rely on the distribution of crystallizability of polymer chains in a dilute solution. Three analytical techniques are based on this approach: TREF, CRYSTAF, and CEF.

All these three techniques fractionate polymers according to the same principle: chains with more "defects" (higher fraction of α-olefin comonomer or higher frequency of stereo- and/or regio-irregularities) have lower crystallization temperatures than chains with fewer defects and can be fractionated from a dilute solution by decreasing the temperature slowly, in either a stirred vessel (CRYSTAF) or a packed column (TREF and CEF). These concepts are illustrated schematically in Figure 2.8.

Chain crystallizability is strongly influenced by how regular the polymer microstructure is. In the case of binary copolymers, the comonomer present in the smallest fraction generally does not crystallize and acts as a chain defect. The crystallization (T_c) and melting (T_m) temperatures, therefore, decrease for polymers with higher comonomer fractions. Typical examples are copolymers of ethylene and α-olefins where the α-olefin molecules form SCBs that are too large to be included in the crystalline lattice, leading to the formation of smaller crystallites with lower T_m and T_c. When the two comonomers crystallize, such as for ethylene/propylene copolymers, the comonomer present in the lower fraction acts as a defect and T_m and T_c pass through a minimum located between the values for

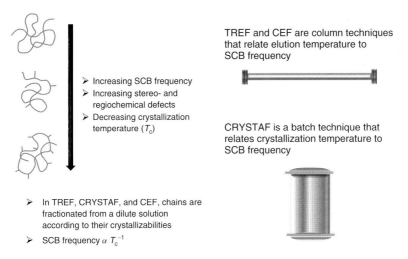

Figure 2.8 The crystallizability of chains in solution determines the fractionation in TREF, CRYSTAF, and CEF.

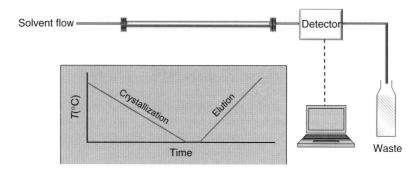

Figure 2.9 Schematic for temperature rising elution fractionation (TREF).

the two semicrystalline homopolymers. A similar phenomenon takes place with isotactic polypropylene. In this case, propylene molecules inserted in a racemic configuration or in 2-1 orientation will decrease T_m and T_c below the values for isotactic polypropylene. A similar rationale applies to syndiotactic polypropylene, but now, the meso insertions are the ones responsible for decreasing T_m and T_c below the values for the stereoregular polymer.

TREF was the first method developed to determine the CCD of polyolefins. This technique can be considered revolutionary because it revealed a microstructural dimension that was very poorly known for polyolefins before its invention. TREF results showed, for instance, that the CCDs of LLDPE resins made with heterogeneous Ziegler–Natta catalysts were bimodal, independent of the process used for their manufacture, while the CCD for LDPE made by free radical polymerization was unimodal, settling the controversy about the multiple-site-type nature of heterogeneous Ziegler–Natta catalysts. TREF has done much to elucidate the nature of coordination catalysts used for olefin polymerization and to develop polyolefin resins with designed properties.

TREF comprises two steps: crystallization and elution (Figure 2.9). In the crystallization step, a very dilute polymer solution (TCB or ODCB is generally the preferred solvent) is transferred at high temperature to a column packed with an inert support. The type of support has minor influence on the fractionation: glass beads, stainless steel shots, and Chromosorb P, among others, have been used in TREF columns. The main role of the support is to provide a place for polymer crystallization to occur and to allow for uniform solvent flow through the column during the elution step. The polymer solution is then cooled slowly, typically from 140 to 120 °C to room temperature. The crystallization step is the most important procedure during fractionation: a slow cooling rate is required for the polymer chains to crystallize near thermodynamic equilibrium, minimizing cocrystallization effects and ensuring good resolution of chains with different comonomer fractions or tacticity. Chains with higher crystallizabilities (and fewer chain defects) crystallize and precipitate first, followed by chains with lower crystallizabilities. When the cooling step is completed, solvent starts flowing through the column at increasing temperatures, removing polymer chains in the reverse order they were precipitated,

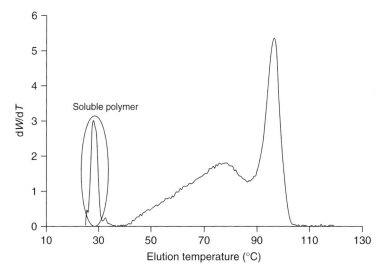

Figure 2.10 TREF profile for an LLDPE resin made with a heterogeneous Ziegler–Natta catalyst.

from least to most crystalline. A mass detector (generally, a temperature-insensitive single-frequency IR detector) is placed at the exit of the TREF column to measure the concentration of chains being eluted as a function of temperature.

A TREF profile for a typical Ziegler–Natta LLDPE is shown in Figure 2.10, where three characteristic regions can be seen: high crystallinity, medium crystallinity, and room-temperature soluble fraction, corresponding to low, medium, and high α-olefin fractions, respectively. A calibration curve is required to convert TREF elution temperature to α-olefin fraction in the copolymer.

Unfortunately, a universal calibration curve for TREF does not exist. TREF calibration curves depend on comonomer type and, to a lesser extent, on comonomer SLD. Calibration curves are generally linear over the whole composition range allowed by the technique, from pure polyethylene to about 10–12% molar fraction of α-olefin. Polyolefins with higher α-olefin content are soluble at room temperature in most solvents used in TREF and cannot be analyzed by this technique, appearing as the *purge* or *room temperature peak* shown in Figure 2.10.[8] TREF is generally calibrated with narrow-CCD standards of known average comonomer fraction, often made with metallocene catalysts, to generate curves such as the ones shown in Figure 2.11. Longer α-olefins are more effective in decreasing the TREF elution temperature, presumably because shorter olefins such as 1-butene, and 1-hexene, to a lesser extent, can be partially included in the polyethylene crystallites, while 1-octene is too bulky and is completely excluded from the amorphous phase. The

---

8) Subambient TREF analysis can only partially overcome this limitation since the melting point of TREF solvents is relatively high; TCB, for instance, freezes at 17 °C.

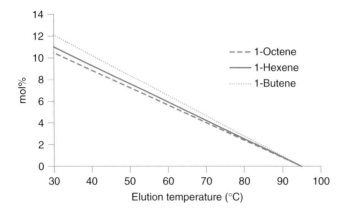

Figure 2.11 Comonomer-type effect on generic TREF calibration curves.

comonomer effect on TREF calibration curves ceases to be important for olefins longer than 1-octene, except for those that form very long SCBs, which may also be able to crystallize.

The linear relation observed in Figure 2.11 is theoretically expected. For random copolymers, Flory showed that

$$\frac{1}{T_m} - \frac{1}{T_m^0} = -\frac{R}{\Delta H_u} \ln x_A \qquad (2.26)$$

where T_m^0 is the melting point of an infinitely long chain of homopolymer A, T_m is the melting temperature of the random copolymer AB, ΔH_u is the enthalpy of melting for the repeating unit A, and x_A is the molar fraction of the crystallizable unit A. This equation can be extended to describe the melting point depression of stereoregular homopolymers due to stereo- or regio-irregular insertions. For instance, in the case of isotactic polypropylene, x_A would represent the molar fraction of meso insertions and $(1 - x_A)$ would then represent the molar fraction of racemic insertions.

For copolymers with small comonomer fractions, such as most industrial polyolefins, the following simplification applies to the terms in the left- and right-hand sides of Eq. (2.26):

$$\frac{1}{T_m} - \frac{1}{T_m^0} = \frac{T_m^0 - T_m}{T_m T_m^0} \simeq \frac{T_m^0 - T_m}{(T_m^0)^2} \qquad (2.27)$$

$$\ln x_A = -(1 - x_A) = -x_B \qquad (2.28)$$

Therefore, Eq. (2.26) assumes the linear form

$$T_m \cong T_m^0 - \frac{R(T_m^0)^2}{\Delta H_u} x_B \qquad (2.29)$$

which shows that the melting temperature, or TREF elution temperature, is a linear function of the fraction of noncrystallizable comonomer units in the copolymer.

Several operation conditions affect TREF results. A slow cooling rate (2–6 °C h^{-1}) is essential to attain adequate resolution, as most of the fractionation takes place during the crystallization step. In fact, initially it was considered that the elution step in TREF was mainly required to recover and to monitor the concentration of the polymer already fractionated during the precipitation step and did little to improve the fractionation quality. However, recent studies with ethylene/α-olefin copolymer blends indicate that the heating during the elution step may lead to recrystallization of the crystals formed during the crystallization stage, enhancing TREF peak resolution as compared to CRYSTAF that does not have this additional step, as described below. TREF performance is relatively insensitive to polymer concentration in the range of 1–4 mg ml^{-1}. Solvent flow rate and heating rate can also affect peak position and breadth, but it has been demonstrated that these variations can be eliminated by operating TREF at constant (heating rate)/(solvent flow rate) ratio, since these conditions ensure that each volume element exiting the column contains polymer chains that were eluted during the same time/temperature interval. Finally, solvent type plays only a minor effect on TREF fractionation, with better solvents leading to lower elution temperatures, as expected. Polymer molecular weight may also influence TREF fractionation, but it only starts significantly affecting TREF peak elution temperatures for M_n values below 6000 g mol^{-1} and can be ignored for most industrial polyolefins.

One of the disadvantages of TREF is the relatively long analysis time required to reach good resolution. For a relatively fast cooling rate of 6 °C min^{-1}, and also using a heating rate of 6 °C min^{-1}, a typical TREF analysis for a Ziegler–Natta LLDPE takes about 24 h. Faster TREF analyses can be performed, but at the cost of lower peak resolution. As discussed above, even though the fractionation takes place during the crystallization step, this information is only retrieved when the chains are eluted from the column. This concept is sometimes illustrated with the "onion skin" analogy shown in Figure 2.12. This analogy illustrates the fractionation mechanism in TREF, but keep in mind that it is only a conceptual representation and does not describe the actual morphology of the polymer fractions deposited onto the TREF packing particles. As the temperature decreases during the crystallization step, the polymer chains are envisioned to precipitate on the surface of the support particles, forming layers of decreasing crystallinity. The subsequent elution step is needed to "peel off" the polymer layers so they can be quantified by the IR detector placed at the exit of the TREF column. In reality, as mentioned above, the elution process seems to enhance TREF resolution, likely due to recrystallization of the crystals formed during the crystallization step. Nonetheless, it is clear that most of the fractionation takes place during crystallization and if the polymer concentration could be monitored during crystallization, the elution step could be eliminated altogether. CRYSTAF was developed as a faster alternative to TREF on this exact premise.

Polymer chains are fractionated by CRYSTAF inside a stirred vessel in the absence of a support. Similar to TREF, a dilute polyolefin solution is slowly cooled from 120 °C to room temperature, causing the precipitation of chains from higher to lower crystallizabilities, but different from TREF, the concentration of polymer

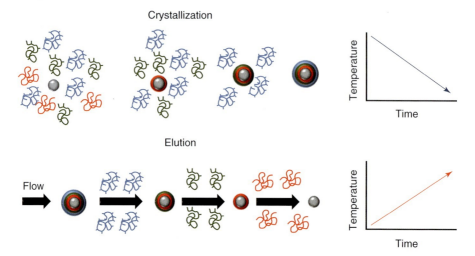

Figure 2.12 Onion skin analogy for TREF.

left in solution is monitored during the crystallization step using an on-line IR detector. Aliquots are taken from the polymer solution at predetermined sampling intervals through a dip tube equipped with a filter (to avoid sampling the polymer crystallites that have already precipitated from solution) and sent to a mass detector that measures the concentration of polymer still remaining in solution. The curve relating polymer solution concentration to crystallization temperature is called the *cumulative distribution*. The *differential distribution* is calculated by taking the first derivative of the cumulative distribution with respect to the crystallization temperature. Both cumulative and differential CRYSTAF curves for a blend of two metallocene ethylene/1-hexene copolymers with different 1-hexene fractions are shown in Figure 2.13. CRYSTAF differential distributions provide essentially the same information as TREF curves, which is not surprising since both techniques rely on the same polymer crystallization mechanism. Similar to TREF, a calibration curve is needed to translate the CRYSTAF differential distribution into a polymer microstructural distribution.

Figure 2.14 compares TREF and CRYSTAF profiles of a heterogeneous Ziegler–Natta ethylene/1-butene copolymer. The profiles have similar shapes, but the CRYSTAF curve is shifted toward lower temperatures because it is measured as the polymer chains crystallize, while the TREF curve is determined as the polymer chains dissolve (melt) and are eluted from the TREF column. The rectangular peak shown at the low T_c end of the CRYSTAF profile is the standard way the soluble peak is represented in CRYSTAF analysis. The area under this peak is proportional to the weight fraction of polymer that remains soluble at room temperature. Even a superficial inspection of Figure 2.14 shows that CRYSTAF and TREF reflect the same microstructural information but that the TREF profile looks a little sharper than the CRYSTAF profile. Today, it is well accepted that TREF has better peak resolution than CRYSTAF for analyses done at similar crystallization rates. This

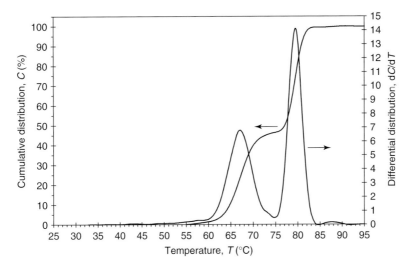

Figure 2.13 CRYSTAF analysis of a blend of two metallocene ethylene/1-hexene copolymers, showing cumulative and differential distributions.

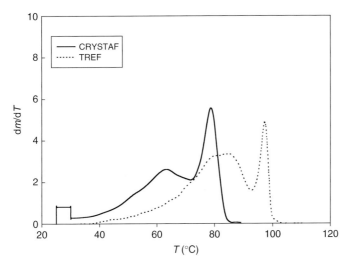

Figure 2.14 TREF and CRYSTAF profiles for an ethylene/1-butene copolymer made with a heterogeneous Ziegler–Natta catalyst.

increased resolution becomes evident in Figure 2.15, which compares TREF and CRYSTAF profiles of a ternary blend of metallocene copolymers. It is clear not only that peak resolution increases for both techniques as the cooling rate decreases from 0.2 to 0.1 °C min^{-1} but also that TREF profiles have the best resolution.

Commercial CRYSTAF instruments come equipped with five crystallization vessels that can be operated in parallel, which is an advantage over most TREF apparatuses. For the commonly used crystallization rate of 6 °C h^{-1}, a typical

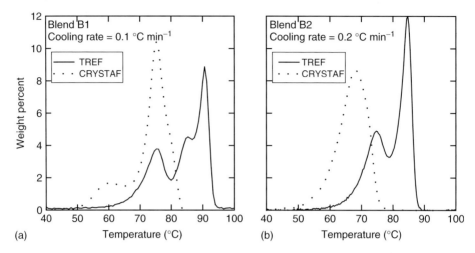

Figure 2.15 TREF and CRYSTAF profiles of a ternary blend of metallocene copolymers measured at two different crystallization rates: (a) 0.1 °C min^{-1} and (b) 0.2 °C min^{-1}.

CRYSTAF analysis only takes about 8–10 h, and five samples may be analyzed in parallel. Despite its lower resolution,[9] CRYSTAF is faster and easier to use than TREF. Several of the comments made above for TREF regarding the effect of solvent type, molecular weight, cooling rate, and calibration curves can be extended to CRYSTAF and do not need to be mentioned again.

More recently, CEF was developed to combine the higher resolution of TREF with the faster analysis time of CRYSTAF. CEF is similar to TREF but uses a much longer column because the solvent flow is not interrupted during the crystallization step. Rather, crystallization takes place under a steady flow of solvent to distribute the polymer populations of different crystallizabilities along the CEF column, as shown in Figure 2.16. The resulting spatial segregation of chains having different crystallizabilities reduces cocrystallization, simultaneously increasing peak resolution and decreasing analysis time. CEF analyses achieve resolutions that are similar to TREF in only 1–2 h, undoubtedly making CEF one of the most attractive analytical techniques for polyolefin CCD determination by crystallizability nowadays.

Figure 2.17 compares the TREF and CEF profiles of a trimodal olefin copolymer measured using a high crystallization rate of 2 °C min^{-1}. CEF peak resolution is much better than that of TREF's. The key to this better performance is the use of the long CEF column where the polymer chains of different crystallizabilities precipitate in separate locations, thus significantly reducing negative cocrystallization effects.

9) It needs to be remarked, however, that CRYSTAF resolution can be increased significantly by decreasing the crystallization rate. Lower crystallization rates increase analysis time, but considering that five samples can be analyzed simultaneously in a typical CRYSTAF instrument, this still gives ample advantage to CRYSTAF when compared to typical single-column TREF instruments.

2.3 Chemical Composition Distribution

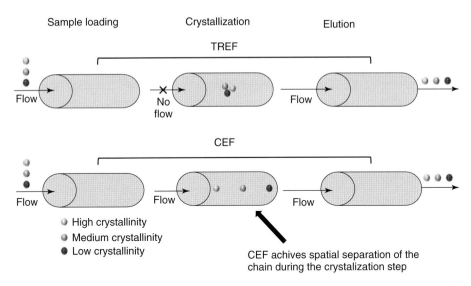

Figure 2.16 Crystallization and elution in TREF and CEF.

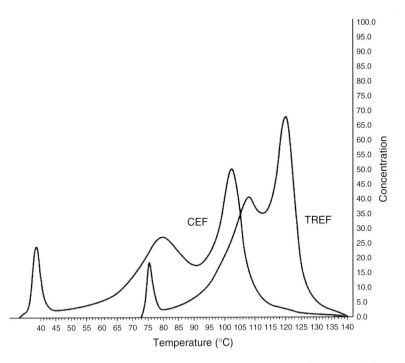

Figure 2.17 TREF and CEF comparison. Analysis conditions: crystallization and elution rates = 2 °C min^{-1}, elution flow rate = 0.2 ml min^{-1}, crystallization flow rate = 0.12 ml min^{-1} (CEF only).

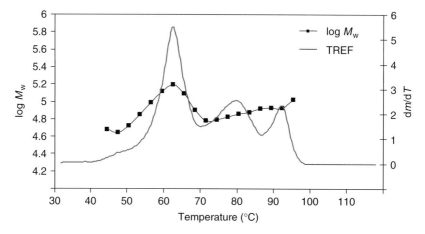

Figure 2.18 TREF profile of a trimodal polyolefin resin showing the correlation between M_w and elution temperature. An on-line laser light scattering detector was used to measure M_w.

In addition to the factors that affect TREF and CRYSTAF, CEF is also significantly influenced by crystallization and elution flow rates. CEF is still a very new technique, for which the operation conditions have not been completely optimized, but one rule seems to be clear: to enhance resolution, it is important to maximize the area in the column used during polymer crystallization. The crystallization flow rate and cooling rate must be matched so that the polymer chains with the highest crystallinities start precipitating as soon as they enter the column and the least crystalline ones precipitate just before they leave the column; this procedure ensures maximum spatial segregation of the chains in the CEF column.

Similar to GPC, multiple detectors may be used with TREF, CRYSTAF, and CEF. A dual length IR detector or an FTIR detector may be connected to these instruments to allow for the direct determination of the average chemical composition or stereoregularity of the polymer, thus eliminating the need for a calibration curve. Viscosity and LS detectors may also be used to estimate average molecular weights across the CCD, as shown in Figure 2.18 for a polyolefin with trimodal CCD. These hyphenated techniques provide a wealth of information that is required for the complete characterization of polyolefins.

In any of the three techniques discussed in this section, polymer chains are assumed to crystallize when the temperature reaches the T_c of its longest crystallizable segment which, in the case of ethylene/α-olefin copolymers, is supposed to be the longest ethylene sequence per chain. Several mathematical models have been proposed for CRYSTAF, TREF, and CEF using this principle. A detailed discussion of these models is beyond the scope of this book (a few simple examples are discussed in Chapter 6), but several references are added at the end of the chapter where these models are discussed in detail.

Finally, DSC has also been used to evaluate the CCDs of polyolefins. DSC principles are well known and are not reviewed here. This method suffers from some inherent limitations but also sheds light on microstructural aspects that

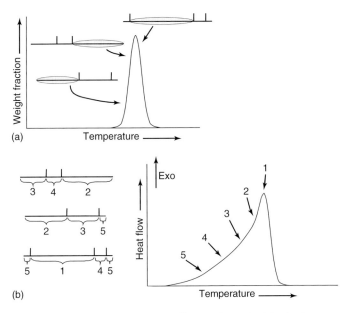

Figure 2.19 TREF/CRYSTAF/CEF profiles are determined by the longest crystallizable sequence in the chain (a), while DSC is sensitive to all crystallizable sequences (b).

might not be apparent in analytical results measured by CRYSTAF and other related techniques. Since most DSC analyses involve melting and crystallization of pure polymer samples, the influence of chain entanglement, crystallization kinetics, and cocrystallization is large due to the high polymer concentration in the melt. Therefore, the translation of the DSC endo- or exotherms into CCDs is not easy and may not be even possible to achieve. Several techniques involving the successive annealing of polymer populations at different temperatures have been developed to enhance DSC resolution, but, despite providing useful additional information on the polymer crystallinity distribution, they cannot be easily quantified.

In addition, DSC responses are proportional to the weight fraction of polymer melting or crystallizing at a given temperature *and* to their enthalpies of phase transition, which makes DSC increasingly less accurate for lower crystallinity fractions and completely insensitive to amorphous components. On the other hand, DSC can detect the crystallization of all crystallizable sequences in the polymer chain, while TREF, CRYSTAF, and CEF respond only (or mostly) to the longest crystallizable sequence per chain, as illustrated in Figure 2.19. DSC is, therefore, sensitive to another dimension of the polymer microstructure and, to a certain extent, more responsive to changes in comonomer SLD.

Chain entanglement and cocrystallization effects can be minimized if the DSC measurements are made from a dilute polymer solution instead of a polymer melt. A highly sensitive DSC instrument is required in this case, since the signal-to-noise ratio is low for diluted polymer solutions. Figure 2.20 shows CRYSTAF and solution DSC crystallization curves of four ethylene/1-hexene copolymers, where

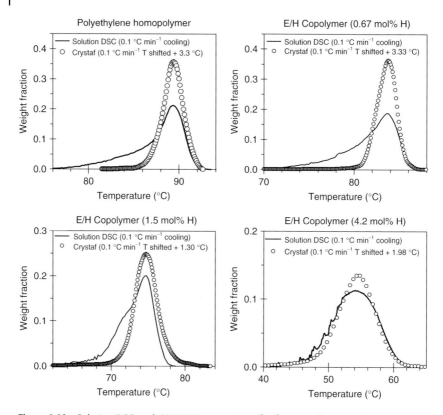

Figure 2.20 Solution DSC and CRYSTAF comparison for four metallocene polymers.

the polymer solution concentrations for the DSC experiments were kept as close as possible to those used in CRYSTAF to facilitate the comparison of the two techniques. The DSC curves are broader than the CRYSTAF profiles, indicating that shorter crystallizable sequences were detected by DSC, as expected from the discussion above. In addition, the curve shapes approximate each other as the 1-hexene content increases, as also expected, since the distribution of crystallizable sequences in the polymer becomes more uniform as the fraction of 1-hexene units increases.

Solution DSC is not a standard analytical technique for polyolefins as TREF, CRYSTAF, and CEF, but it is an interesting and relatively easy to run alternative method for polyolefin characterization that complements the information provided by these better established procedures.

2.3.2
High-Performance Liquid Chromatography

HPLC has been used to characterize polymers for many years, but its application to polyolefins is relatively recent due to the high temperatures required and the relatively narrow range of solvents available for their analysis. Since the polyolefin chains should be in solution to avoid crystallization effects (which would make the procedure similar to TREF), temperatures in excess of 100 °C are generally required. At these high temperatures, differences in interaction between polymer chains in the mobile phase and the stationary phase are expected to be minimal, limiting the applicability of HPLC to polyolefins. In a typical HPLC experiment, the polymer sample is first dissolved in a good solvent and injected into a mobile phase of low solvent strength, causing it to precipitate or adsorb onto the HPLC column packing. The quality of the solvent mixture (solvent mixtures with a higher quality have higher ratios of good solvent to poor solvent) in the mobile phase is then increased gradually, generally by mixing a good and a poor solvent in different proportions, redissolving the precipitated polymer, or desorbing the adsorbed polymer, into the mobile phase. One of the advantages of HPLC over techniques that rely on polymer crystallization is that it can be used to analyze the CCD of samples with low crystallinity (even amorphous samples such as the ethylene/propylene rubber component in impact polypropylene) that would be soluble at room temperature and appear as a soluble peak in TREF, CRYSTAF, and CEF. Cocrystallization effects are also not a concern, since no crystallization takes place during HPLC analysis.

For instance, TCB is a good solvent for polypropylene and polyethylene, while ethylene glycol monobutylether (EGMBE) is a good solvent for polypropylene and a nonsolvent for polyethylene. Therefore, ethylene/propylene copolymers can be fractionated by composition using a TCB/EGMBE gradient based on a precipitation/redissolution mechanism in a HPLC column packed with silica: propylene-rich chains are eluted at low TCB/EGMBE ratios; ethylene-rich fractions, at high TCB/EGMBE ratios.

The choice of column used for HPLC fractionation is critical. According to several recent investigations, Hypercarb® columns, made of porous graphitic carbon, are considered to be among the best choices for polyolefin fractionation by CCD based on the adsorption/desorption mechanism. A TCB/1-decanol mobile phase has been used to fractionate polypropylenes by tacticity and ethylene/α-olefin copolymers by composition using this type of column.

Because the solvent composition varies during the gradient HPLC analysis, RI and IR detectors commonly used with other polyolefin analytical techniques are not easily applicable. Instead, an evaporative light scattering detector (ELSD) is often used, which may be seen as a limitation of this technique, since the detector signal is nonlinear with respect to concentration and solvent composition. For the same reason, VISCs and LS detectors cannot be easily coupled to the HPLC instrument, which further reduces the amount of information that can be obtained in a single analysis.

Very recently, high-temperature temperature gradient interaction chromatography (TGIC) was proposed to eliminate most of these limitations. TGIC also uses Hypercarb columns as the stationary phase; however, instead of relying on a good-solvent/poor-solvent gradient, it uses a temperature gradient to promote the fractionation, but at temperatures high enough to prevent crystallization from happening. Since a constant composition solvent is used (isocratic solvent), all

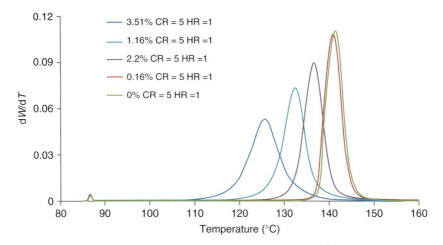

Figure 2.21 TGIC profiles for five metallocene ethylene/1-hexene copolymers. CR, cooling rate °C min^{-1}; HR, heating rate °C min^{-1}.

detectors commonly used with GPC, CRYSTAF, TREF, and CEF can be employed with TGIC.

TGIC can be run in instruments that are similar to TREF and CEF, with or without solvent flow during the adsorption step. The TGIC analysis consists of (i) introducing the polymer in solution at high temperature, (ii) adsorbing the sample onto the column support by reducing the temperature, either with (dynamic cooling, similar to CEF) or without (static cooling, similar to TREF) flow, and (iii) desorbing and eluting the sample under solvent flow. Dynamic cooling may be preferred because it minimizes the risk of column plugging and multilayer adsorption effects. From the limited results published thus far, it seems that molecular weight effects are not very important in TGIC analysis, but further experiments are required to clarify this point.

Figure 2.21 shows the TGIC profiles for five ethylene/1-hexene copolymers made with a metallocene catalyst. It is clear that the polymers can be fractionated according to their average comonomer fractions and that the CCDs can be estimated from the TGIC profiles using the calibration curve shown in Figure 2.22.

It is interesting to notice that the peak positions in TGIC are relatively close, likely due to the small interaction differences between the polyolefin and the stationary phase at high temperatures. As a consequence, the TGIC profiles of Ziegler–Natta LLDPE resins look different from their TREF or CEF profiles since the peak corresponding to the population with low α-olefin content is not so clearly separated from the intermediate crystallinity peak, as shown in Figure 2.23.

TGIC is a very new polyolefin characterization technique, and several factors that may affect its operation still need to be thoroughly investigated, but it may change the way polyolefins are analyzed in the future. CEF and TGIC are the most important contributions to CCD analysis made after the development of TREF in the 1980s and CRYSTAF in the 1990s.

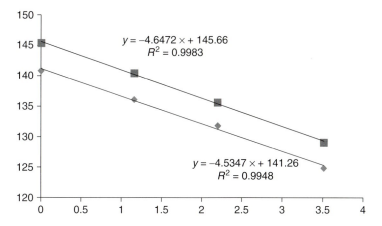

Figure 2.22 TGIC calibration curve for ethylene/1-hexene copolymers measured at two heating rates: 1 °C min^{-1} (lower curve) and 3 °C min^{-1} (upper curve).

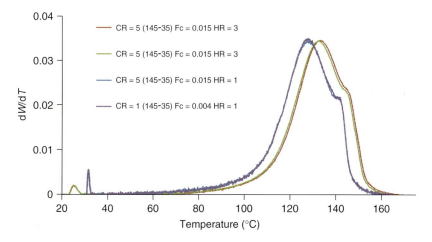

Figure 2.23 TGIC profiles for LLDPE resin made with a heterogeneous Ziegler–Natta catalyst. CR, cooling rate °C min^{-1}; HR, heating rate °C min^{-1}; Fc, flow rate during cooling stage ml min^{-1}.

2.4
Cross-Fractionation Techniques

In the previous sections, we discussed how an FTIR detector could be added to an SEC instrument to determine the average comonomer composition across the MWD or, alternatively, how an LS detector could be combined with CEF to measure M_w across the CCD. These multidetector setups provide a wealth of microstructural information but fall short of giving the complete description of the joint distribution of molecular weight and chemical composition (MWD × CCD) that is required to unequivocally characterize a polyolefin.

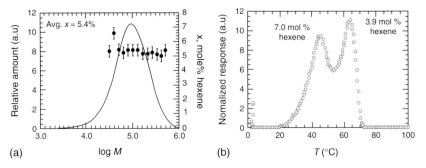

Figure 2.24 (a) SEC-FTIR results and (b) TREF profile for a blend of two metallocene copolymers with different 1-hexene fractions and same MWD.

An example of such a limitation is presented in Figure 2.24, where the SEC-FTIR results for a blend of two metallocene ethylene/1-hexene copolymers (50/50 wt%) with different 1-hexene fractions but same MWDs is depicted. By examining the plot in Figure 2.15a, we could justifiably argue that the polyolefin was made by a single-site catalyst in a well-stirred reactor, since the MWD is narrow, with polydispersity index close to 2.0, and the average 1-hexene fraction seems to be independent of molecular weight. The TREF profile shown in the same figure, however, tells a different story, showing that this material is, in reality, a blend of two copolymers with distinct 1-hexene incorporations. Since their MWDs superimpose, the on-line FTIR measures the average composition for the two polymer samples for each SEC molecular weight cut, $(3.9 + 7.0\%)/2 = 5.45\%$, giving the false impression that the sample was made with a single metallocene catalyst.

This phenomenon can be visualized more clearly in the simulated results shown in Figure 2.25. This plot was created using the Stockmayer bivariate distribution that is discussed at length in Chapter 6. For now, it suffices to say that it describes the joint MWD × CCD[10] of linear polymers made with metallocene catalysts. It is clear that the two distributions superimpose completely in the MWD plane but are seen as distinct populations in the CCD plane.

Even though the polymer blend discussed above can be considered an oddity, this example highlights how the use of different characterization techniques is fundamental to uniquely define the microstructures of polyolefins.

If the joint MWD × CCD is required, a CCD-sensitive technique must be combined with an MWD fractionation method in a single *cross-fractionation* instrument. Several alternative configurations have been used to separate polyolefins by molecular weight and chemical composition, such as TREF-SEC, SEC-TREF and, more recently, HPLC-SEC.

The first and still most widely used cross-fractionation instrument uses the TREF-SEC combination. Figure 2.26 shows the joint MWD × CCD for the resin

---

10) In Figure 2.25, the chain length distribution (CLD), measuring the number of monomer units (r) in the chain, is shown in place of the MWD. The CLD can be easily transformed into the MWD through multiplication by the average molar mass of the repeating unit and renormalization.

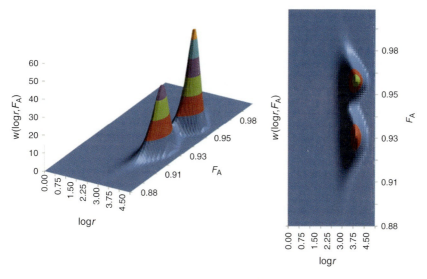

Figure 2.25 MWD × CCD simulation for the polymer blend shown in Figure 2.25 (r = chain length and F_A = fraction of ethylene in the copolymer).

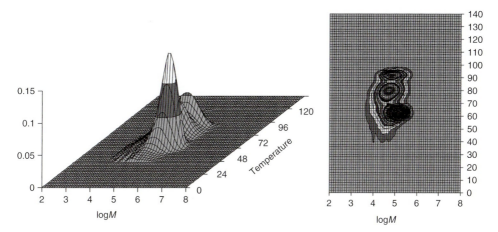

Figure 2.26 TREF-GPC cross-fractionation results for the polyolefin resin shown in Figure 2.18.

depicted in Figure 2.18. During the analysis, polymer fractions are collected by TREF in predetermined elution temperature intervals following a stepwise temperature increase program and then injected into the SEC columns to measure their MWDs. In effect, the joint MWD × CCD is cut across its CCD dimension by TREF before each narrow-CCD cut is sent for MWD determination by SEC. Figure 2.26 demonstrates more convincingly than any long discussion the importance of cross-fractionation for the proper understanding of polyolefins.

The SEC-TREF configuration reverses the cross-fractionation order discussed above, first separating the polymer molecules by molecular weight and then analyzing the narrow-MWD fractions by TREF. In principle, the joint MWD × CCD measured by SEC-TREF should coincide with that measured by TREF-SEC, but in practice, difference may arise because neither does SEC fractionate polymers purely on the basis of their molecular weights nor does TREF separate them based only on their CCD, as explained in the two previous sections in this chapter.

If TREF is chosen as the first fractionation step, cocrystallization may cause chains with different comonomer fractions to elute at the same temperature and be considered as having the same average comonomer fraction in the joint MWD × CCD. Polymer chains with very low molecular weight may also elute at temperatures that are lower than those expected for their actual comonomer contents. Therefore, TREF-SEC is not the best configuration for resins where significant cocrystallization may be expected or where a polymer fraction of very low molecular weight is present. The use of CEF, and perhaps TGIC, in place of TREF may ameliorate some of these practical limitations.

On the other hand, if SEC is chosen as the first separation step, microstructural characteristics, other than chain length, that affect the hydrodynamic volume of the chains (such as wide distribution of SCBs and presence of LCBs) may also reduce the resolution, since chains with same retention time may not have the same molecular weight. Adding additional detectors to the SEC instrument to detect SCB and LCB may correct some of these discrepancies but may also make the analytical procedure and data interpretation more complex.

Unfortunately, there are no systematic comparative studies between TREF-SEC and SEC-TREF in the literature, likely because of the time-consuming nature of these experiments, but we can be certain that neither has the configuration that solves all problems. New cross-fractionation developments involving CEF-SEC and TGIC-SEC (in any order) may shed some more light on this interesting subject in the near future.

2.5
Long-Chain Branching

We have already seen in Section 2.2.1 how SEC coupled with either a VISC or an LS detector can be used to estimate the LCB frequency of polyolefins. Despite being a powerful technique, this method relies on the Zimm–Stockmayer equations and requires that assumptions be made regarding LCB topology (for instance, trifunctional vs tetrafunctional branching points, and random vs comb branching).

Another method frequently used to quantify LCB in polyolefins is ^{13}C NMR. In principle, ^{13}C NMR can provide an absolute LCB measurement, but is not without limitations. First, ^{13}C NMR cannot distinguish between alkyl branches longer than hexyl; therefore, branches with six or more carbons (C_6^+) are counted as LCBs by ^{13}C NMR. Second, owing to the relatively low LCB frequency of most

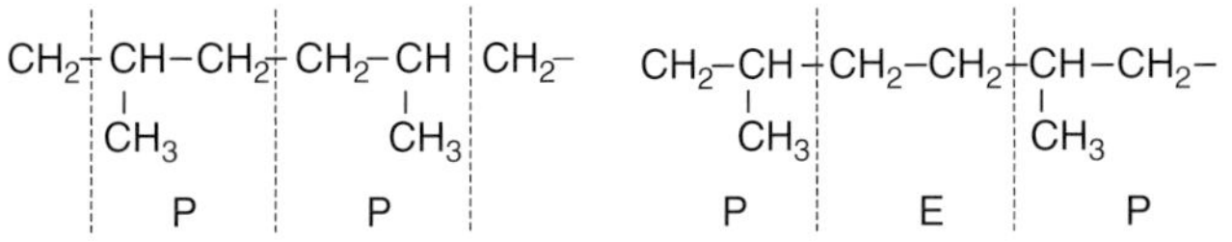

CH$_2$–CH–CH$_2$–CH$_2$–CH–CH$_2$–   CH$_2$–CH–CH$_2$–CH$_2$–CH–CH$_2$–
　　|　　　　　　|　　　　　　　　　　|　　　　　　　　|
　　CH$_3$　　　　CH$_3$　　　　　　　 CH$_3$　　　　　　CH$_3$

　　P　　　　　　P　　　　　　　　　　P　　　　　E　　　P

Figure 2.27 2-1 Propylene/1-2 propylene (PP) and 1-2 propylene/ethylene/2-1 propylene (PEP) sequences form the same chain structural unit and cannot be distinguished by ^{13}C NMR analysis.

polyolefins made with coordination catalysts, the peak signal-to-noise ratio may be low, increasing the uncertainty of the LCB frequency estimate.

The second limitation is gradually becoming less relevant as more powerful NMR spectrometers become available and as better analytical techniques for LCB determination are developed. Currently, LCB in polyethylene can be detected at levels of 1 LCB per 10 000 carbon atoms.

The first limitation is more interesting because it involves two important concepts: (i) How long is an LCB? and (ii) Can a mathematical model help unravel the C_6^+ signal measured by ^{13}C NMR? Before we attempt to answer these questions, we first review how LCBs, or C_6^+ branches, are quantified by ^{13}C NMR.

In contrast to polymers having functional groups, the monomer identity is lost in polyolefins. A polyethylene is just a long methylene group chain. In ethylene/α-olefin copolymers where α-olefin inversions may occur, the sequence descriptions may not be unique. For instance, in ethylene/propylene copolymers, a propylene-propylene sequence cannot be distinguished from a propylene-ethylene-propylene sequence (Figure 2.27).

It is up to the analyst to translate a given carbon resonance identification to the precise monomer sequence in which it resides, on the basis of the analyst's knowledge of the polymerization process. Therefore, in addition to sequence identification (for copolymers), a carbon atom nomenclature is needed. An efficient nomenclature system for polyolefins was proposed by Carman and Randall. Methylene (CH$_2$) carbons located along the backbone of an ethylene/α-olefin copolymer chain are identified by a pair of Greek letters, indicating the location of the nearest methine (CH) carbons in either direction. The Greek letter α indicates that a methine carbon is bonded to a methylene carbon of interest. Two Greek letters, $\alpha\alpha$, indicate that the identified methylene carbon is sandwiched between two methine carbons. The Greek letter β indicates that a methine carbon is two carbons away from the carbon of interest, and so forth. Since neighboring carbon contributions to chemical shifts seldom exceed four carbons, a methine carbon four or more carbons away from the methylene carbon of interest is indicated by δ^+. This terminology is illustrated in Figure 2.28.

LCBs can be clearly identified by ^{13}C NMR in LDPE prepared by free radical polymerization in high pressure reactors because these materials are highly branched. The measurement of LCBs in polyolefins made with coordination catalysts is more elusive because of their relatively low frequency, but it has been also demonstrated in several publications. As described in more detail in Chapter 6, the type of LCB in polyethylenes made with coordination catalysts is called a *"Y" type*, which gives

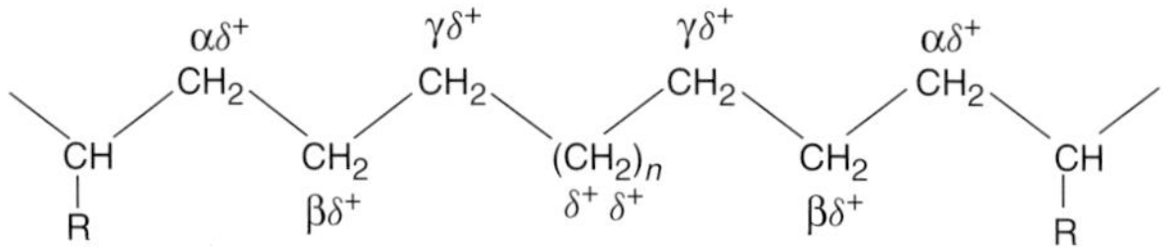

2 Polyolefin Microstructural Characterization

Figure 2.28 ^{13}C NMR nomenclature convention for methylene carbon atoms.

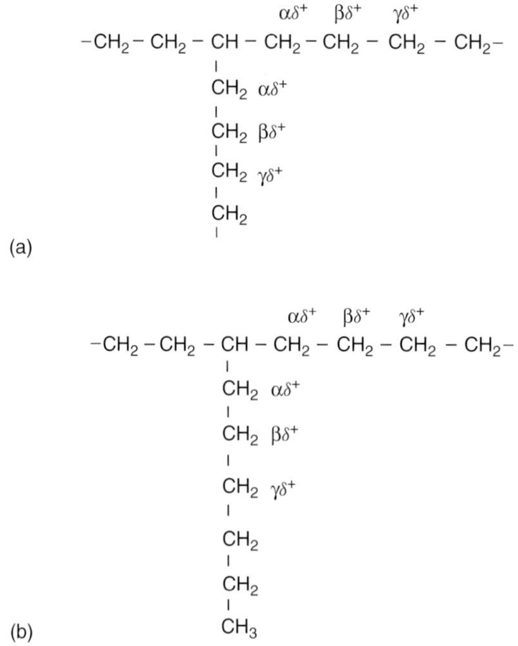

(a)

$$-CH_2-CH_2-\underset{|}{CH}-\overset{\alpha\delta^+}{CH_2}-\overset{\beta\delta^+}{CH_2}-\overset{\gamma\delta^+}{CH_2}-CH_2-$$

with side chain: $CH_2\,\alpha\delta^+$ — $CH_2\,\beta\delta^+$ — $CH_2\,\gamma\delta^+$ — CH_2

(b)

$$-CH_2-CH_2-\underset{|}{CH}-\overset{\alpha\delta^+}{CH_2}-\overset{\beta\delta^+}{CH_2}-\overset{\gamma\delta^+}{CH_2}-CH_2-$$

with side chain: $CH_2\,\alpha\delta^+$ — $CH_2\,\beta\delta^+$ — $CH_2\,\gamma\delta^+$ — CH_2 — CH_2 — CH_3

Figure 2.29 C_6^+ branches in polyethylene: (a) Y LCB in polyethylene and (b) ethylene/octene structural sequence.

rise to methine, $\alpha\delta^+$, $\beta\delta^+$, and $\gamma\delta^+$ carbons, corresponding to those observed in ethylene/1-octene copolymers, as depicted in Figure 2.29.

Since NMR cannot distinguish among branches longer then C_6, ethylene/1-octene copolymers serve as good models for LCB in polyolefins. Detection of these resonances in polyethylene where no 1-octene or longer α-olefins were added during the polymerization is a strong indication of the presence of LCBs.

For polyethylenes made with coordination catalysts, generally methyl and vinyl chain ends are dominant, as illustrated in Figure 2.30.

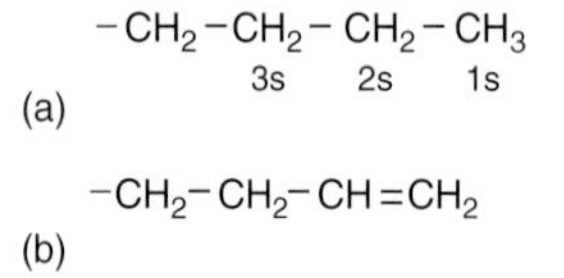

(a) $-CH_2-CH_2-CH_2-CH_3$
       3s   2s   1s

(b) $-CH_2-CH_2-CH=CH_2$

Figure 2.30 Chain ends in polyethylene: (a) saturated chain end and (b) allylic chain end.

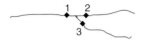

Figure 2.31 Schematic LCB in a polyethylene molecule.

Using these definitions, we can calculate the average number of LCB per polymer molecule, B_n, using the information available in ^{13}C NMR spectrum, as explained below.

First, assume that α is the intensity of the three carbon atoms in the schematic LCB shown in Figure 2.31. The carbon atoms at locations 1, 2, and 3 are equivalent and have the same peak positions and intensities in ^{13}C NMR. Since only one of them is in the LCB, while the other two are in the main chain, the number of LCBs in the sample is proportional to

$$n_{LCB} \propto \frac{\alpha}{3} \tag{2.30}$$

The number of moles of polymer chains is equal to half the number of main chain ends (excluding the LCB chain ends) in the sample. The total number of chain ends is proportional to the sum of the intensities of allylic (a) and saturated (s) end group carbons

$$n_{\text{total chain ends}} \propto s + a \tag{2.31}$$

Therefore, the number of main chain ends (excluding LCB chain ends) is proportional to

$$n_{\text{main chain ends}} \propto s + a - \frac{\alpha}{3} \tag{2.32}$$

Consequently, the number of polymer chains in the sample is given by,

$$n_{\text{chains}} \propto \frac{1}{2}\left(s + a - \frac{\alpha}{3}\right) \tag{2.33}$$

The average number of LCB per chain can now be calculated by the ratio $n_{LCB}/n_{\text{chains}}$

$$B_n = \frac{\frac{\alpha}{3}}{1/2\left(s + a - \frac{\alpha}{3}\right)} = \frac{2\alpha}{3(s+a) - \alpha} \tag{2.34}$$

Frequently, LCB frequencies are expressed as the number of LCBs per 1000 carbon atoms (λ). This average can be also determined from ^{13}C NMR intensities using the simple equation

$$\lambda = \frac{\frac{\alpha}{3}}{I_t} \times 1000 \tag{2.35}$$

where I_t is the ^{13}C NMR intensity for all the carbons in the polymer sample.

Figure 2.32 shows a typical ^{13}C NMR spectrum of a polyethylene made with a single-site catalyst, where the LCB peak at about 38.24 ppm is indicated.

We can now return to the two questions asked before this brief description on how to quantify branching points in polyethylene by ^{13}C NMR. As explained above,

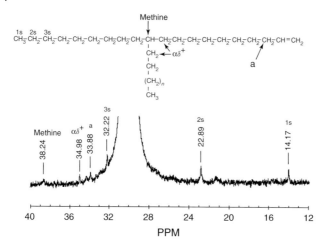

Figure 2.32 ^{13}C NMR spectrum of a polyethylene containing LCBs made with a single-site catalyst.

all C_6^+ branches will be counted as LCBs in Eqs. (2.34) and (2.35), but how long is an LCB? Some of the answers suggested for this question include the following:

1) An LCB has length comparable to the main chain.[11]
2) The minimum length of an LCB is equal to the critical entanglement molecular weight which, for polyethylene, has been reported in the range 2100–5200 (150–370 carbon atoms).
3) The length of an LCB is at least twice the molecular weight between entanglements, about 180 carbon atoms for polyethylene.

Neither of these answers is entirely satisfactory. This question does not have a clear answer because it is an ill-posed question. A better question is how long does an LCB need to be to affect the rheological behavior of a polyolefin? In this case, rheological measurements that depend on the critical entanglement molecular weight and average length between entanglement points can provide a more precise answer, as suggested in (2) and (3) above.

From a chain topology point of view, another way to answer this question is to consider how SCBs and LCBs are formed in ethylene/α-olefin copolymers made with coordination catalysts: SCBs are made by the incorporation of α-olefin comonomer added to the reactor with the specific intention of lowering its density and melting temperature, but LCBs are formed via a mechanism of macromonomer

---

11) This seems like a good answer, until we realize that in a branched material, it is not easy, and often not even possible, to determine which chain is the main chain, unless we are dealing with simple comb structures. In addition, the "main" chain will also have a wide distribution of sizes.

incorporation that is discussed at length in Chapters 3 and 6. These LCBs have a length distribution that is the same as the distributions for the "main chains" and therefore can be considered LCBs according to answer (1) proposed above. In fact, the LCB size distribution can be described precisely with a relatively simple model that is explained in Chapter 6. This model can also be used to calculate the fraction of LCBs that are above the entanglement molecular weight and find out how many branches would be considered LCBs according to definition (2), for instance.

Finally, it needs to be mentioned that ^{13}C NMR can only measure average branching frequencies. In this regard, SEC analysis coupled with a VISC or an LS detector may give more information on LCB topology than ^{13}C NMR, since it permits the determination of the LCB frequency as a function of molecular weight.[12]

Because even a very small fraction of LCBs can have a very large impact on the zero-shear viscosity and shear thinning of polyolefins, many rheological methods have been developed to try to quantify LCB in polyolefins. These methods are very sensitive to the presence of LCBs but are not without drawbacks either. The rheological behavior of polyolefins is outside the scope of this book.

Further Reading

There is a vast literature on room-temperature and high-temperature SEC dating back to 1960s. An excellent recent book was published by Striegel, A.M., Yau, W.W., Kirkland, J.J., and Bly, D.D (2009) *Modern Size-Exclusion Liquid Chromatography: Practice of Gel Permeation and Gel Filtration Chromatography*, 2nd edn, John Wiley & Sons, Inc., Hoboken, which includes many aspects not covered in this chapter, such as column resolution and band broadening, as well as the effect of several other experimental variables and techniques.

TREF and CRYSTAF have also been the subject of several reviews.

Wild, L. (1990) *Adv. Polym. Sci.*, **98**, 1.

Monrabal, B. (1994) *J. Appl. Polym. Sci.*, **52**, 491.

Anantawaraskul, S., Soares, J.B.P., and Wood-Adams, P.M (2005) *Adv. Polym. Sci.*, **182**, 1.

Soares, J.B.P. and Anantawaraskul, S. (2005) *J. Polym. Sci., Part B: Polym. Chem.*, **43**, 1557.

CEF, a much more recent analytical technique, has not been the subject of many scientific publications yet (Monrabal, B., Sancho-Tello, J., Mayo, N., and Romero, L. (2007) *Macromol. Symp.*, **257**, 71; Al-ghyamah, A.A. and Soares, J.B.P. (2011) Crystallization elution fractionation of polyolefins made with metallocene catalysts. *Macromol. Symp.*, in print), but this situation is bound to change very soon because of its significant advantages over TREF and CRYSTAF.

Excellent review articles have been written on high-temperature liquid chromatography techniques for polyolefin analysis recently.

Macko, T., Brüll, R., Zhu, Y., and Wang, Y. (2010) *J. Sep. Sci.*, **33**, 3446.

Macko, T., Brüll, R., Alamo, R.G., Thomann, Y., and Grumel, (2009) *Polymer*, **50**, 5443.

Pasch, H., Malik, M.I., and Macko, T (2011) High-temperature fractionation of polyolefins. *Adv. Polym. Sci.*, in print.

12) Experimental limitations in both techniques make absolute comparisons difficult to make in this case. It is always advisable to use both techniques for a better description of the LCB frequencies in the polymer. SEC-VISC, SEC-MALLS, and ^{13}C NMR are better seen as complementary, not alternative, techniques.

This class of techniques holds the promise for faster polyolefin analyses that are free of cocrystallization effects and can be extended to low-crystallinity or amorphous samples but, as discussed above, are still in its infancy for polyolefins. Among them, high-temperature thermal gradient interaction chromatography (HT-TGIC) has attracted considerable interest due to its simplicity, since it uses a single solvent, allowing the use of conventional mass detectors such as refractometers and IR detectors.

Cong, R., deGroot, W., Parrott, A., Yau, W., Hazlitt, L., Brown, R., Miller, M., and Zhou, Z. (2011) *Macromolecules*, **44**, 3062.

There is a vast literature on the use of DSC for polyolefin analysis, but only a few papers follow the solution DSC approach described here for chemical composition determination.

Sarzotti, D.M., Soares, J.B.P., Simon, L.C., and Britto, J.D. (2004) *Polymer*, **45**, 4787.

TREF-SEC cross-fractionation has been practiced for a couple of decades, generally in home-made equipment that were relatively hard to operate and require long analysis time. More recently, a commercial TREF-SEC unit has been available, considerably simplifying the use of this very useful instrument.

Ortin, A., Monrabal, B., and Sancho-Tello, J (2007) *Macromol. Symp.*, **257**, 13.

The reverse fractionation, SEC-TREF, has also been discussed in the literature.

Yau, W. (2007) *Macromol. Symp.*, **257**, 29.

Finally, an extensive overview of hyphenated analytical techniques for polymers has been published by Pasch, H. (2000) *Adv. Polym. Sci.*, **157**, 1.

3
Polymerization Catalysis and Mechanism

> [..] from so simple a beginning endless forms most beautiful and most wonderful have been, and are being, evolved.
> *Charles Darwin (1809–1892)*

> The historian of science may be tempted to exclaim that when paradigms change, the world itself changes with them.
> *Thomas Khun (1922–1996)*

3.1
Introduction

The development of coordination catalysts for olefin polymerization started in the early 1950s with the discovery of Ziegler–Natta and Phillips catalysts and continues to these days with the refinement of the original systems and the discovery of early and late transition metal single-site catalysts. This area has had a rich history of gradual improvements on existing catalysts, punctuated by discoveries that revolutionized the polyolefin manufacturing industry, as illustrated in the approximate timeline shown in Figure 3.1.

Each new breakthrough discovery created new opportunities for polyolefins without eliminating the catalysts and processes that preceded them, a true testimony to the resilience and vitality of the olefin polymerization technologies: the processes for making low-density polyethylene (LDPE) by free radical polymerization were not rendered obsolete neither when Ziegler and Natta discovered that $TiCl_3$ complexes could be used to make high-density polyethylene (HDPE), linear low-density polyethylene (LLDPE), and isotactic polypropylene at mild polymerization conditions nor when Hogan and Banks developed their remarkable chromium oxide catalyst to produce HDPE; similarly, the advent of single-site early (metallocene) and late transition metal catalysts did not herald the demise of Ziegler–Natta or Phillips catalysts – rather, new high-performance polyolefins became available with the development of these catalysts. Instead of substituting the previous technologies, each new catalyst family allows the manufacture of polyolefins with distinct microstructural characteristics that are adequate for different applications.

Polyolefin Reaction Engineering, First Edition. João B. P. Soares and Timothy F. L. McKenna.
© 2012 Wiley-VCH Verlag GmbH & Co. KGaA. Published 2012 by Wiley-VCH Verlag GmbH & Co. KGaA.

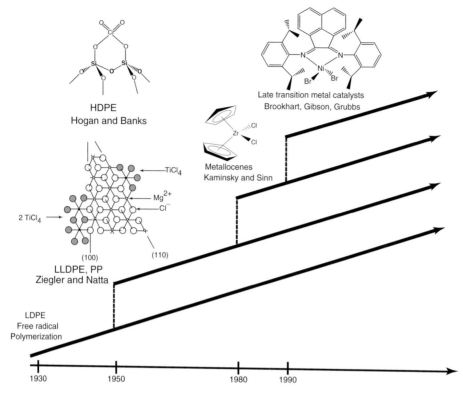

Figure 3.1 Evolution of coordination catalysts for olefin polymerization. The arrows indicate incremental improvements of catalysts and polymerization processes, while the vertical dashed lines are "discontinuities" (paradigm shifts) caused by main discoveries in the field.

For instance, an LLDPE made with a Ziegler–Natta catalyst differs substantially from the one synthesized by a metallocene; they share an acronym and an average density range, but little more than that. The versatility of the polyolefin industry resides in the ability to make polymers with widely varying mechanical and rheological properties from simple monomers, due largely to the different types of catalysts used in their manufacture.

The four major olefin polymerization catalyst families are depicted in Figure 3.1: Ziegler–Natta, Phillips, metallocene, and late transition metal catalysts. The first three are used commercially, while late transition metal catalysts still have to find industrial applications. The main characteristics of these catalysts are listed in Table 3.1.

Phillips and Ziegler–Natta complexes were the first coordination catalysts used for olefin polymerization. Both catalysts created a revolution in the plastic industry when they were discovered in the early 1950s and are, to this day, the workhorses of the polyolefin industry. They make polymers with nonuniform microstructures, characterized by broad and sometimes bimodal molecular weight distributions (MWDs) and CCDs. Heterogeneous Ziegler–Natta and Phillips catalysts have

Table 3.1 Main characteristics of coordination catalysts for olefin polymerization.

Type	Physical state	Examples[a]	Polymer type
Ziegler–Natta	Heterogeneous	$TiCl_3$, $TiCl_4/MgCl_2$	Nonuniform
	Homogeneous	VCl_4, $VOCl_3$	Uniform
Phillips	Heterogeneous	CrO_3/SiO_2	Nonuniform
Metallocene	Homogeneous	Cp_2ZrCl_2	Uniform
	Heterogeneous	Cp_2ZrCl_2/SiO_2	Uniform
Late transition metal	Homogeneous	Ni, Pd, Co, Fe with diimine and other ligands	Uniform

[a]There is a huge variety of coordination catalysts for polymerization, especially metallocene and late transition metal complexes. The examples provided in this table simply illustrate some of the most common types.

more than one type of active site, each making polyolefin populations with different average properties. Consequently, the polymer made with these catalysts can be considered a blend of polymers with different M_n, M_w, average comonomer content, and regioregularity and stereoregularity. There is no consensus regarding the number of site types on these catalysts, although some numerical techniques, such as MWD and CCD deconvolution, are routinely used to estimate it, as detailed in Chapter 6; also, there is no general agreement on what makes them behave differently, even though several hypotheses, such as differences in oxidation states, coordination with cocatalysts and/or donors, and transition metal clustering degree, have been proposed. Ziegler–Natta catalysts can also be homogeneous (soluble in the reaction medium). Contrary to their heterogeneous counterparts, soluble vanadium-based Ziegler–Natta catalysts may have only one site type, making polyolefins with more uniform properties.

Two families of single-site catalysts are also included in Table 3.1: metallocene and late transition metal catalysts. They make polyolefins with narrow MWD and CCD that can be described, ideally, with Flory and Stockmayer distributions. These microstructural distributions are the subject of much discussion in Chapter 6, but it suffices to say now that they obey a single set of polymerization kinetic constants or probabilities. These catalysts are usually soluble in the reaction medium but can be supported onto organic or inorganic carriers. This is commonly done for metallocenes to permit their use in industrial reactors designed to operate with heterogeneous Ziegler–Natta or Phillips catalysts. Similar procedures can be followed to anchor late transition metal catalysts onto a variety of supports. When metallocene catalysts are supported, their polymerization behavior may be affected – notably, the polymer yield may be reduced, the polymer molecular weight averages may increase, the MWD and CCD may become slightly broader, and the catalyst stereoselectivity and regioselectivity may change – but they are still considered single-site catalysts because they produce polymers with properties that are considerably more uniform than those made with Phillips or heterogeneous Ziegler–Natta catalysts.

Most of the research activities on catalyst development aims at achieving either better polymerization activity or enhanced polymer microstructure control (and preferably both), but other factors are also important for olefin polymerization catalysts. What then are the requirements of a good olefin polymerization catalyst?

1) **High stability and productivity**: Catalyst activity must be high enough to permit the production of several kilograms of polymer per gram of catalyst per hour; typical activities for polyethylene and polypropylene modern industrial catalysts vary from 10 to 100(kg polymer) · (g catalyst · h)$^{-1}$.

2) **Good polymer microstructural control**: Ziegler–Natta and Phillips catalysts, because of their multiple-site nature, fare less well than metallocene and late transition metal catalysts in this respect. There is a wealth of information (published and confidential) on how to enhance microstructural control by catalyst design, donor usage, support technology, and polymerization conditions.

3) **Production of polymer with good processability**: Since polyolefins must be processed at high throughput rates, their flowability in extruders and other polymer melt-processing equipment has a huge impact on how successful a polyolefin grade will become. Polyolefin processability is commonly controlled by changing the breadth of the MWD or the average long-chain branching frequency of the polymer.

4) **Production of polymer with controlled particle size distribution, morphology, and high density**: These are requirements that influence reactor operation and postreactor processing. Polymer particles with relatively narrow size distribution and high density are required to minimize reactor fouling and to facilitate the transport from the reactor through monomer/diluent recovery sections, and, finally, to pelletizing extruders or polymer powder storage.

In this chapter, we discuss how catalyst selection influences the factors mentioned in items (1) and (2). Microstructural control is discussed at length in Chapters 5 and 6, while properties related to particle morphology are the subject of Chapter 7. The important topic of polyolefin processability falls outside the scope of this book.

3.2
Catalyst Types

3.2.1
Ziegler–Natta Catalysts

In its broadest definition, Ziegler–Natta catalysts are composed of a transition metal salt of metals from groups IV to VII[1] and a metal alkyl of a base metal from

---

1) Rigorously speaking, these metal salts, such and TiCl$_3$ and TiCl$_4$, should be called *precatalysts*, since they are not active for polymerization until reacted with a *cocatalyst*. For convenience, however, they are often simply called *catalysts*. This is the convention we will adopt in this book.

groups I to III, known as the *cocatalyst or activator*. The cocatalyst is required to form the active species in the catalyst in a two-step process involving alkylation and reduction of the transition metal centers. The preferred cocatalysts are alkyl aluminum compounds such as trimethyl aluminum (TMA), triethyl aluminum (TEA), and diethyl aluminum chloride (DEAC). Ziegler–Natta catalysts can be heterogeneous or homogeneous.

Even though the above definition is rather broad, not all combinations are equally efficient. The most common type of heterogeneous Ziegler–Natta catalyst is $TiCl_4$ supported on $MgCl_2$ or SiO_2. This description, however, is deceptive in its simplicity: there are various ways to synthesize these catalysts to guarantee high activity, good control of molecular weight, comonomer incorporation, stereoselectivity and regioselectivity, and adequate polymer particle morphology, as attested by the vast scientific literature and numerous patents on Ziegler–Natta catalysis.

Homogeneous Ziegler–Natta catalysts are generally (but not exclusively) vanadium based and are used to produce ethylene-propylene-diene (EPDM) elastomers. They make polymers with uniform microstructures: narrow MWD and CCD, and polydispersitys (PDIs) close to the theoretical value of 2.0. Vanadium-based catalysts used in solution processes with short average reactor residence times (in the order of a few minutes) at relatively high temperatures (above 140 °C) generally have high deactivation rates. Homogeneous Ziegler–Natta catalysts are an attractive choice for EPDM manufacture because they do not make polymers with the high crystallinity peak characteristic of heterogeneous Ziegler–Natta catalysts. (The high crystallinity fraction has a negative impact on the properties of most elastomers.) Despite its economic importance in the rubber industry, homogeneous vanadium Ziegler–Natta catalysts are not covered further in this book.

The history of the development of heterogeneous Ziegler–Natta catalysts is fascinating. Perhaps no system is more illustrative of this progress than catalysts for isotactic polypropylene production. Table 3.2 classifies the stages of this development into four generations, although divisions into more generations have also been proposed. In our opinion, the four-generation classification suffices because it breaks the catalyst development stages into four main discoveries: the use of $TiCl_3$ activated with DEAC (first generation), the modification of $TiCl_3$ with donors to increase catalyst stereoselectivity (second generation), the development of $MgCl_2$ as a support for $TiCl_3$ and the adoption of internal and external donors for increased stereoselectivity (third generation), and, finally, the development of catalysts (mainly for liquid propylene polymerization processes) with precisely controlled morphology that do not require pelletization because they produce spherical polymer particles with diameters in the range of a few millimeters (fourth generation).

The original $TiCl_3$ catalyst had relatively low activity and poor stereoselectivity, requiring the removal of atactic polypropylene and catalyst residues (deashing) from the isotactic polypropylene product. Several postreactor unit operations were needed to obtain polypropylene with the required isotacticity index and acceptable catalyst residue content. Polypropylenes made with modern heterogeneous Ziegler–Natta

Table 3.2 Evolution of heterogeneous Ziegler–Natta catalysts for propylene polymerization.

Generation	Catalyst	Yield (kg PP per g catalyst)	Isotacticity index[a]	Process steps
First	δ-TiCl$_3$/DEAC	2–4	90–94%	Deashing and atactic polypropylene extraction
Second	δ-TiCl$_3$/isoamylether/ AlCl$_3$/DEAC	10–15	94–97	Deashing
Third	MgCl$_2$/ester/TiCl$_4$/TEA/ ester	15–30	95–97	No purification
Fourth	MgCl$_2$/ester/TiCl$_4$/TEA/ PhSi(OEt)$_3$	>100[b]	>98	No purification, no extrusion/ pelletization

[a] Measured as the weight percentage of polymer soluble in boiling heptanes.
[b] In liquid (bulk) polypropylene.

catalysts, on the other hand, exit the polymerization reactor with an insignificant amount of catalyst residues and negligible atactic polypropylene fraction because they are made with catalysts that have very high activities and isospecificity. For this reason, they do not require postreactor purification steps. Some catalysts, such as the ones used in the Spheripol process (Chapter 4), even produce large spherical polypropylene particles that do not require pelletization. As a consequence of these advances in catalyst technology, modern polypropylene and polyethylene manufacturing processes are very streamlined; they basically consist of one or more reactors in series, recovery systems for diluent (in the case of slurry and solution but not for liquid propylene and gas-phase processes) and unreacted monomers, and extruders for making polyolefin pellets.

These advances were possible because of two main findings: the discovery of MgCl$_2$ as an ideal support for TiCl$_4$ (TiCl$_4$ and MgCl$_2$ form a mixed crystal where the TiCl$_4$ active sites are easily accessible to the monomer) and the use of electron donors, such as ethers and esters, that selectively poison or modify the specific sites responsible for the formation of atactic polypropylene.

The first and second generation Ziegler–Natta catalysts were composed of crystalline TiCl$_3$ in four different geometries: hexagonal (α), fibrous (β), cubic (γ), and a mixed hexagonal-cubic form (δ). Three of these forms, α, γ, and δ, have high stereoselectivities. The δ-TiCl$_3$ complex is the most active for propylene polymerization and is obtained as porous particles with diameters varying from 20 to 40 μm (*secondary particles*), resulting from the agglomeration of even smaller TiCl$_3$ particles with diameters in the range 0.03–0.04 μm (*primary particles*). The controlled fragmentation of these catalyst particles during polymerization was one of the major challenges during the development of effective heterogeneous Ziegler–Natta catalysts. The morphology of δ-TiCl$_3$ particles had a pronounced effect on the development of single-particle mathematical models used to explain

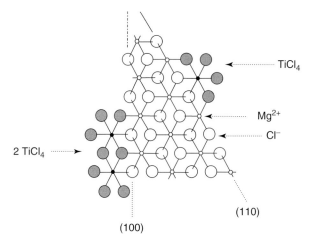

Figure 3.2 Structure of a generic TiCl$_4$/MgCl$_2$ Ziegler–Natta catalyst.

intraparticle mass and heat transfer resistances during olefin polymerization, particularly on the well-established *multigrain model*, as discussed in Chapter 7.

The use of electron donors (Lewis bases) during polymerization increased the stereoselectivity and productivity of TiCl$_3$, leading to the second generation Ziegler–Natta catalysts. Catalysts belonging to these two first generations were formed by crystalline TiCl$_3$, with most of the potentially active sites located inside the catalyst crystals where they could not promote polymerization because they were inaccessible to the monomer molecules; not surprisingly, their productivities were relatively low, and a postreactor deashing step was required to remove the catalyst residues imbedded in the polymer particles. The lower stereoselectivities of first and some second generation catalysts also demanded a postreactor step to extract the atactic polypropylene fraction from the final product.

The elimination of these two shortcomings was among the main driving forces behind the development of the third generation Ziegler–Natta catalysts, which came about when TiCl$_4$ was supported on porous MgCl$_2$ particles, as schematically represented in Figure 3.2. Third generation TiCl$_4$/MgCl$_2$ catalysts have very high activities and stereoselectivities and do not require the expensive and cumbersome postreactor steps for atactic polypropylene removal and catalyst residue deashing. Further improvements on catalyst particle morphology control (with the development of larger catalyst particles that form spherical polyolefin particles that do not require pelletization) and donor technology led to the fourth generation Ziegler–Natta catalysts.

Donors are classified as *internal* or *external* donors. Internal donors are added to the system during the catalyst-preparation step, while external donors are only introduced into the polymerization reactor. Electron donors are supposed to control the TiCl$_4$ distribution on the (100) and (110) faces of the MgCl$_2$ surface, as illustrated in Figure 3.3. Ti$_2$Cl$_8$ species coordinate with the (100) faces through dinuclear bonds to form the isospecific polymerization sites, while the electron

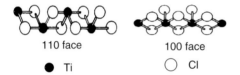

Figure 3.3 Lateral faces of a TiCl$_4$/MgCl$_2$ Ziegler–Natta catalyst.

donor molecules tend to coordinate with the nonstereospecific (more acidic) sites on the (110) faces. When aromatic monoesters and diesters are used as internal donors, the addition of the alkyl aluminum cocatalyst causes the partial removal of the internal donors; therefore, the use of external donors is essential to maintain the high stereoselectivity level of these catalysts.

Table 3.3 lists the main steps in the development of electron donors for Ziegler–Natta catalysts. Initially, aromatic monoesters, such as ethyl benzoate (EB), were used as internal donors to increase the isotacticity index from 40 to 60%. Later on, in addition to their use as internal donors, aromatic monoesters were also used as external donors, increasing the isotacticity index to 95%. Furthermore, the isotacticity index was increased to 97–99% with the use of aromatic diesters as internal donors (di-iso-butylphthalate, DIBP), and silanes as external donors (n-propyltrimethoxysilane, NPTMS). Later studies showed that very high isotacticity indices of 97–99% could be obtained in the absence of external donors when 1,3-diethers were used as internal donors, presumably because they coordinate strongly with the (110) faces and cannot be removed by the addition of the alkyl aluminum cocatalyst.

A typical TiCl$_4$/MgCl$_2$ catalyst for olefin polymerization is prepared in four main steps: digestion, activation, washing, and drying. The digestion step includes the reaction of an organomagnesium (MgOR) compound, TiCl$_4$, and an internal electron donor (if the catalyst is to be used for propylene polymerization, but not necessarily for ethylene polymerization where stereoselectivity is not a concern) in a chlorinated organic solvent. In this step, the active TiCl$_4$ is dispersed in the precursor porous surface, forming the MgCl$_2$ crystal and TiCl$_3 \cdot$ OR. The latter is removed by further addition of TiCl$_4$ and solvent in the activation step. Then, the formed catalyst is washed using a volatile organic compound in the washing step. Finally, the catalyst is obtained as a free-flowing powder after the volatile organic

Table 3.3 Summary of electron donor development of propylene polymerization catalysts.

Internal donor	External donor	Isotactic index (%)
Aromatic monoesters (EB)	–	60
Aromatic monoesters (EB)	Aromatic monoesters (methyl p-toluate)	95
Aromatic diesters (DIBP)	Silanes (NPTMS)	97–99
Diethers (1,3-diether)	–	97–99

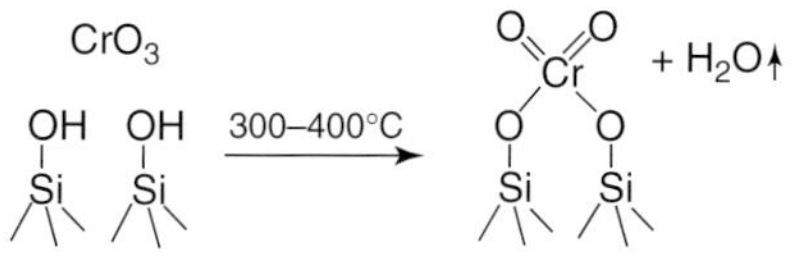

Figure 3.4 Structure of a chromium oxide Phillips catalyst.

compound is evaporated using hot nitrogen in the drying step. Several other techniques are used for synthesizing catalysts for olefin polymerization, including emulsion, and spray drying approaches for the production of $MgCl_2$ particles with finely controlled morphology and particle size distribution, but their description is beyond the scope of this book. For more information, see the references cited in the section titled Further Reading.

3.2.2
Phillips Catalysts

Phillips catalysts are generally made by the impregnation of chromium compounds, such as CrO_3, onto SiO_2 (Figure 3.4), followed by calcination in dry air at high temperature. They differ from Ziegler–Natta catalysts in several aspects:

1) They are not activated by alkyl aluminum cocatalysts but by calcination at high temperatures (200–900 °C). The thermal activation step helps attach the Cr species to the silica (200–300 °C) through reactions with surface silanol groups and eliminate neighboring silanol groups (>500 °C).
2) The thermal treatment used during activation affects polymerization activity, polymer MWD, and long-chain branch (LCB) formation.
3) Hydrogen is not an effective chain transfer agent; rather, the MWD is affected by support porosity and pore volume.
4) Long induction times are very common before the polymerization starts taking place.

Phillips catalysts also have lower reactivity ratios toward α-olefin incorporation and therefore are not used to produce LLDPE. However, they are excellent catalysts for HDPE and are responsible for approximately 40% of the market for this resin. HDPE made with Phillips catalysts have a very broad MWD, often with PDIs of 10 or higher, which can be controlled by support selection and calcination conditions before the polymerization. Such broad MWDs point to the existence of several types of active sites on the surface of Phillips catalysts.

There is significant evidence that HDPE made with Phillips catalysts have trace amounts of LCB that imparts excellent melt strength properties to these resins. The absence of H_2 during polymerization coupled with significant β-hydride elimination during polymerization are conditions that favor LCB formation via the same mechanism that is operative with metallocenes, as explained in Section 3.4.

3.2.3
Metallocenes

Metallocene catalysts have been known for a long time before they started being used industrially for the production of polyolefins – Natta himself published papers where he used metallocene catalysts for olefin polymerization. When activated with common alkyl aluminum compounds, such as TMA and TEA, they have very poor polymerization rates and deactivate rapidly via bimolecular reactions. It was the fortuitous discovery by Kaminsky and Sinn in the early 1980s that methylaluminoxane (MAO), a product of the controlled hydrolysis of TMA, could be used to activate and stabilize metallocene precursors that took metallocenes from academic laboratories to industrial reactors, creating in the process a new revolution in the polyolefin world. For the first time in their history, HDPE, LLDPE, and isotactic polypropylene having uniform properties could be produced under industrially relevant conditions.

Metallocene catalysts are called *sandwich compounds* because they are composed of a transition metal atom "sandwiched" between two cyclopentadienyl or cyclopentadienyl-derivative rings, as depicted in Figure 3.5. The two rings may

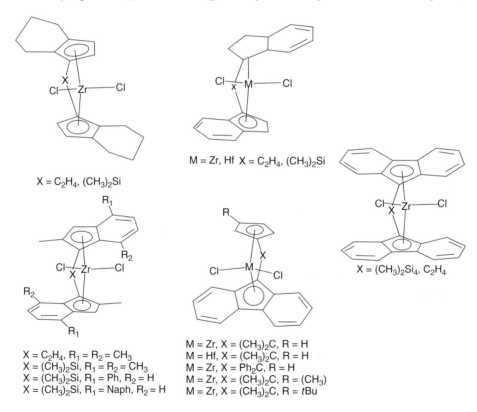

Figure 3.5 Different metallocene catalyst structures.

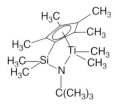

Figure 3.6 A half-sandwich metallocene or constrained geometry catalyst (CGC).

be connected through bridges of different types (*ansa* metallocenes), altering the ligand–metal–ligand angle (bite-angle). By altering the electronic and steric environment around the active sites, it is possible to modify the accessibility and reactivity of the active sites and produce polyolefins with different microstructures, although it is not always easy to predict a priori the result of a given ligand modification. Figure 3.5 shows only an insignificant number of metallocene structures that have been used for olefin polymerization. The combination of different ligand types is essentially infinite. The type of transition metal atom also affects the catalyst behavior, but the most common choices are Zr, Ti, and Hf.

Figure 3.6 depicts an example of a monocyclopentadienyl catalyst, also called *half sandwich* or constrained geometry catalyst (CGC). As for regular metallocenes, the CGC ligands can also be varied to synthesize catalysts with different polymerization activities and abilities to control polymer molecular weight and composition. This family of catalysts has very high reactivity ratios for α-olefin incorporation, allowing the easy copolymerization of ethylene and α-olefins. This is often attributed to the absence of a second cyclopentadienyl ring, creating an active site that is more easily accessible to bulkier α-olefin comonomers, but it is also a strong function in the electronic environment around the transition metal site. As a consequence, polymer chains that contain a terminal vinyl group can be copolymerized with ethylene and other α-olefins to form LCBs. Vinyl-terminated polymer chains are known as *macromonomers* and may be formed in the reactor through several mechanisms that are discussed in Section 3.4.

Why is the introduction of LCBs important for these single-site polymers? Metallocene polyolefins have very good mechanical properties because of their narrow MWD, but, unfortunately, their narrow MWDs also make them more difficult to process than Ziegler–Natta and Phillips resins with similar melt indices and densities. Broad MWD resins have a low-molecular-weight component that acts as a lubricant, increasing shear thinning and enhancing processability. Since this component is absent in metallocene polyolefins, they are much harder to process at high throughputs, having lower critical shear rates at the onset of surface melt fracture. LCBs, even when present at frequencies lower than one LCB/C atoms, have a large impact on shear thinning and increase processability considerably, besides improving polymer melt elasticity.

Another important characteristic of metallocene catalysts is that their stereoselectivity is determined by their ligand configuration. Figure 3.7 shows how different ligand symmetries can be used to produce atactic, isotactic, and syndiotactic polypropylene. Compared with the control of stereoselectivity for heterogeneous Ziegler–Natta catalysts by internal and external electron donor addition, the

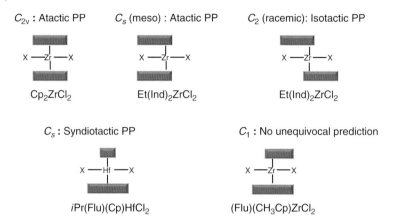

Figure 3.7 Ligand symmetry and stereoregular control for propylene polymerization. Cp, cyclopentadienyl; Ind, indenyl; Flu, fluorenyl.

ligand-design approach available with metallocene catalysts seems a lot more elegant and rational. Unfortunately, although they excel in stereoregularity control, most metallocenes are still not as effective as heterogeneous Ziegler–Natta catalysts for polypropylene regioselectivity (1-2 vs 2-1 insertions). As a consequence, metallocene polypropylenes may be highly isotactic but still contain enough 2-1 insertion defects to adversely affect their melting points. This has been one of the main reasons slowing down the industrial implementation of metallocene catalysts for isotactic polypropylene manufacturing. On the other hand, metallocenes are the only catalysts that can produce syndiotactic polypropylene at commercially acceptable rates.

Besides the three classic polypropylene configurations – isotactic, syndiotactic, and atactic – polypropylene chains with novel structures can also be made with other metallocene types, as illustrated in the Fischer projections in Figure 3.8. Although

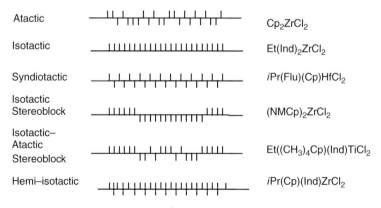

Figure 3.8 Some types of polypropylene chain configurations produced with metallocene catalysts. Cp, cyclopentadienyl; Ind, indenyl; Flu, fluorenyl; NM, neomenthyl.

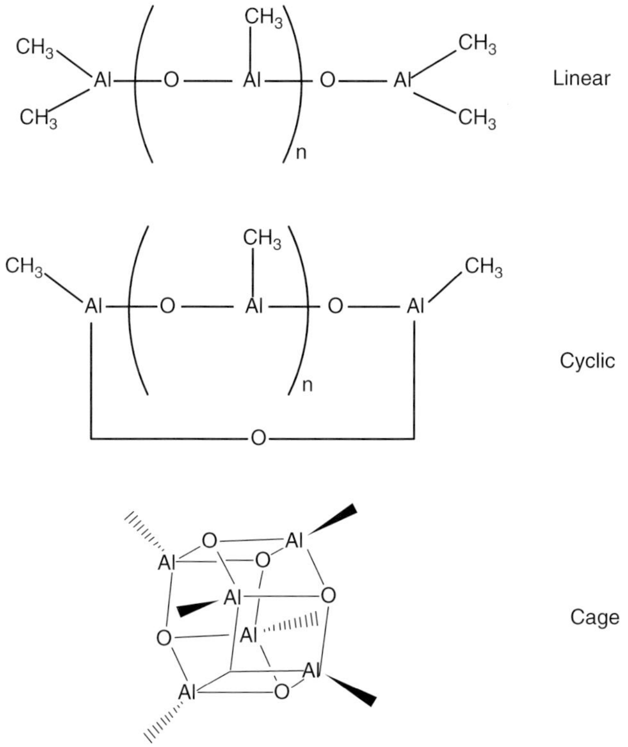

Figure 3.9 Proposed MAO structures.

these new polypropylene types are mostly laboratory curiosities (the isotactic–atactic polypropylene has very interesting thermoplastic elastomeric properties), they illustrate well how ligand design can affect the structure of polymers made with metallocene catalysts.

As mentioned previously in this section, the finding that MAO could activate and stabilize metallocene catalysts was the landmark discovery in transforming them from laboratory curiosities to a multimillion dollar industrial reality. MAO is an oligomeric compound with the degree of oligomerization varying approximately from 6 to 20. There are several different MAO grades, made following different synthetic pathways, most of which are proprietary. Despite its importance, the structure of MAO is still not completely established. Figure 3.9 shows some structures that have been proposed for MAO, including linear, cyclic, and cage structures. The oligomeric MAO form exists in equilibrium with its TMA monomer, with different MAO grades having distinct TMA fractions. It is likely that this residual TMA also influences the final cocatalyst performance when combined with the metallocene, although this effect can be compensated for by adjusting the reactor conditions during polymerization.

Other alkyl aluminums can also be reacted with water to produce equivalent aluminoxanes: for instance, TEA and water produce ethylaluminoxane (EAO),

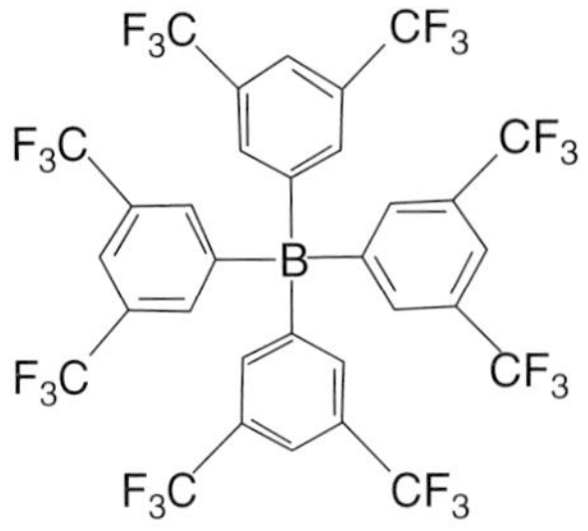

Figure 3.10 Tetrakis(3,5-bis(trifluoromethyl)-phenyl)borate, an effective counter anion for metallocenes used in solution polymerization.

t-butyl aluminum and water produce *t*-butylaluminoxane (*t*BAO), and so on. MAO, however, seems to be the most effective of all aluminoxane cocatalysts tried with metallocenes.

In general, a large excess of MAO is needed to achieve high catalyst activity – ratios in excess of 1000 aluminum atoms to transition metal atoms are commonly reported – if the metallocene is to be used as an unsupported catalyst. It has been claimed that this large excess of the bulky MAO cocatalyst is needed to shield the active sites from one another and prevent bimolecular deactivation reactions. As for Ziegler–Natta catalysts, the metallocene active site is a cation associated with an anion that is capable of stabilizing the transition metal cation and is labile enough to be displaced by the coordinating monomer. Fortunately, a much lower ratio (in the order of a few hundreds) is required when the catalyst is supported on SiO_2 and other carriers, significantly decreasing the MAO requirements for catalysts used in slurry or gas-phase processes.

Industrially, unsupported metallocenes are only used in solution processes at polymerization temperatures above the polymer melting temperature to prevent the polymer from precipitating inside the reactor. The high MAO/transition metal ratios required to achieve adequate polymerization activities under these conditions would render impractical the use of metallocenes in solution processes. Fortunately, other bulky, noncoordinating anions, such as tetrakis(3,5-bis(trifluoromethyl)-phenyl)borate (Figure 3.10) are also very effective cocatalysts for metallocenes and have the added advantage of being required in nearly stoichiometric amounts. These cocatalysts are used to replace MAO in industrial solution processes (although MAO or similar compounds may still be required during the alkylation step of the catalyst, as described in Section 3.4).

Perhaps the most important reason leading to the industrial implementation of metallocene catalysts, in addition to their high activity and excellent microstructural control, is that they could be easily adapted to existing industrial olefin polymerization processes. The transition from Ziegler–Natta or Phillips catalysts to metallocenes is sometimes called *drop-in technology* exactly to indicate that the new catalysts can simply be dropped into the existing reactors. Of course, reality is often not as simple as catchy phrases may imply, but the fact remains that metallocenes can be used with industrial processes designed to operate with Ziegler–Natta or Phillips catalysts without prohibitively expensive adjustments.

All olefin polymerization processes that are described in Chapter 4 can be operated commercially with metallocene catalysts. Metallocenes can be used directly in solution processes, but need to be supported to be used in slurry and gas-phase processes. In the latter case, the support of choice is SiO_2. Some supporting techniques used for metallocenes are briefly reviewed in Section 3.3.

3.2.4
Late Transition Metal Catalysts

Late transition metal catalysts for olefin polymerization were first discovered by Brookhart and researchers from DuPont in the early 1990s, leading to the longest patent in DuPont's history, nicknamed the 500/500/500 patent for having approximately 500 pages, 500 examples, and 500 claims. These complexes are much less oxophilic than Ziegler–Natta, Phillips, or metallocene catalysts, allowing the copolymerization of olefins and polar comonomers such as vinyl acetate and methyl metacrylate. Since these important commodity polymers are currently manufactured using free radical polymerization, with all the inconveniences that this process entails (more aggressive polymerization conditions and presence of side reactions such as chain scission and chain transfer to polymer), it was expected that a new industrially relevant synthetic route to ethylene/polar comonomer copolymers had been finally discovered. Unfortunately, even though the copolymerization of ethylene, α-olefins, and several polar comonomers are feasible with late transition metal catalysts, they still cannot compete industrially with the existing free radical polymerization process.

Some late transition metal catalysts, such as the Ni-diimine catalyst shown in Figure 3.11, have an intriguing property called *chain walking*: when polymerizing ethylene, the active center can move away from the chain end and "walk" on the polymer backbone, leading to the formation of short-chain branches (SCBs) in the absence of α-olefin comonomers. By varying the polymerization temperature and monomer pressure, it is possible to make polymers with densities varying from those of high to ultralow-density polyethylene and, in fact, to that of completely amorphous, ethylene-propylene-like elastomers.

The chain-walking mechanism is illustrated in Figure 3.12. If monomer insertion takes place when the Ni catalyst is placed at the end of the polymer molecule, the chain grows by one monomer unit without the formation of an SCB. However, some late transition metal catalysts can move along the chain backbone through

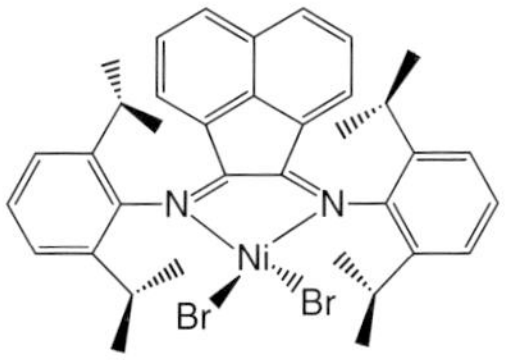

Figure 3.11 A Ni-diimine late transition metal catalyst.

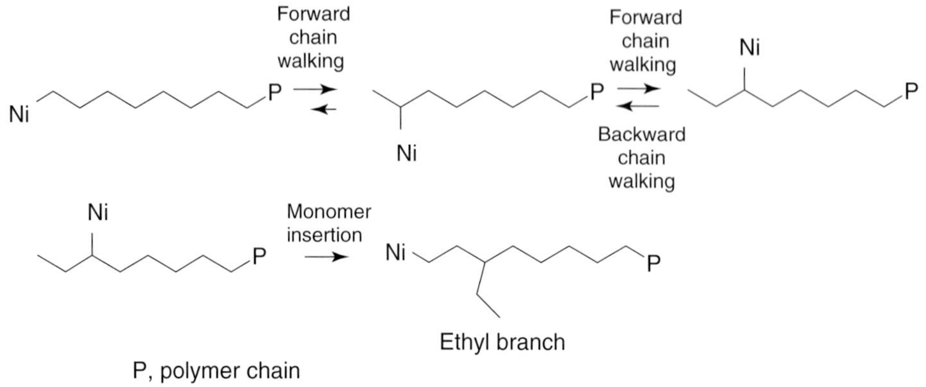

Figure 3.12 Chain-walking mechanism with a late transition metal catalyst.

isomerization reactions, as indicated in the first reaction step in Figure 3.12, where the catalyst moved from carbon 1 to carbon 2. After moving away from the chain end, the catalyst may either move forward to carbon 3 or back to carbon 1. This back-and-forward movement will take place until a monomer is inserted into the growing polymer chain. In Figure 3.12, an ethylene molecule is inserted when the active site is placed on carbon 3, forming an ethyl branch. If the insertion had taken place when the Ni site was bonded to carbon 2, a methyl branch would have resulted. Longer SCBs are formed as the active site moves farther away from the chain end.

It should be apparent from this discussion that LCBs are not likely to be formed by the chain-walking mechanism, as this would require that the active site moved away from the chain end by hundreds of carbons. This, however, is extremely improbable, as the likelihood of backward and forward isomerization reactions should be the same as soon as the catalyst site moves a few carbon atoms away from the chain end, approximating a random walk on the polymer backbone.

As expected, high ethylene concentrations in the reactor will increase the probability that the monomer will add to the chain before the catalyst has time to move away from the chain end, decreasing the SCB frequency of the polymer. Likewise, increasing the polymerization temperature increases chain walking and SCB formation frequency, as illustrated in Figure 3.13. The chain-walking mechanism seems to favor the formation of methyl branches, likely because the active site is in a more stable configuration when bonded to the β-carbon in the polymer chain, but chain walking assumes a near random pattern for the higher carbon in the chains. Table 3.4 shows the detailed SCB distribution for the polymers shown in Figure 3.13. Owing to ^{13}C NMR limitations, hexyl and longer branches are detected as a single signal and reported under the hexyl+ column.

The literature on late transition metal catalysts is very large and keeps growing as new complexes are developed. Several families have been developed such as Brookhart (Ni and Pd), Gibson–Brookhart (Co and Fe), and Grubbs (neutral Ni) catalysts. Some of their properties are very enticing: these catalysts are much less

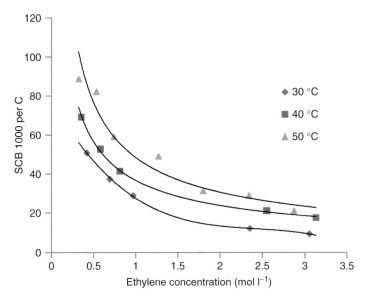

Figure 3.13 Overall short-chain branching frequencies as a function of polymerization temperature and ethylene concentration for the polymerization of ethylene with a Ni-diimine catalyst (1,4-bis(2,6-diisopropylphenyl))-acenaphthenediimine-dichloronickel(II)/trimethylaluminum) in a semibatch reactor.

Table 3.4 SCB distribution of polymers shown in Figure 3.13.

Polymerization conditions		SCB per 1000 C by ^{13}C NMR						
T (°C)	E(mol l^{-1})	Methyl	Ethyl	Propyl	Butyl	Pentyl	Hexyl+	Total
30	0.42	34.0	3.2	2.3	2.3	1.5	7.7	50.9
30	0.69	27.4	2.4	2.2	1.4	0.9	3.4	37.5
30	0.97	21.4	2.6	1.7	1.4	0.4	1.5	29.0
30	2.35	10.1	0.9	0.5	0.3	0.0	0.2	12.1
30	3.05	8.0	0.4	0.3	0.2	0.0	0.3	9.3
40	0.35	43.7	4.2	3.7	3.5	2.8	11.2	69.2
40	0.58	36.2	3.5	3.1	2.8	1.4	5.7	52.8
40	0.81	32.8	2.0	1.2	1.0	0.7	3.9	41.6
40	2.55	16.1	1.6	1.2	0.9	0.4	0.9	21.2
40	3.13	14.2	1.0	0.7	0.5	0.2	1.1	17.7
50	0.32	56.3	5.7	4.9	4.4	4.1	13.4	88.8
50	0.53	47.9	6.3	5.9	4.5	4.4	13.4	82.3
50	0.74	40.0	4.1	3.0	2.7	2.6	6.8	59.2
50	1.27	31.7	4.5	3.2	2.2	2.0	5.6	49.2
50	1.80	25.7	1.6	1.2	0.9	0.4	1.7	31.5
50	2.34	22.3	2.0	1.2	1.1	1.0	1.5	29.0
50	2.87	19.0	0.7	0.5	0.3	0.2	0.5	21.1

sensitive to polar compounds and can be used to copolymerize olefins with polar monomers such as acrylates and methyl acrylates. Fascinating as they are, these catalysts have not found commercial applications yet and are not discussed any further. The interested reader is referred to the reviews cited at the end of this chapter.

3.3
Supporting Single-Site Catalysts

Since slurry and gas-phase reactors (which are responsible for about 70% of the total industrial polyolefin production) require morphologically uniform catalyst particles that can be continuously fed to the reactor, soluble metallocene catalysts must be fixed onto insoluble carriers before they can be used in these processes. It is important that the catalyst structure, activity, comonomer reactivity, and stereoselectivity be maintained after supporting. Moreover, the catalyst should not leach from the support during polymerization to avoid reactor fouling.

The main advantages of supported catalysts over their homogeneous counterparts, besides becoming compatible with slurry and gas-phase processes, are as follows

1) The cocatalyst/catalyst ratio required to reach maximum activity is lower for supported catalysts than for homogeneous catalysts.
2) The average polymer molecular weight generally increases on supporting.
3) The MWD may broaden on catalyst supporting, facilitating processability (at the cost of poorer mechanical properties).
4) Some supported catalysts can be activated by common alkyl aluminum compounds in the absence of more expensive MAO.
5) Different single-site catalysts can be incorporated onto the same support to produce polymers with tailored MWD and CCD.

The key factors when selecting a support for an olefin polymerization catalyst are its chemical composition, surface characteristics, morphology (surface area and distribution of particle size, pore size, and pore volume), and mechanical strength. Silica is the most extensively used support for metallocene catalysts because it has several properties that make it attractive as a catalyst support: it is relatively chemically inert; it is stable at high temperatures; and it can be synthesized with several pore sizes, volumes, and surface areas. In addition, silica is inexpensive, ideal for the production of commodity polymers such as polyolefins. For the same reasons, SiO_2 is also preferred as a support for Cr oxide Phillips catalysts and even for some $TiCl_4$ Ziegler–Natta catalysts with finely controlled morphology for gas-phase processes.

The chemical properties of SiO_2 are governed by the presence of silanol groups on its surface that can be changed with appropriate thermal or chemical treatments. Figure 3.14 illustrates the thermally induced change of silica surface groups from silanol to siloxane. The hydroxylated surface is hydrophilic and easily absorbs

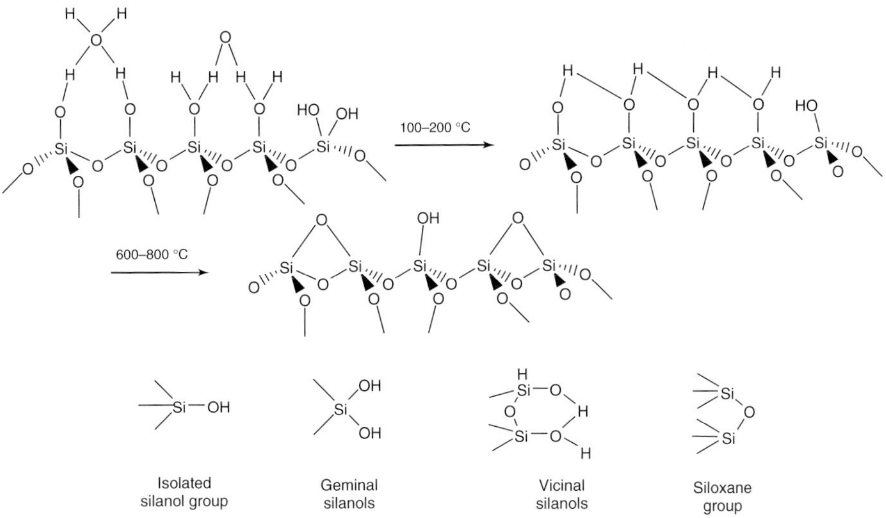

Figure 3.14 Schematic representation of SiO$_2$ dehydration.

moisture from air. This physically absorbed water can be desorbed by raising the temperature to 200 °C. The number of hydroxyl groups decreases continuously as the temperature is raised, until at a temperature of 600–800 °C an almost completely dehydroxylated silica surface is left.

Hydroxyl groups on the silica surface may also be treated with a dehydroxylating agent such as hexamethyldisilazine (HMDS) to form other functional groups; this treatment effectively eliminates surface hydroxyl groups, as illustrated in Figure 3.15.

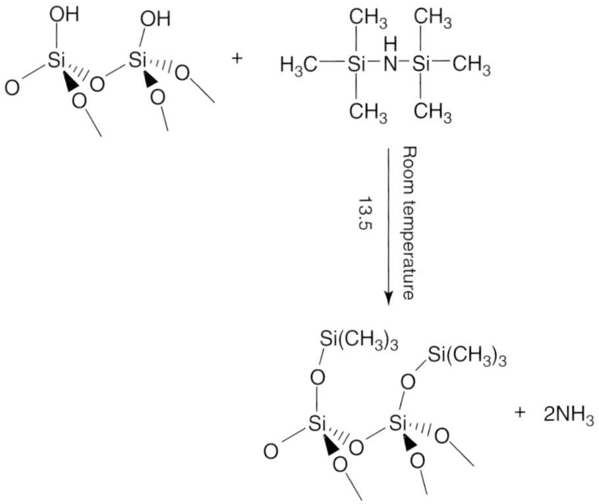

Figure 3.15 Modification of SiO$_2$ with HMDS.

3 Polymerization Catalysis and Mechanism

Numerous supporting techniques have been used to anchor early and late transition metal catalysts on SiO_2 carriers. We are not attempting to review all possible alternatives herein, but to cover some of the main techniques, which are classified as follows:

1) Physical adsorption methods
   a. SiO_2/metallocene/cocatalyst
   b. SiO_2/cocatalyst/metallocene
   c. SiO_2/(metallocene + cocatalyst).
2) Covalent attachment methods
   a. SiO_2–OH/metallocene/cocatalyst
   b. Si–O–Si/metallocene/cocatalyst.

Physical adsorption methods are very common because of their simplicity. Method 1(a) consists in adsorbing the metallocene on the support and then

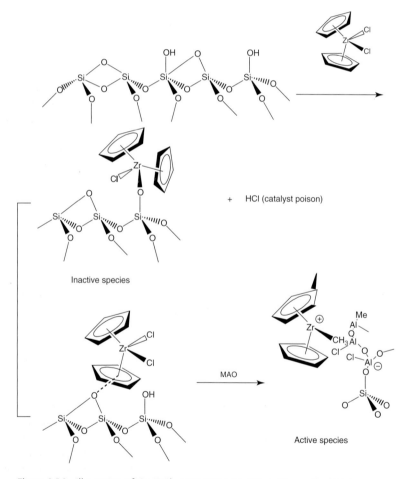

Figure 3.16 Illustration of Cp_2ZrCl_2 supporting on SiO_2 using method 1(a).

3.3 Supporting Single-Site Catalysts

reacting it with a cocatalyst such as MAO. The hydroxyl density on the silica surface decreases with increasing temperature and duration of the thermal treatment, as represented in Figure 3.14, increasing the density of siloxane groups that leads to active immobilized metallocene sites (Figure 3.16). However, metallocenes may also react with hydroxyl groups that remain on the silica surface after the thermal treatment, producing inactive sites and significantly decreasing catalyst activity. In addition, catalyst poisons such as HCl may be formed during supporting, further reducing catalyst activity when this method is applied.

In method 1(b), the SiO_2 support is first treated with an alkyl alumoxane, followed by washing, drying, and impregnation with an appropriate metallocene. The SiO_2/MAO/metallocene/MAO system, commonly referred to as *sandwich* structure on account of the position of the metallocene between two MAO layers, is a variation of this technique. Instead of contacting SiO_2 with aluminoxane, TMA or a mixture of alkyl aluminums can be reacted directly with the water adsorbed onto the silica surface, generating *in situ* aluminoxanes, as shown Figure 3.17. When successful, this method has the advantage of not using the more expensive MAO; in addition, TMA is a much smaller molecule capable of penetrating into the smaller pores of the SiO_2 support and likely creating a more uniform catalyst support for metallocene deposition.

Finally, in the last physical adsorption technique, method 1(c), the metallocene and the cocatalyst are precontacted before supporting. Because the metallocene is activated in solution, this procedure may increase the number of active sites and lead to highly active catalysts. This method also has the benefit of reducing the number of reaction steps required during supporting.

All the variations on the physical adsorption methods summarized above assume that the reagents and the support material onto which they adsorb are suspended or dissolved in a solvent. A gas-phase variation of these procedures is also possible, where gaseous TMA, water vapor, and metallocene are impregnated onto SiO_2 in a fluidized bed reactor. This method reduces the preparation costs of supported

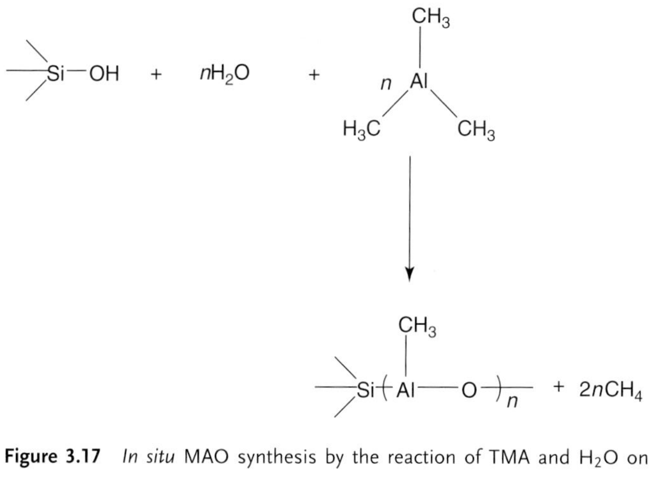

Figure 3.17 *In situ* MAO synthesis by the reaction of TMA and H_2O on the SiO_2 support.

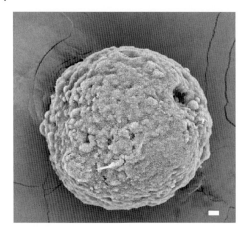

Figure 3.18 SEM image of a polyethylene particle made with a Ni-diimine catalyst supported on SiO$_2$ using method 2(b) (scale bar = 400 nm).

catalysts because MAO, generated and supported *in situ*, achieves higher loads than in slurry processes.

Despite their relatively simple preparation, metallocenes supported through physical adsorption are often extracted by diluents during polymerization in slurry reactors. (This, evidently, poses less of a problem for gas-phase reactors.) These solubilized sites make polymer particles with poor morphology that cause reactor fouling in slurry processes.

Supporting methods that involve the covalent bonding of the catalyst to the support surface (methods 2(a) and 2(b)) overcome the disadvantage of physical adsorption supporting methods, but may be more complex and expensive to implement. The scanning electron microscopic (SEM) image of a polyethylene particle made with a late transition metal catalyst supported on SiO$_2$ using method 2(b) shown in Figure 3.18 illustrates how effective these techniques can be in producing spherical, high bulk-density polymer particles.

Method 2(a) is very attractive because it promotes strong chemical bonding between support and catalyst. A functional group must be added to the metallocene ligand to react with other functional groups present on the support surface. Metallocenes with chemically tethered groups are strongly supported and cannot be appreciably leached from the support surface, avoiding reactor fouling. Figure 3.19 illustrates a common method for supporting metallocenes on silica, where surface hydroxyl groups provide the anchoring sites for the functionalized metallocene complexes. Unfortunately, other side reactions may also take place during supporting (for instance, between other metallocene ligands such as chlorine), leading to the formation of inactive sites and other by-products such as water and hydrochloric acid that act as catalyst poisons.

Method 2(b) is a modification of method 2(a), where siloxane groups on the silica surface react with functional groups on the metallocene ligands to create a

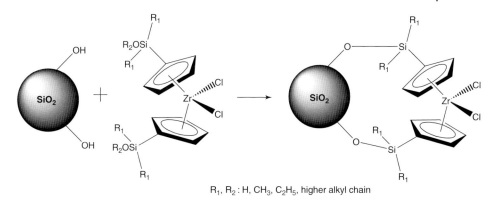

R_1, R_2 : H, CH_3, C_2H_5, higher alkyl chain

Figure 3.19 Illustration of functionalized Cp_2ZrCl_2 supporting on SiO_2 using method 2(a).

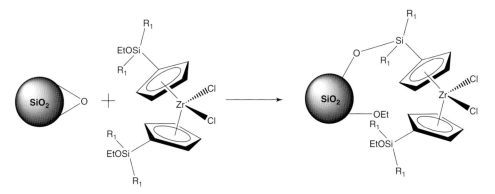

R_1, H, CH_3, C_2H_5, higher alkyl chain

Figure 3.20 Illustration of functionalized Cp_2ZrCl_2 supporting on SiO_2 using method 2(b).

supported metallocene. This method is attractive because fewer by-products are formed during catalyst supporting (Figure 3.20).

In general, supported metallocenes are less active than their homogeneous counterparts. This may be due to the proximity of the catalyst to the support surface. Significant electrical and/or sterical effects caused by the presence of the support around the active sites may decrease the monomer insertion frequency or deactivate some of the catalytic sites. Spacer groups may be used to improve catalyst activity because they help reduce electronic and steric effects between support and catalyst, as illustrated in Figure 3.21.

Not only the order of reactants but also the supporting conditions influence the activity of the resulting catalyst and the microstructure of the polymer made by it. This makes comparison of the relative performances of supported metallocene catalysts prepared by different research groups difficult and the conclusions drawn from it of uncertain merit. Fairly extensive investigations regarding the effects of silica activation temperature, grafting temperature, and solvent type on the amount

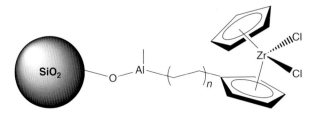

Figure 3.21 Metallocene catalyst supported with the help of a spacer group to minimize the interactions between support and active site.

of catalyst supported and activities were explored but are not discussed herein. More references on this subject can be found in the section titled Further Reading at the end of the chapter.

3.4
Polymerization Mechanism with Coordination Catalysts

What all catalysts discussed in Section 3.2 have in common, despite their widely different chemical structures and properties, is their basic polymerization mechanism. The mechanism of olefin polymerization with coordination catalysts has been studied extensively since the discovery of Ziegler–Natta and Phillips catalysts. Some of the steps in this mechanism are very well established and constitute what we call the *fundamental model* for polymerization kinetics in this chapter. Some important phenomena that take place during olefin polymerization, however, are not described by the fundamental model. For instance, it is commonly observed that hydrogen increases the rate of propylene polymerization, while decreasing the rate of ethylene polymerization with heterogeneous Ziegler–Natta catalysts and other coordination catalysts. Similarly, the addition of an α-olefin at small concentrations accelerates the overall polymerization rate of coordination catalysts, the so-called comonomer effect. We come back to these phenomena at the end of this section, and propose some models to quantify their behavior in Chapter 5.

Figure 3.22 shows the main steps in the catalytic cycle of the fundamental model. An active site (C^*) is formed when a catalyst (C) reacts with a cocatalyst molecule (TEA or MAO, for instance). This reaction is generally very fast. The first monomer insertion produces a living polymer chain of length 1 ($P_{1,i=0}^*$). The subscript i describes the number of LCBs in the chain; we will ignore it for now and assume that there are no LCBs in the chain ($i = 0$). The polymer chain then grows through successive monomer insertions (innermost cycle), its chain length increasing to $2, 3, 4, \ldots, r$ until a chain transfer reaction occurs. For the sake of simplicity, we included only two types of transfer reaction in Figure 3.22: transfer to hydrogen (second cycle) and β-hydride elimination (third cycle). When a hydrogen molecule reacts with a living polymer chain, one hydrogen atom bonds to the active site, forming a metal hydride site (P_H^*), while the other hydrogen atom is transferred to the end of the living polymer chain, creating a dead polymer chain with chain length

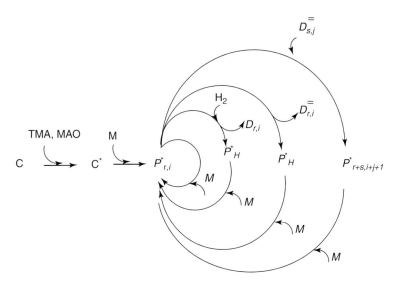

Figure 3.22 Coordination polymerization mechanism (fundamental model): C, catalyst; TMA and MAO, cocatalysts; C*, active site; M, monomer; $P_{r,i}^*$, living chain with length r and i LCBs; P_H^*, metal hydride site; H_2, chain transfer agent (hydrogen); $D_{r,i}$, dead chain with saturated chain end; and $D_{r,i}^=$, dead chain with vinyl chain end (macromonomer).

r and a saturated chain end ($D_{r,i=0}$). The same metal hydride site is created during β-hydride elimination, but now the hydrogen atom is abstracted from the second carbon in the living polymer chain (counting from the transition metal–carbon bond), generating a dead polymer chain with a vinyl terminal group ($D_{r,i=0}^=$). As indicated in the diagram, metal hydride sites are also active for polymerization and initiate another polymer chain through a monomer insertion step. For linear polymerization, this is all there is to the fundamental model, plus a few additional chain transfer steps, such as transfer to monomer and cocatalyst, that are discussed later in this section.

When the proper catalyst is used under the right polymerization conditions, vinyl-terminated dead polymer chains may be further polymerized with the living chains, forming LCBs. Accordingly, vinyl-terminated chains are often called *macromonomers*, as they can be considered to be very long α-olefins. The outermost cycle of Figure 3.22 illustrates this step: macromonomers with chain length s and j LCBs, when reacting with living polymer chains of length r and i LCBs, generate a living polymer chain of length $r + s$ having $i + j + 1$ LCBs. For instance, a linear ($j = 0$) macromonomer of chain length $r = 1000$ reacting with a linear ($i = 0$) living polymer of chain length $s = 1500$ will create a living polymer chain with $i + j + 1 = 1$ LCB and length $r + s = 2500$.

So far, we have mentioned several times that the cocatalyst is required to activate the catalyst, but have not proposed a mechanism for this reaction. Let us now explore this reaction in Scheme 3.1, where L indicates a ligand – cyclopentadienyl groups for metallocenes, Cl or Mg atoms for Ziegler–Natta catalysts, for instance, – A is the

3 Polymerization Catalysis and Mechanism

Scheme 3.1 Catalyst activation with cocatalyst.

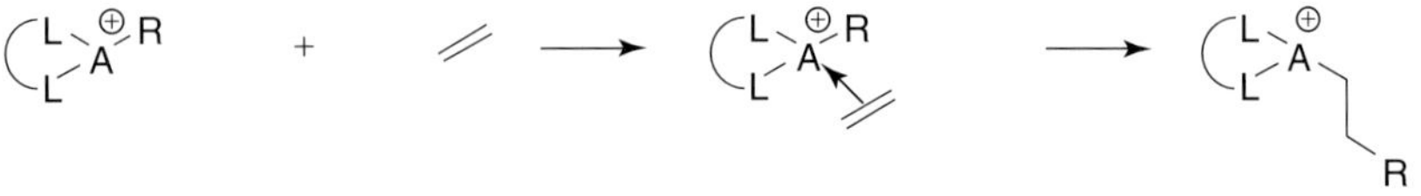

Scheme 3.2 Chain initiation.

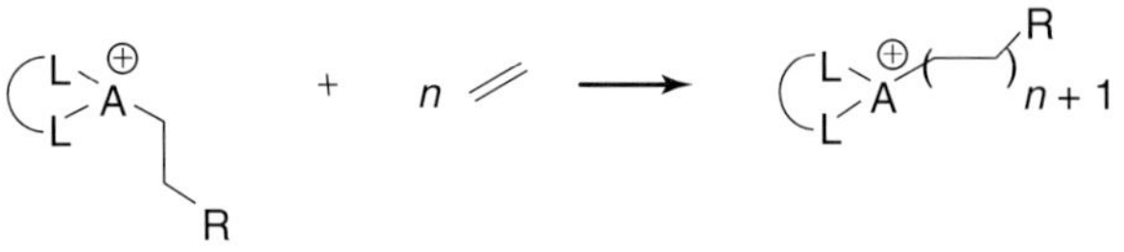

Scheme 3.3 Monomer propagation.

transition metal active center, and X are halogens, generally chlorine atoms. The cocatalyst (AlR$_3$, where R is an alkyl group) acts as an alkylating and reducing agent, extracting two halogen atoms from and transferring one alkyl group to the catalyst. The resulting active site is positively charged and the cocatalyst product (AlR$_2$X$_2^-$) is a noncoordinating anion required to stabilize the catalyst. The electron-deficient site is now ready to attract the π-electrons in the olefin double bond, initiating the polymerization. The alkyl group transferred to the active site will become one of the polymer chain ends.

After the catalyst is activated by reaction with the cocatalyst, the first monomer insertion takes place, as illustrated in Scheme 3.2. This step is called *initiation*.

Further monomer insertion steps lead to monomer propagation, as shown in Scheme 3.3.

It is common to differentiate between the first (initiation) and the subsequent monomer insertion steps (propagation) by assigning them different reaction rate constants. In practice, however, it is seldom possible to quantify this difference.

Hydrogen is the most commonly used chain transfer agent during olefin polymerization. Metallocene and most late transition metal catalysts are very sensitive to hydrogen, while Ziegler–Natta catalysts require a much larger hydrogen concentration to achieve a comparable molecular weight reduction. The molecular weight of polymers made with Phillips catalysts, on the other hand, is regulated by the thermal treatment of the catalyst and support selection, instead of by using hydrogen as a chain transfer agent. Chain transfer to hydrogen is described in Scheme 3.4.

The metal hydride site formed after transfer to hydrogen can also initiate other polymerizations according to the reaction in Scheme 3.5.

Scheme 3.4 Chain transfer to hydrogen.

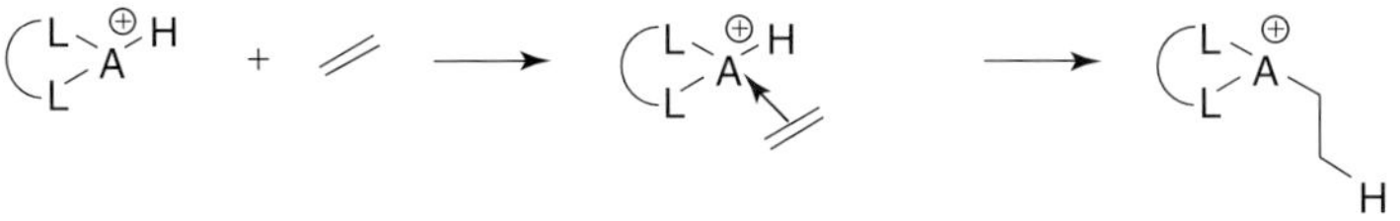

Scheme 3.5 Initiation of a metal hydride site.

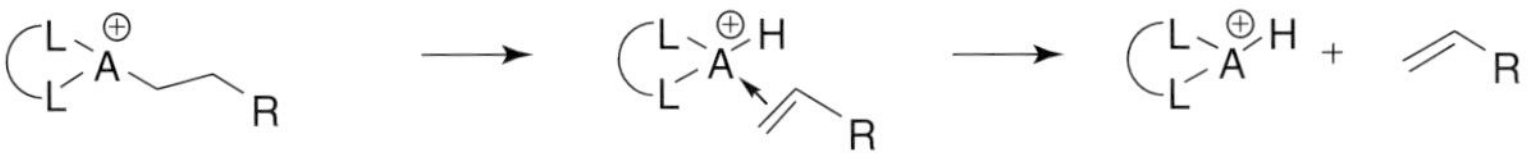

Scheme 3.6 β-Hydride elimination.

The active species shown in Scheme 3.5 differs from the equivalent one in Scheme 3.2 only by the hydrogen atom at the end of the chain. This small difference has been linked to polymerization kinetic effects such as the decrease in the overall ethylene polymerization rate observed in the presence of hydrogen. Mathematical models to describe this phenomenon are presented in Chapter 5.

β-Hydride elimination takes place following the elementary step in Scheme 3.6.

For ethylene polymerization, a vinyl-terminated chain is formed after β-hydride elimination, but for propylene polymerization, a vinylidine-terminated chain is produced instead. Vinyl-terminated chains are more reactive than vinylidine-terminated chains and more likely to form LCBs. It is, however, possible to produce vinyl-terminated polypropylene chains if the catalyst favors β-methyl elimination. In this case, vinyl-terminated polypropylene macromonomers are produced according to Scheme 3.7.

In the presence of an adequate catalyst, these macromonomers can be reincorporated to form an LCB, in a reaction step that is analogous to monomer propagation, as illustrated in Scheme 3.8.

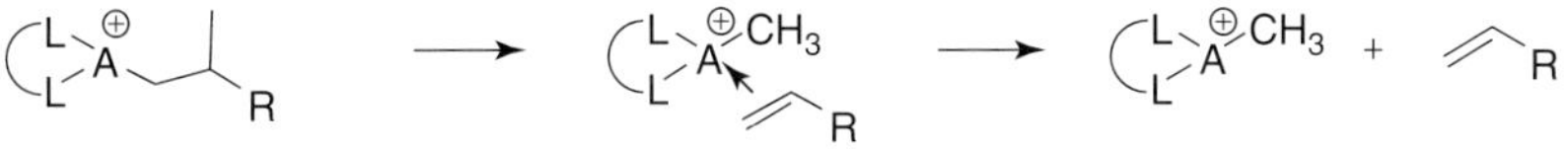

Scheme 3.7 β-Methyl elimination.

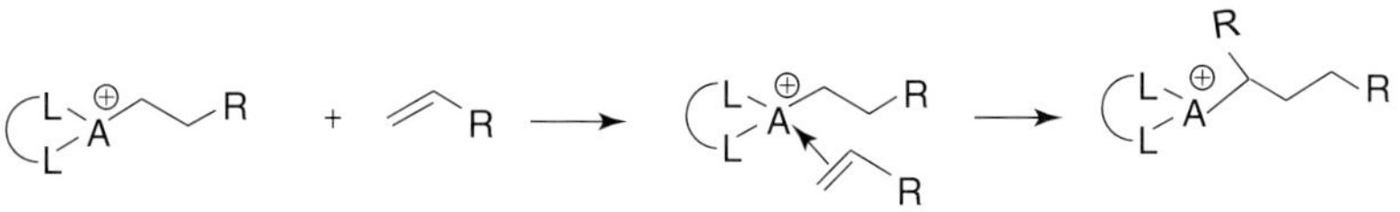

Scheme 3.8 LCB formation by macromonomer incorporation.

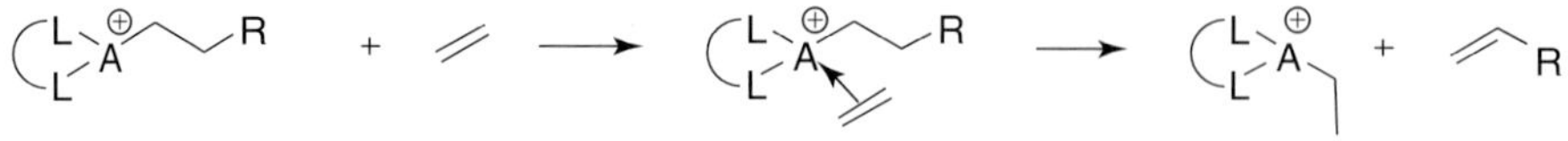

Scheme 3.9 Chain transfer to ethylene.

Scheme 3.10 Chain transfer to cocatalyst.

Chain transfer to monomer is also usually observed during olefin polymerization. In this case, a dead chain (vinyl-terminated, in the case of polyethylene) and a living chain of length 1 are formed, as shown in Scheme 3.9.

Finally, besides being a reducing and alkylating agent for the catalyst, the cocatalyst may also act as a chain transfer agent, as indicated in Scheme 3.10.

Our discussion above was restricted to homopolymerization reactions, but the same polymerization steps are valid for copolymerization. For a given catalyst system, the propagation and chain transfer rates during copolymerization depend on the type of monomers being added to the polymer chain and attached to the polymer chain end. It is easy to understand why the chain-end type affects the polymerization rate, since for the monomer to be inserted into the carbon–metal bond it needs to interact with the last monomer molecule inserted into the chain. This mechanism is called the *copolymerization terminal model*;[2] it is illustrated for ethylene/1-butene polymerization in Scheme 3.11. Each one of the four steps is assumed to have a different propagation rate.

Likewise, the rates of β-hydride elimination, transfer to hydrogen, and cocatalyst may be also affected by the type of monomer at the chain end. However, with the proper mathematical transformation, the equations derived to describe homopolymerization can also be used for copolymerization. This modeling technique, called the *method of pseudo constants*, is described in Chapter 5.

Stereoisomerism must be considered when asymmetric monomers, such as propylene, are polymerized. In this case, the orientation of the monomer coordinating at the active site will determine if the insertion generates meso (*m*) or racemic (*r*) placements (stereoisomerism), and 1-2 or 2-1 insertions (regioisomerism). Scheme 3.12a illustrates a meso placement of a propylene molecule in a polypropylene chain. If all the insertions have this orientation, an isotactic polypropylene chain would be produced. In practice, since no catalyst is perfectly selective toward meso insertions, we call *isotactic polypropylene* those polymers wherein most of the insertions are meso. Contrarily, Scheme 3.12b

---

2) Higher order copolymerization models propose that the next to last monomer (penultimate model) or the second next to last monomer (pen-penultimate model) also affect polymerization rates, but there is little evidence that this added complexity is justifiable for olefin polymerization with coordination catalysts.

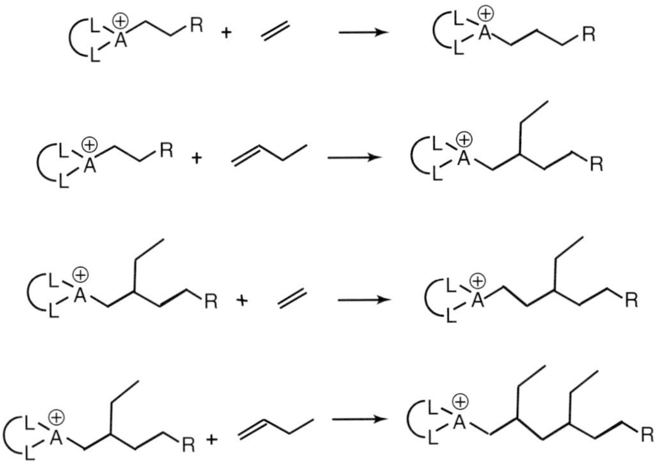

Scheme 3.11 Terminal model of copolymerization.

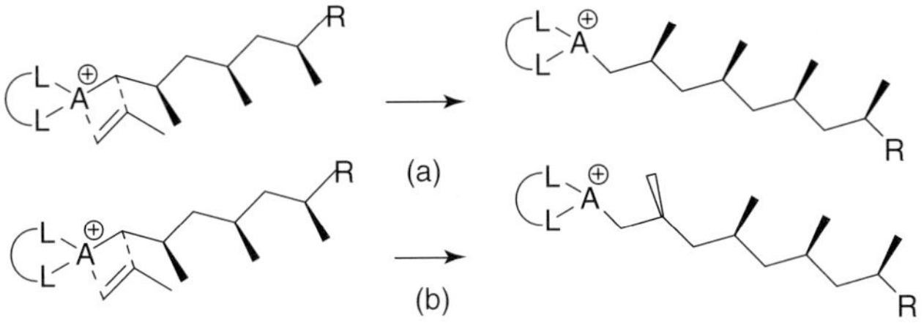

Scheme 3.12 Stereoselectivity during propylene polymerization: (a) meso insertion and (b) racemic insertion.

shows a racemic insertion; a syndiotactic polypropylene chain is produced when all (or most) monomer insertions are racemic.

Scheme 3.13a shows a 1-2 meso propylene insertion into a growing polypropylene chain, while a 2-1 meso insertion is illustrated in Scheme 3.13b. Most modern heterogeneous Ziegler–Natta catalysts favor 1-2 meso insertions, leading to highly stereoregular and regioregular polypropylenes with high crystallinities and melting points. Electron donors are commonly used to regulate these two modes on insertion. Metallocene catalysts are also capable of reaching very high isotacticity levels, but are not as regioselective as most heterogeneous Ziegler–Natta catalysts. As a consequence, isotactic polypropylene made with metallocene catalysts often has lower melting temperatures due to the presence of regiodefects in their backbones.

These four modes of insertion are possible for any α-olefin, but they are generally of concern only for polypropylene, because of its commercial importance.

The most widely accepted mechanism of olefin polymerization was proposed by Cossee and is illustrated in Scheme 3.14 for a Ziegler–Natta active site. Since it does not explicitly include the cocatalyst, it falls into the category of *monometallic mechanisms* (although the cocatalyst is implicitly assumed to alkylate and reduce the

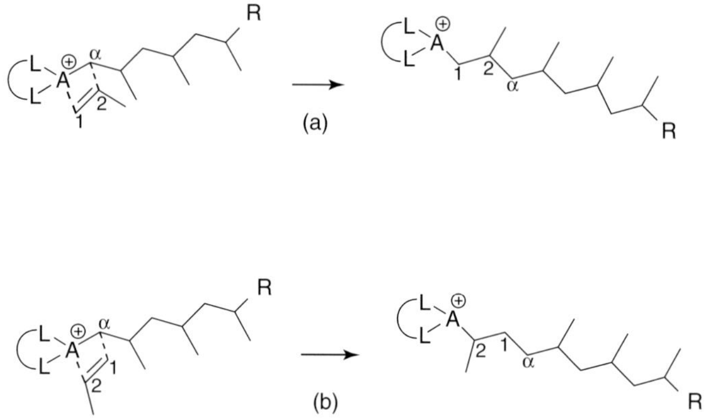

Scheme 3.13 Regioselectivity during propylene polymerization: (a) 1-2 insertion and (b) 2-1 insertion.

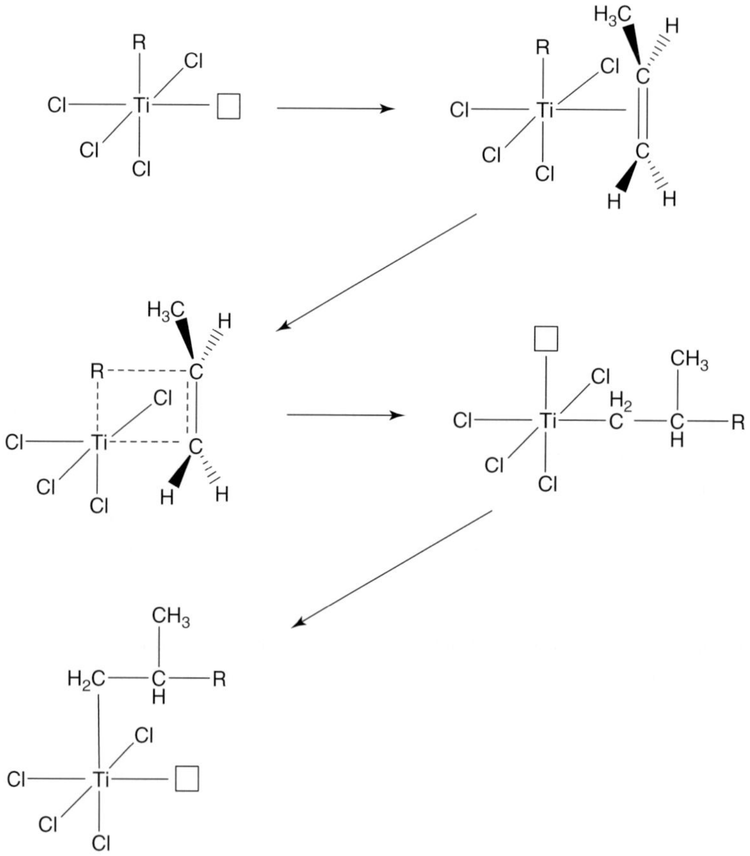

Scheme 3.14 Cossee's mechanism.

catalyst, as explained above). The active site is formed by a Ti atom bonded to four Cl atoms, an alkyl group or polymer chain (R), and a coordination vacancy (□). Before insertion, the incoming olefin molecule becomes coordinated at the vacant site with the double bond parallel to an octahedral axis. After the formation of a four-centered transition state, the polymer chain migrates to the position previously occupied by the coordinating monomer, as indicated in Scheme 3.14 (*chain migratory insertion mechanism*). Even though the resulting site is able to coordinate another monomer molecule at the newly created vacancy, Cossee proposed that the polymer chain skipped back to its original position to ensure that the coordinating monomer always experienced the same steric constraints, as required for the production of isotactic polypropylene. This last step has been considered a weakness in Cossee's mechanism. Some alternative explanations such as the *trigger mechanism*, where monomer insertion can take place only with the aid of another incoming olefin, have been proposed to eliminate the need for the back skip. *Bimetallic mechanisms*, taking into account the presence of the cocatalyst, have also been proposed to explain olefin polymerization with Ziegler–Natta catalysts, but the simpler monometallic mechanism suggested by Cossee's is still considered the most general description of these systems.

All polymerization steps discussed above have kinetic rate constants that are defined for single-site metallocene or late transition metal catalysts. Multiple-site heterogeneous Ziegler–Natta and Phillips catalysts can be considered as having several different types of active sites, each following the polymerization steps described above, but with different rate constant values. What differentiates one site type from the other in a Ziegler–Natta catalyst is not the polymerization mechanism they follow but rather the values of the rate constants for the same elementary steps. We see in Chapters 5 and 6 how this concept can be extended to model the polymerization rate and microstructure of polyolefins made with Ziegler–Natta and Phillips catalysts.

Finally, we should mention a few complicating effects observed during olefin polymerization with coordination catalysts that were not included in any of the mechanistic steps described above: (i) the comonomer effect, (ii) the hydrogen effect, and (iii) the catalyst/cocatalyst ratio effect.

The comonomer effect is one of the best known phenomena in coordination polymerization. In general, the rate of homopolymerization with a given catalyst decreases in the order ethylene > propylene > 1-butene > ... > higher α-olefins. However, when ethylene is copolymerized with small fractions of α-olefins for the production of LLDPE, the rate of copolymerization may be significantly higher than the rate of ethylene homopolymerization. This observation cannot be explained with the reactions listed in Scheme 3.11 alone. A few alternative mechanisms for this behavior are explored in Chapter 5.

Besides acting as a chain transfer agent, hydrogen can also influence the polymerization rate significantly. Hydrogen most often increases the rate of propylene polymerization; this effect is well documented and its explanation is now widely accepted. We saw before that the 1-2 insertions are favored during propylene polymerization with heterogeneous Ziegler–Natta catalysts, but 2-1 insertions may also take place, generating a regiodefect on the chain. It is easy to see that the active site that is formed after a 2-1 insertion is sterically hindered because the methyl group is closer to the transition metal, as shown in Scheme 3.13b. It is reasonable to expect that this site will have a lower propagation rate than sites terminated with 1-2 insertions. Hydrogen, being a small molecule, is assumed to react more easily with 2-1 terminated sites, freeing up these "dormant" species for further propagation reactions. Contrarily, hydrogen generally decreases the rate of ethylene polymerization; there is no consensus in the literature regarding a general theory for the effect of hydrogen on ethylene polymerization, but some hypotheses have been advanced to explain this phenomenon, as discussed in Chapter 5.

Figure 3.23a shows the hydrogen effect on the polymerization rate of ethylene with a heterogeneous Ziegler–Natta catalyst in a semibatch reactor. The polymerization starts without hydrogen, which is only injected in the reactor after

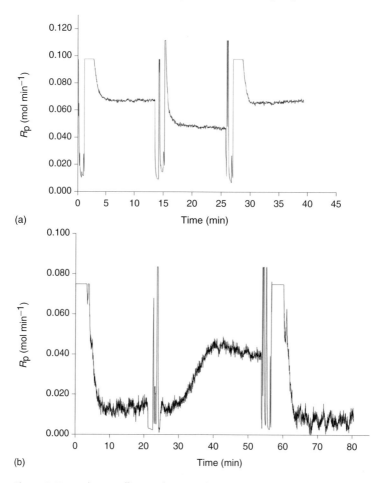

Figure 3.23 Hydrogen effect on the rate of polymerization with a heterogeneous Ziegler–Natta catalyst for: (a) ethylene and (b) propylene.

approximately 15 min. After a short transient period, the monomer uptake rate settles to a lower value, illustrating the negative effect hydrogen has on the ethylene polymerization rate. However, if the reactor gas contents are vented off and the reactor is pressurized again with pure ethylene, the monomer uptake rate is restored to its original value before hydrogen is introduced in the reactor, showing that the hydrogen effect is completely reversible for this catalyst.

This behavior is in clear contrast to that for propylene polymerization with the same catalyst, as shown in Figure 3.23b. When hydrogen is introduced after approximately 25 min of polymerization, the monomer uptake curve shows a period of rapid acceleration to a new higher polymerization rate with a slight, but clear, catalyst decay profile. Venting hydrogen off after about 55 min of polymerization leads to a new monomer uptake rate, with values slightly lower than the ones before hydrogen was introduced in the reactor, likely because the catalyst underwent some degradation during the period hydrogen was present in the reactor. Nonetheless, the hydrogen effect on the polymerization rate seems to be clearly reversible in this case as well.

Scheme 3.1 shows that the cocatalyst alkylates and reduces the catalyst to form the active sites, but, in reality, this is an oversimplification of the cocatalyst roles. The cocatalyst also scavenges polar impurities, stabilizes the active sites, and, if present in too large an excess, reduces the polymerization rate. In general, if we plot the polymerization rate as a function of the cocatalyst/catalyst ratio, the curve will initially increase up to a maximum value and then start decreasing, as illustrated in Figure 3.24. At the same time that the cocatalyst changes the polymerization rate, it also acts as a chain transfer agent, affecting molecular weight averages and MWD width, as shown in Figure 3.25. The effect of the cocatalyst on the different polymerization rates is difficult to quantify with a theoretical model and is mostly done with empirical correction factors.

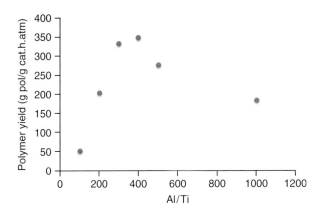

Figure 3.24 Effect of Al/Ti molar ratio on polymer yield for ethylene polymerized with a heterogeneous Ziegler–Natta catalyst and TEA for 1 hour at 80 °C, P_H, 2 atm; $P_{ethylene}$, 5 atm in a slurry reactor.

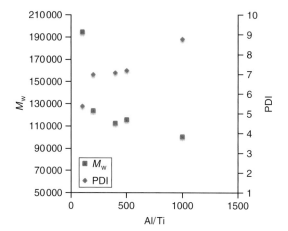

Figure 3.25 Effect of Al/Ti molar ratio on M_w and PDI for ethylene polymerized with a heterogeneous Ziegler–Natta catalyst and TEA for 1 hour at 80 °C, P_H, 2 atm; $P_{ethylene}$, 5 atm in a slurry reactor.

Further Reading

Several excellent review articles are available covering different aspects of coordination polymerization catalysts and mechanisms. Ziegler–Natta catalysts have been the subject of several reviews in the past.

Galli, P. (1994). *Prog. Polym. Sci.*, **19**, 959.Soga, K. and Shiono, T. (1997) *Prog. Polym. Sci.*, **22**, 1503.

Galli, P. and Veccellio, G. (2001) *Prog. Polym. Sci.*, **26**, 1287.

The microstructural details of polypropylene made with Ziegler–Natta and metallocene catalysts has also been extensively revised.

Busico, V. and Cipullo, R. (2001) *Prog. Polym. Sci.*, **26**, 443.

Several phenomena that affect the behavior of Phillips catalysts have been investigated in several papers.

Smith, P.D. and McDaniel, M.P. (1990) *J. Polym. Sci., Part A: Polym. Chem.*, **28**, 3587.

McDaniel, M.P., Collins, K.S., Benham, E.A., and Cymbaluk, T.H. (2008) *Appl. Catal. A: Gen.*, **335**, 180.

McDaniel, M.P., Collins, K.S., Benham, E.A., and Cymbaluk, T.H. *Appl. Catal. A: Gen.*, **335**, 252.

McDaniel, M.P. and Collins, K.S. (2009) *J. Polym. Sci., Part A: Polym. Chem.*, **47**, 845.

McDaniel, M.P., Collins, K.S., and Benham, E.A. (2007) *J. Catal.*, **252**, 281.

An authoritative issue of *Chemical Reviews* was published in 2000 covering in great detail several topics on olefin polymerization with single-site catalysts.

Gladysz, J.A. (2000) *Chem. Rev.*, **100**, 1167.

Ittel, S.D., Johnson, L.K., and Brookhart, M. (2000) *Chem. Rev.*, **100**, 1169.

Alt, H.G. and Köppl, A. (2000) *Chem. Rev.*, **100**, 1205

Coates, G. (2000) *Chem. Rev.*, **100**, 1223.

Resconi, L., Cavallo, L., Fait, A., and Piemontesi, F. (2000) *Chem. Rev.*, **100**, 1253.

Hlatky, G.G. (2000) *Chem. Rev.*, **100**, 1347

Fink, G., Steinmetz, B., Zechlin, J., Przybyla, C., and Tesche, B. (2000) *Chem. Rev.*, **100**, 1377.

Chen, E.Y.X. and Marks, T.J. (2000) *Chem. Rev.*, **100**, 1391.

Angermund, K., Fink, G., Jensen, V.R., and Kleinschmidt, R. (2000) *Chem. Rev.*, **100**, 1457.

Boffa, L.S. and Novak, B.M. (2000) *Chem. Rev.*, **100**, 1479.

4
Polyolefin Reactors and Processes

> Any sufficiently advanced technology is indistinguishable from magic.
> *Arthur C. Clarke (1917–2008)*
>
> It worked! (said after witnessing the first atomic detonation).
> *J. Robert Oppenheimer (1904–1967)*

4.1
Introduction

As we saw in Chapter 1, the importance of polyolefins stems from the tremendous variety of products and properties that can be made with these materials. In order to obtain such a broad range molecular weight, composition, and branching distributions, it is necessary to use a variety of catalyst types and a broad range of polymerization conditions. At the risk of oversimplifying, it is for this reason that so many different reactors and processes are used to make polyolefins; some are simply more effective and economical than others for producing polymers with a given range of properties.

It is almost impossible to discuss reactors for polyolefin production without describing the entire polymerization process. There is a strong relationship between the process, the reactor configuration, and the catalyst used for polymerization. Advances in one component often enable or stimulate the development of the other components of the process. The reactor is the heart of the process, and the process is designed to complement the various attributes of a particular reactor setup. While the reactor is the heart of the process, the catalyst is the heart of the reactor.

Modern processes for polyolefin production are extremely efficient at producing large quantities of polymer. World-scale plants have grown from 80 kt per year in 1980 to more than 750 kt per year for the newest plants being built now. Despite these changes in production capacity, the basic process for making polyolefins has changed little since it was first commercialized in the 1960s. Figure 4.1 is an illustration of the basic blocks of such a process. What has changed from the basic 1960s process design is that now each block has much better efficiency, higher

Polyolefin Reaction Engineering, First Edition. João B. P. Soares and Timothy F. L. McKenna.
© 2012 Wiley-VCH Verlag GmbH & Co. KGaA. Published 2012 by Wiley-VCH Verlag GmbH & Co. KGaA.

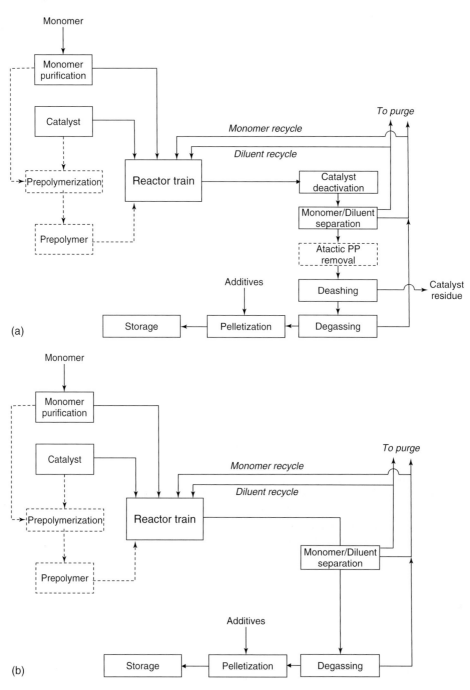

Figure 4.1 Schema of polyolefin plant using (a) pre- and (b) post-fourth generation Zielger–Natta catalysts.

throughput and requires less capital investment. Advances in catalyst technology allowed for the design of more efficient reactors, higher catalyst yields, and the elimination of some of the intermediate steps, such as catalyst deactivation and deashing.

Broadly speaking, polyolefins can be made in solution, in slurry, or in gas-phase reactors. In solution reactors, both the catalyst and the polymer are soluble in the reaction medium. They are used to produce most of the commercial EPDM rubbers and some polyethylene resins. Autoclave, tubular, and loop reactors can be used in solution processes. In slurry- and gas-phase reactors, the polymer is formed around heterogeneous catalyst particles, as discussed in detail in Chapter 7. Slurry processes can be subdivided into diluent and bulk. In diluent processes, an inert diluent (typically a C_4–C_6 alkane) is used to suspend the particles while gaseous (ethylene and propylene) and liquid (higher α-olefins) monomers are fed to the reactor. On the other hand, only liquid propylene monomer is used in the bulk process. Polyethylene and polypropylene can be produced in diluent reactors, while bulk reactors are restricted to polypropylene and its copolymers. Slurry processes involve the use of autoclaves or loop reactors. Gas-phase reactors are also used to polymerize ethylene, propylene, and higher α-olefins. They can be classified into fluidized bed reactors (FBRs) and stirred bed reactors. A gaseous stream of monomer and nitrogen fluidizes the polymer particles in FBRs, while mechanical stirring is responsible for suspending the polymer particles in gas-phase stirred bed reactors. Gas-phase stirred bed reactors are further subdivided into horizontal and vertical reactors.

Several polymerization processes use only one reactor, but, as discussed in Chapter 1, two or more reactors, often of different types, can also be operated in series to produce polyolefins with more complex microstructures. This variety of reactor configurations is unique to polyolefins among all commodity and specialty polymers. It also indicates how important polymer reaction engineering is for the manufacture of polyolefins.

It is useful to separate the discussion on processes into those for polyethylene and those for polypropylene, as the requirements for these two polymers are different and have led to similar, but not identical, processes. A qualitative overview of the various reactor types is presented first, followed by descriptions on how each reactor configuration is used in different polymerization processes throughout the world.

4.2
Reactor Configurations and Design

All commodity polyolefin processes, and thus reactors, are continuous processes for reasons essentially linked to economics. This means that it is necessary to consider residence time distributions (RTDs) in order to fully understand the impact of a reactor choice on the range of products that can be made in it. Given the high production rates mentioned above, it is extremely uneconomical to run batch or semibatch processes where up to 50% of the process time is consumed by steps such as emptying, cleaning, and refilling the reactor.

One of the main concerns during the design and control of polymerization reactors is how to remove the heat of reaction efficiently, since polymerizations are highly exothermic reactions. One could reasonably claim that a significant driving force for the advances in reactor design over the years, from the original simple stirred autoclaves to more sophisticated loop technologies, has been the need to efficiently remove the heat of reaction during the polymerization. To illustrate the importance of heat removal, let us consider a small-scale polyethylene plant producing a very modest 200 000 tonnes of polyethylene per year in a two-reactor train. The enthalpy of reaction for ethylene polymerization is on the order of 3600 kJ kg^{-1}. Assuming approximately 8000 operating hours per year, the total heat generation rate for the reactor train, $\dot{Q}$, is on the order of

$$\dot{Q} = 2 \times 10^5 \text{ t per year} \times \frac{1 \text{ year}}{8000 \text{ h}} \times \frac{1 \text{ h}}{3600 \text{ s}} \times 3600 \text{ kJ kg}^{-1}$$
$$= 2.5 \times 10^4 \text{ kJ s}^{-1} = 25 \text{ MW} \qquad (4.1)$$

In other words, a conservative estimate tells us that one reactor in the train will be generating heat at a rate of over 10 MW – an impressive amount of energy to evacuate (and recover). Clearly, heat removal will be a concern in the design and operation of polyolefin reactors, regardless of the process used. The enthalpy of reaction of propylene polymerization is on the order of 2400 kJ kg^{-1}. This means that for a given bed volume and production rate, it is necessary to remove about 50% more heat in a polyethylene process than in a polypropylene process. So, while heat transfer remains a concern for polypropylene, it is critical in polyethylene.

In polymerization reactors, it is also imperative to be able to produce polymer that can be easily separated from unreacted monomer, catalyst residues, and other by-products, while assuring that polymer properties remain on target for a given grade. These stringent requirements led to the design of several reactor types, each with their own advantages and disadvantages. As mentioned above, this variety of reactor configurations is unique to polyolefins among all commodity and specialty polymers; for example, almost all emulsion polymers are produced in stirred autoclaves. Each of these different reactor configurations is discussed below.

4.2.1
Gas-Phase Reactors

Gas-phase processes were the last of the three major families of polymerization processes to be developed. They offer an economical and energy-efficient alternative to liquid-phase polymerization. Since the monomer is in the gas phase, separating the polymer from the unreacted monomer is relatively easy. There is no need to flash off large amounts of liquids, a step with high energy requirement; this represents a significant cost reduction with respect to slurry processes using hydrocarbon diluents. An extended product range is also theoretically possible with gas-phase reactors, as there is no solubility limit for hydrogen and monomers in the reaction medium, resulting in products with higher melt flow index and increased comonomer content. In solution and slurry reactors, on the other hand, the low solubility of comonomers and hydrogen in the reaction medium limits the molecular weight and comonomer incorporation that can be achieved in these processes. However, the reality is more complex, and the flexibility of a process in terms of grade changes will depend on the size of the reactors as well as their RTDs.

In addition, while heat transfer is always an issue with exothermic reactors, it is particularly problematic in gas-phase processes, given the poor thermal characteristics of gaseous species. For this reason, "per pass" conversion of monomer tends to be lower in gas-phase reactors than in slurry processes, and special steps must be taken to enhance heat transfer to keep gas-phase reactors economically competitive in terms of productivity. Such steps include the injection of small amounts of liquid components below their dew points or the use of inert gas-phase compounds with higher heat capacities than nitrogen.

Four types of gas-phase reactors are used in industry: FBRs, which are used for both polyethylene and polypropylene, and reactor configurations used essentially for polypropylene and its copolymers, including vertical stirred bed reactor (VSBR), horizontal stirred bed reactor (HSBR), and the multizone circulating reactor (MZCR). As mentioned above, the enthalpy of polymerization is approximately 2400 kJ kg^{-1} for polypropylene and 3600 kJ kg^{-1} for polyethylene. This difference is responsible for the fact that FBRs are the only type of reactor used to make polyethylene in gas-phase processes.

4.2.1.1 Fluidized Bed Gas-Phase Reactors

A schema of an FBR is shown in Figure 4.2. The reactor is essentially an empty cylinder with a distributor plate near the bottom and a disengagement zone at the top. Polymerization takes place in the zone between the distributor plate and the freeboard zone. Economic necessity dictates that they be quite large and operated continuously. The height of the reaction zone is typically on the order of 10–15 m (but can be even higher), and the height-to-diameter ratio of this part of the bed is between 2.5 and 5. The disengagement zone at the top of the reactor is at least two times wider than the reactor bed.

Fresh feed and recycled gases are injected below the distributor plate. As the name suggests, its principle role is to distribute the components in the powder bed in an appropriate manner. The design of the plate is not trivial. The holes in the plate that allow the gases to pass must orient the gases (and eventually liquid droplets) in order to promote fluidization. They must also be configured to prevent the particles in the reactor from settling in the injection zone and blocking the arrival of gases, as well as to stop the powder from falling through the plate into the feed zone below. Both would prove catastrophic to the operation of the FBR, as a reduction in the gas flow rate into the reactor would lead to loss of heat removal capacity and melt down of the polymer inside.

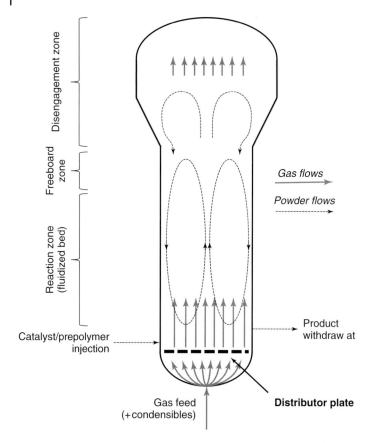

Figure 4.2 Sketch of a fluidized bed reactor for olefin polymerization.

The reaction zone begins above the distributor plate. Here, catalyst (or occasionally prepolymerized catalyst[1]) is injected just above the plate, and polymer particles are also withdrawn at a similar bed height. Both the feed and withdraw streams are pulsed rather than continuous. Despite the fact that catalyst injection rates are on the order of grams per second, care must be taken when designing the injectors. Improper injector design can lead to plugging of the feed lines because of the formation of polymer; it is imperative to avoid backflow of the monomer into the feed lines at the end of each pulse. In addition, it might be preferable to feed the catalyst using propane than nitrogen because propane has a higher heat capacity and helps avoid local hot spots in the feed zone.

Product withdrawal also requires some precision. The powder is discharged under the influence of gravity through a series of at least two chambers. The frequency of the opening and closing of the valves to the withdraw chambers must

---

1) Prepolymerization refers to the practice of producing a small amount of polymer on a fresh catalyst particle under mild conditions before injecting it into the main reactor train. See Chapter 7 for a discussion of the importance of prepolymerization.

be carefully controlled to maintain a constant bed level and to reduce the amount of gas leaving the reactor. In addition, it is important to prevent both any direct connection between the reactor and the downstream degassing section of the plant and the chambers being plugged by the formation of any undesired polymer.

The gas flow rate into the reactor zone is set in order to fluidize the growing polymer particles. Typically, the superficial gas velocity (volumetric flow rate of gas divided by the cross-sectional area of the reaction zone) is on the order of $0.5-1 \text{ m s}^{-1}$, and the relative gas-particle velocity is in the vicinity of two to eight times the minimum fluidization velocity. In order to get such high flow rates through the bed, one typically uses a recycle ratio upward of 50, with low per pass conversions (2–30%). The recycled gas is recovered at the top of the bed, compressed and cooled before being fed back to the bottom of the reactor. The relative gas-particle velocities are in fact higher for FBRs than for the stirred bed and recirculating reactors described below, and this implies that FBRs have the best heat removal capacity of any of the gas-phase reactors. This is the reason why FBRs are the only reactors used to make polyethylene in the gas phase.

The fluidization of the powder bed is not a trivial process. One of the reasons for this is that the continuous stirred tank reactor (CSTR)-like residence time of the solid phase means that particle size distribution (PSD) in the bed can be quite broad: going from several tens of microns if fresh catalyst is injected into the reactor up to several hundreds of microns (or even millimeters) for the final powder. The minimum fluidization velocity for an FBR varies as the square of the particle diameter, so care must be taken to maintain proper gas flow rates in this case. Thus, proper fluidization requires robust process control and proper design of catalyst particles. One of the many reasons for a prepolymerization step to be included in front of an FBR process would be to avoid the presence of too many fine, highly active catalyst particles that could easily be blown out the top of the reactor.

The denser reaction zone is separated from the upper disengagement zone by the freeboard zone. Here, the void fraction approaches one, and the velocity of most particles will go below the value for minimum fluidization and they will fall back into the bed. However, some particles are small enough that they get blown out of main bed and through the freeboard zone. Therefore, the diameter of the reactor is increased by a factor of at least two in the *disengagement zone*. This causes the superficial gas velocity to decrease even further to prevent (or at least minimize) the fines from being carried away by the fluidizing gas. Without the disengagement zone, the gas flow through the bed would have to be reduced, which would impose a serious limitation on the maximum reaction rate for reasons related to heat transfer. Ideally fine particles will fall out of the gas phase and back into the main bed of the reactor along the sloped walls of the disengagement section. However, it is possible that the particles that fall out of suspension in this part of the reactor are deposited on the reactor wall, where they build up to form chunks or sheets that can occasionally fall back into the reactor bed.

The gas and powder phases obviously have different RTDs (in fact, this is often the case with gas-phase reactors). In the simplest representation of flow in the reactor, the gas phase will flow upward from the distributor plate in a plug-flow-like

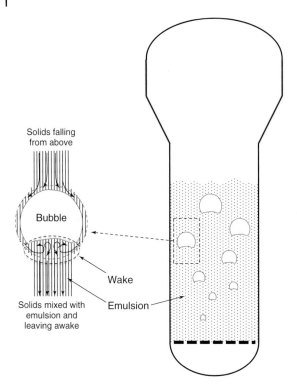

Figure 4.3 Schema of gas and solid flows in a bubbling FBR.

manner. The powder tends to be blown up the center of the reactive bed and falls down the outside walls, forming a recirculation zone. As this recirculation takes at most 30–60 s to complete a full cycle, and the average residence time of the powder phase is on the order of 30–180 min, the RTD of the powder phase is likened to that of a CSTR.

While this might be true of a smoothly fluidized bed, the FBRs used in polyolefin production tend to be operated in a bubbling regime (Figure 4.3), so the reality of the situation is not quite so simple. In a bubbling FBR, the gas will be distributed between a bubble phase made entirely of the gaseous components and an emulsion phase that behaves like a fluid and is made up also of gas and polymer/catalyst particles. The solid particles are found in the emulsion and in the wake trailing the bubbles. The gas phase will probably not be in true plug flow because the bubbles rise faster than the emulsion phase, and also some gaseous components will be entrained in the polymer particles as they fall back down into the bed. The particles tend to mix in the wake region behind the bubbles, so one can consider that, at least locally, the emulsion phase is well mixed. However, since the bubbles will grow as they move up through the bed, causing the local conditions to change, it is not uncommon to see the emulsion phase of the reactor modeled as a cascade of CSTRs with recirculation streams that exchange mass with the bubble phase.

It is also possible that the particles become segregated in the bed, with the average particle size decreasing as the bed height increases. Given that the minimum fluidization velocity decreases as particle size decreases, it is not illogical that this occurs. The net result of bed segregation is that the smaller particles will have a slightly longer time to grow before descending in the bed, making the PSD of the powder at the outlet slightly narrower than that one would obtain if the solid phase residence time were equivalent to a perfect CSTR.

Pressures and average temperatures in the bed are set primarily as a function of the desired polymer properties and production rates. The upper temperature limit is determined by the polymerization kinetics, by the evolution of molecular weight distribution (MWD), and ultimately, by the softening point of the polymer. In the case of polyethylene, average softening temperatures are close to 90 °C for linear low-density polyethylene (LLDPE) or upward of 110 °C for high-density products. The thermal sensitivity of the catalyst will also play a role in the choice of reactor temperature, with certain catalysts for propylene polymerization being used at 70–80 °C despite the much higher melting point of polypropylene because stereospecificity and activity can decrease at higher temperatures.

FBRs can be highly efficient reactors. Per pass conversions are typically on the order of 5% (but can be as high as 30% depending on the heat removal efficiency), but overall conversions are usually greater than 95%. Despite their efficiency, there are several drawbacks associated with the operation of FBRs. Stable temperature control in the reactor is extremely important; all localized hot spots must be minimized. Hot spots will cause the polymer particles to melt and stick together, forming chunks and lumps. Static electricity can also be generated in the reactor and can lead to similar problems. Small chunks and lumps can cause transport and processing problems downstream, while large ones can block the outlet of the reactor. The latter condition creates an extremely dangerous situation and the reactor has to be shut down right away. If it is not shut down properly, the content of the entire reactor may melt together to form a big polymer lump such as the one shown in Figure 4.4.

Figure 4.4 Example of the result of poor control over fluidization and heat transfer in an FBR. (Reproduced with permission from the Big Pellet Club.)

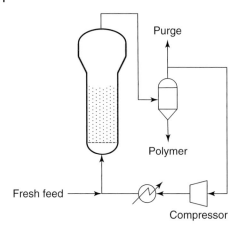

Figure 4.5 Schema of feed/recycle configuration for an FBR.

Efficient heat removal is critical and can be achieved by different means. The reactor itself is effectively adiabatic since only a very small fraction of the heat generated by polymerization can be removed through the reactor walls. In the absence of condensed cooling (see below), transferring energy to the gas phase is the only means of eliminating the heat generated by the polymerization. Increasing the flow rate of the gas can improve the relative gas-particle velocity and thus the heat transfer coefficient between the particle and the continuous gas phase. This is much easier in an FBR than it is in the other gas-phase reactors presented in this section. There are obviously limits to this option because the higher the flow rates, the lower the per pass conversion, the higher the recycle rate (and associated compression costs), and the greater the risk of entrainment of small particles out of the disengagement zone. In addition, chilled feeds to the reactor can improve the sensible heat removal from the reactor. As shown in Figure 4.5, the recycled gas phase is compressed and then passed through an external heat exchanger. Another tactic in "dry mode" is to use inert gases with high heat capacity. An inert gas such as nitrogen is added to control the partial pressure of the reactive gases. However, the heat capacity of nitrogen is lower than that of an inert such as propane (0.029 kJ/(mol K)$^{-1}$ for nitrogen and 0.075 kJ (mol K)$^{-1}$ for propane at 1 bar and 25 °C [1]), so some producers will use the latter as a process gas, allowing the gas phase to absorb more heat as it passes through the reactor.

In some processes, further heat removal can be achieved using what is called a *condensed mode operation* (or *supercondensed*, depending on the amount of liquid used). This consists of including condensable material in the recycle stream. The recycle stream can be cooled below the dew point of the condensable material, which then vaporizes when it heats up in the reactor. The condensable material can be a monomer (in the case of propylene), a comonomer such as 1-hexene (in the case of some LLDPE grades), or an inert alkane such as isopentane. While the benefits outweigh the costs, adding an alkane in polyethylene production will increase the levels of residual hydrocarbons in the reactor effluents, thereby increasing the load (and cost) on downstream separation and purification steps. Changing the composition of the condensable phase can be used to tailor its dew point. This method of heat removal through the latent heat of vaporization is extremely efficient and can significantly increase reactor throughput. It can also give producers more flexibility in terms of the range of products that can be made in the reactor because of the increased rate of polymerization due to the comonomer effect when making LLDPE. A reactor operating in the condensed mode can more than double the throughput of a reactor operating in the normal noncondensed mode. It should be noted that care must be taken not to destabilize the FBR during condensed mode operation. Regardless of whether the liquid droplets are sprayed into the reactor under the distributor plate, or near the catalyst injection zone, most of the liquid will vaporize and thus expand in the first 1–2 m of the bed. The increased gas flow rate in the bed due to the expansion of the liquid droplets must be accounted for by turning down the flow rate of the lighter components such as ethylene and nitrogen. One must also avoid pooling of liquid below the distribution plate, or the presence of large droplets in the bed.

Static charges can also be a significant problem in gas-phase reactors in general and in FBRs in particular. Polymer particles are poor conductors of electricity, and static electricity generated by the intense mixing and high velocity of the circulating monomer may cause fine particles to stick to the reactor walls. In FBRs, the heat transfer coefficient at the wall is small since there is little or no mixing action. The fine particles that are stuck to the wall can continue to polymerize and, because of poor heat transfer, melt to form sheets and chunks. Polymer sheets can block the outlet of the reactor if they fall off the wall, requiring immediate shutdown of the reactor. Finally, FBRs are quite large, meaning long residence times on the order of hours. This implies that grade changes can be problematic, with the formation of significant amounts of off-spec material.

4.2.1.2 Vertical Stirred Bed Reactor

A vertical stirred gas-phase reactor for polyolefin production was first developed by BASF and put into operation in 1967. The original reactor configuration is basically that of a stirred autoclave but with a bottom-mounted helical stirrer shown in Figure 4.6. Mechanically, this is a fairly complex agitation system that moves the particles in the angular and axial directions. The stirrer, which is just covered by the powder phase, is designed to convey the particles up the reactor wall and let them fall down through the center of the bed; just the opposite of what we commonly see in an FBR. This is the first type of reactor to have been used for gas-phase polymerizations. Initially, the production capacity of these reactors was limited by the shaft guidance system in larger reactors. However, with the advent of high activity catalysts, one can expect production rates of over 200 kt per year from a single VSBR of 80 m^3.

Similar to the fluidized bed gas-phase reactor, heat is removed by circulating monomer gas through an external heat exchanger. Part of the gas is condensed and returned to the reactor as liquid, and part of the gas is cooled and recycled to the bottom of the reactor. It is important to keep the temperature in the reactor

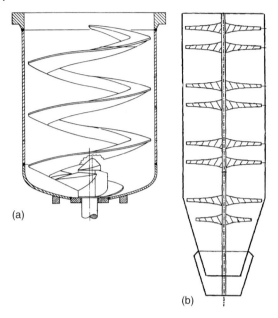

Figure 4.6 Examples of vertical stirred bed reactor. (a) A "traditional" stirred tank reactor according to US Patent 4188132 to BASF. (b) A reactor from US Patent 7 459506 to Novolen Technology Holdings BV. (*Source*: Used with permission from patent holder.)

above the dew point of the returning gas to prevent pooling of liquid monomer at the bottom of the reactor. Liquid pooling can cause uncontrolled polymerization, forming lumps and chunks that can block the outlet of the reactor. The liquid monomer is used to absorb the heat of the reaction via condensed mode cooling. The dynamics of the powder phase in the reactor approaches that of a CSTR, but some plug flow reactor (PFR) characteristics also exist because of the slow speed of agitation in this reactor configuration.

A newer reactor configuration plus agitation system is shown in Figure 4.6b and is described in a patent to Novolen Technology Holdings. This reactor configuration, while still a stirred bed reactor, is somewhat different from the earlier version. The agitation system is composed of pairs of radially curved blades (one set curved in one direction, and the other in the opposite sense) mounted on a central shaft, forming together a "segment." In the figure, the reactor contains three segments, each of which corresponds roughly to a CSTR. In the patent, the inventors claim that it is preferable to have at least 12 segments (probably more) and a height-to-inner diameter ratio of 9–10. Catalyst suspended in liquid propylene is injected into the top of this reactor, and the powder flows downward under the influence of gravity and expansion of the particles as they grow. In this configuration, Novolen claims that the reactor behaves like a cascade of a large number of CSTRs; if enough CSTRs are put in series, they can theoretically behave in a manner similar to a PFR. The advantage of this configuration (which is also exploited in the horizontal stirred beds discussed below) is that the RTD is quite different from that of a single CSTR, allowing one to make a different range of products than can be produced in the latter. Propylene is fed from the bottom of the reactor at a rate that keeps the overall gas-particle velocity at 50–80% of the minimum fluidization velocity, and liquid propylene feeds can be injected into any or all of the segments, thereby allowing one to control (to a certain extent) the local concentration of monomer, reaction rate, and heat evacuation locally (this cannot be done in a reactor with an RTD approaching that of a CSTR). It is also claimed that barrier fluids can be injected at different points in order to "separate" the segments or, in certain instances, to degas the powder. It is not clear whether or not this reactor system is actually used commercially (although the example in US Patent 7459506 employs a reactor 16 m high with a volume of 12 m^3), but this demonstrates that there is certain interest in controlling the RTD of a reactor in order to make different products. For instance, in a cascade of this type, the RTD will be narrow, so most of the particles spend exactly the same amount of time in the reactor, and as a result, the PSD is narrow. By controlling the composition at different points in the reactor (something that would be very difficult to do in an FBR, for instance), one can make products with a different arrangement of chain architectures and sizes in a single reactor. As seen below, this can be of interest for making different types of polypropylene products.

4.2.1.3 Horizontal Stirred Gas-Phase Reactor

Amoco first patented a horizontal stirred gas-phase reactor for polypropylene production in 1976 [2]. This reactor setup has now developed into the Innovene process and the Japan Polypropylene Horizone process. The reactor is a horizontal cylindrical shell 10–15 m long and 1.5–4 m in diameter, with proprietary agitators mounted on a central axis as shown in Figure 4.7. Catalyst and cocatalyst are injected at one end of the reactor, and powder is withdrawn from the other end. The powder produced by the polymerization can be moved through the reactor in two ways. In one configuration, gentle mechanical agitation of the pitched blades push the powder through the reactor. However, if the blades have no pitch, they will have no net transport effect and the particles are moved through the bed via the expansion of the particles. In this second configuration, each of the paddle-like blades will have local forward and back-mixing effects locally, but the net transport of the powder will be zero.

Monomer is fed to the reactor at different points along its axis. Gaseous components, including propylene, hydrogen, and eventually ethylene, are fed from below the bed, and a quench liquid is sprayed onto the surface of the powder higher up in the reactor. The gaseous components flow upward through the bed at a rate carefully controlled to avoid fluidizing the bed. Unreacted gas will be drawn out through the large stacks located along the top of the main cylinder.

In its original configuration, the reactor was separated into different zones by a series of weirs with polymer being moved from one zone to another by the agitator, thereby ensuring that the reactor was divided up into different zones. More modern versions of this reactor with quite complex agitators no longer require weirs.

Since the relative gas-particle velocity in this reactor is lower than in an FBR, convective heat transfer is not sufficient to remove the heat generated by the

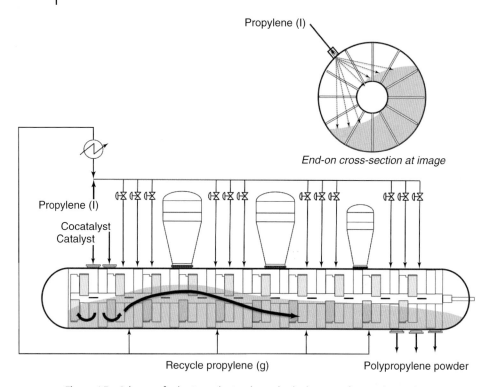

Figure 4.7 Schema of a horizontal stirred powder bed reactor for a polypropylene process. The end-on cross-sectional image shows the quench liquid (here liquid propylene) being sprayed onto the top surface of the bed. The powder bed shows a relatively complex flow pattern.

polymerization, so other means must be found to do so. This is the reason for the injection of the liquefied components in the top part of the reactor. The quench is typically liquid propylene monomer. Note that choosing the rate of liquid monomer feed can be a real challenge as there will be trade-offs between polymer properties determined by the relative composition of the gas phase and heat transfer. It is, therefore, occasionally necessary to combine the liquid monomer and solvent to correctly balance the feed of gaseous components. Per pass conversions upward of 25% are achieved with this method of cooling, compared to only 5–15% if gaseous feeds are used. Using liquefied monomer also greatly reduces the heat load removed by the recycled monomer gas, and thus, the gas velocity can be kept much lower resulting in lower capital as well as operating costs. Uniform mixing of the powder is crucial, as renewing the powder surface exposes hot particles from within the bed to the spray of quench liquid.

The spacing of the catalyst and cocatalyst injection points appears to be a critical issue with this reactor; too close and one encounters lump formation in the feed zone, probably due to overheating, too far apart and the catalyst is not properly activated, causing productivity and quality to suffer. It appears that a certain amount

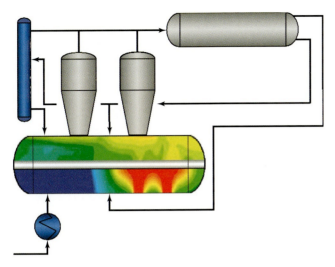

Figure 4.8 Hydrogen concentration (blue, lowest; red, highest) in an HSBR reactor. (*Source*: Reproduced with the consent of INEOS.)

of back mixing in the powder phase at the inlet is required for the uniform activation of the catalyst particles.

As with other gas-phase processes, one needs to consider the RTD of the gas and solid phases separately. This reactor operates essentially in cross-flow. Recall that the gaseous components are fed in at different points along the bottom of the reactor. While there will be some axial transport of the reactive species in the powder, the unreacted gases flow through the bed and out of the stacks at the top in such a way that there is very little mixing of gases from the different injection points. This means that it is possible either to maintain a very uniform gas-phase composition in the reactor or to have a gradient of composition in the axial direction. For instance, the schema in Figure 4.8 shows a computational fluid dynamics (CFD) simulation of an HSBR from INEOS, with different levels of hydrogen concentration at different parts of the bed, leading to the production of a bimodal resin with a Ziegler–Natta catalyst in a single reactor.

The RTD of the powder phase in the HSBR is very different from that of the FBR and the original VSBR (Figure 4.6a). In the HSBR, the RTD is commonly approximated as a cascade of three to five ideal CSTRs. While reality is more complex than this [3], it is reasonable to say that the powder phase has a "PFR-like" RTD (like the more recent reactor proposed in Figure 4.6b). This combination of RTD is unique among commercially licensed polyolefin reactors. Since the RTD of the powder is narrower than in most other gas-phase reactors, the polymer particles at the outlet to the reactor will be more homogeneous in terms of their composition, MWD, and PSD than one would obtain from an FBR. In addition, the ability to exert control of the gas-phase composition along the axis of the reactor also represents an extra tool for the polymer reaction engineer. The narrow RTD also permits faster grade transition times, less off-spec material formation during grade transition, and higher rubber contents in impact polypropylene resins than other technologies.

On the other hand, the reactor can be delicate to operate. As with most gas-phase processes, chunking and agglomeration of the particles can be an issue, as well as cleaning the reactor can be difficult since it is necessary to dismantle it, and slowly and carefully withdraw the agitator shaft, which has a very small clearance with the reactor wall.

4.2.1.4 Multizone Circulating Reactor

The MZCR pictured in Figure 4.9 is the latest reactor configuration developed by Basell and was first commercialized in the early 2000s as the basis of the Spherizone technology. This reactor is similar to the riser-tube reactor used in fluidized catalytic cracking. It consists of two interconnecting reaction zones, a riser and a downer, separated at the top by a cyclone. In the first zone (the riser),

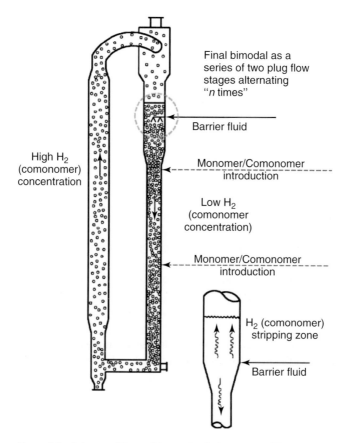

Figure 4.9 Schema of the multizone circulating reactor. (*Source*: Reprinted from Chemical Engineering Science, M. Covezzi, G. Mei, The Multizone Circulating Reactor Technology, 4059–4067, Copyright (2001), with permission from Elsevier.)

the gas velocity is high enough that this section is a fast-fluidized bed. The particles are blown into the cyclone at the top of the reactor, which separates the particles from the gas phase of the riser. The particles then pass into the "downer" section that behaves like a moving packed bed where the particles move simply by the pull of gravity. According to Covezzi and Mei [4], the particle density in the downer is close to 90% of the bulk density of the final powder, whereas it is less than half that level in the riser. Product withdrawal is done at the bottom of the downer in order to reduce as much as possible the amount, and therefore cost, of recycling the gas phase. Much like the slurry loop reactor discussed below, the recycle ratios in the reactor are very high, so the particles move around with a residence time approaching that of a CSTR.

A barrier fluid is injected in the lower portion of the cyclone. The barrier fluid, for instance propane near or below the dew point, is denser than the gas phase leaving the riser and is intended to help stop the entrainment of the lighter gases in the riser, in particular, hydrogen and/or ethylene, into the downer. Barrier fluid is fed in slight excess with respect to the theoretical interstitial volume of the downer in order to maintain a total downward flow of gas in this section of the reactor. The result of this separation of the particles and evacuation of the gas phase leaving the riser means that the two zones can be operated under very different polymerization conditions to produce polymers with different compositions in each zone. The riser usually operates with a hydrogen concentration 2–4 orders of magnitude higher than that in the downer, meaning that it is possible to produce a resin with a bimodal MWD in a single continuous reactor using a standard olefin polymerization catalyst – something that is impossible with a standard FBR or VSBR. In addition, the growing polymer particles circulate between the two different reaction zones, building up layers of different composition in each successive pass. This means that chains with different architectures will be more intimately mixed at the outlet to this reactor. In other words, as was the case with the HSBR, manipulating the RTD of the different phases, it is possible to produce polymers with very different properties. The holdup in each leg of the reactor can also be varied, giving an extra level of control over the properties of the final polymer.

The maximum amount of polymer that can be accommodated in the reactor will be limited by its heat removal capacity. As in the FBR, heat can be evacuated from the riser by chilling the feeds (above or below the dew point) to take advantage of both sensible heat removal and condensed mode operation. This section of the reactor is close to isothermal. The relative gas-particle velocity in the downer zone is much lower than that in the riser, so this leg is operated under pseudoadiabatic conditions. This factor, coupled with the much higher packing of the particles, implies that convective heat transfer to the fluid phase will be rather poor in this zone. Some sensible heat removal will help control the temperature here, but a good portion of the energy evacuation will be accomplished by adding liquid feeds. Since the packing density is relatively high and the entire bed flows down under the influence of gravity, it is possible to add a fair amount of liquid to this section of the reactor. However, given the nature of the bed, it is highly probable that, as in the case of the FBR, a prepolymerization step will be needed to avoid particle agglomeration.

4.2.2
Slurry-Phase Reactors

Slurry processes for olefin polymerization can be broken down into two broad categories: diluent and bulk. Diluent processes use a diluent (C_3H_8–C_6H_{14}) that is a nonsolvent for polyolefins to suspend the polymer particles. The isoforms of butane and hexane are preferred to the normal isomers, as the former are poorer solvents. Although the diluent does not directly affect the polymerization, it has been shown that different diluents might change catalyst behavior, probably due to electronic interaction with the active sites. Gaseous monomers and hydrogen are continuously bubbled through the diluent. Liquid α-olefin comonomers, diluent, catalysts, and cocatalyst are continuously fed to the reactor. The diluent approach is the only means of making polyethylene in slurry since ethylene cannot be liquefied economically. Alternatively, liquefied propylene can be fed to the reactor in the bulk process. Except for this difference, all other conditions are similar to the diluent process. The advantage of using slurries rather than gas phase can be found in far better heat transfer capacity when the particles are suspended in a liquid. In addition, maintaining a fixed pressure and using the heat of polymerization to boil the liquid monomer in the bulk reactors allow one to precisely control the temperature. Better heat transfer means that higher specific reaction rates can be tolerated, and higher rates imply that the reactor residence times (and thus volumes) can be shorter for given production rates than those of gas-phase reactors. As a result, grade changes can be faster in slurry, with less off-spec material.

On the other hand, slurry processes require the purchase, purification, removal, and recycling of the solvent, which can impose extra costs and unit operations. In addition, amorphous polyolefins are soluble in hydrocarbons (obviously, the longer the hydrocarbon, the more soluble the amorphous polymer will be). This can lead to polymer chains leaching from the particles and significant reactor fouling. Furthermore, this also implies that certain products, in particular, ethylene–propylene rubber, a component of high impact polypropylene, cannot be made efficiently in slurry reactors. With Ziegler–Natta catalysts, transfer to hydrogen generally controls the molecular weight. Since the solubility of hydrogen in the diluent is not very high, there is less flexibility for controlling molecular weight with this type of reactor than in the gas phase. This is not as important for chromium catalysts, where molecular weight control is achieved via support treatment, but can become a limiting factor with Ziegler–Natta catalysts. Metallocenes are generally very sensitive to the presence of hydrogen and are therefore less influenced by this reduced solubility than Ziegler–Natta catalysts.

There are only two choices of reactor configurations for slurry-phase olefin polymerization reactors: autoclave or loop reactors. Both reactor configurations date from the beginning of commercial olefin polymerization. Most first-generation olefin polymerization processes used autoclave reactors, with the exception of

Figure 4.10 An eight-leg loop reactor. (*Source*: Image courtesy of LyondellBasell.)

Phillips, which was the first company to introduce a loop reactor for olefin polymerization. Both configurations operate as CSTRs – the autoclaves for obvious reasons and the loops such as the one picture in Figure 4.10 for their very high recirculation ratios.

4.2.2.1 Autoclaves

Autoclaves are perhaps the simplest type of reactor to start-up and to run. No particular difficulties are associated with their construction, and it is said that processes relying on stirred autoclaves can run for more than two years with no maintenance shutdown. In polyolefin production, this reactor is usually designed to operate as a CSTR and is used in slurry, bulk, and solution processes. The autoclave provides a relatively uniform reaction medium with proper stirring. However, their main disadvantage (with respect to loop reactors) is the available heat transfer surface area per unit volume. Commercial autoclaves will have surface-to-volume ratios of approximately 5 $m^2\ m^{-3}$ (loop reactors can have over 100 $m^2\ m^{-3}$). As the heat of polymerization is evacuated via a cooling jacket (with eventually a condenser), reaction rates and solid contents must be kept lower than those in loop reactors. Temperature control can be improved in diluent slurry processes by operating the reactor at the boiling point of the diluent/polymer mixture and using an external condenser to return the vapor phase as a subcooled liquid.

Many modifications to the basic stirred autoclave design have been made to improve its heat transfer characteristics; those commonly used in polyolefin production processes include the use of external coolers, overhead condensers, or internal cooling coils.

When external coolers are used, a portion of the polymer slurry in the reactor is circulated through one or more external heat exchangers to remove the heat of polymerization. This is an efficient method of heat removal but puts stringent requirements on the morphology of the polymer particles being produced. Polymer fines are undesirable, as they tend to deposit on the heat exchanger walls and require frequent shutdowns to clean out. An example of a process that uses stirred autoclaves with external coolers is the Mitsui polyethylene process.

A very efficient alternative for heat removal is to use overhead condensers. This modification uses the latent heat of evaporation of the monomer (propylene) to remove the heat of polymerization. Monomer is evaporated in the reactor, condensed in the overhead condenser, and the cooled liquid monomer is returned to the reactor. This design works well for propylene polymerization but is naturally not a good option for ethylene because of its much lower boiling point. Overhead condensers are used in the El Paso bulk polypropylene process.

Finally, internal cooling coils are also used to increase the heat transfer area inside the reactor, but they are generally less efficient than the other two methods mentioned above, as they are often subject to fouling.

4.2.2.2 Slurry Loop Reactors

Loop reactors are used to make at least 50% of all commercial polyolefins. Many different loop reactor configurations are used in industrial processes. The loop can be in either a vertical position (Phillips and Spheripol processes) or a horizontal position (USI process). Modern loop reactors can have up to 12 legs (the reactor in Figure 4.10 has 8 legs), and each leg can be as much as 60 m tall in the more common vertical position configuration. A typical loop reactor for a 250 kt per year plant will have a volume on the order of 100 m^3. Nonetheless, they are extremely simple in conception since they consist of empty tubes surrounded by cooling jackets. As noted above, they have very high ratios of heat transfer area per unit volume, allowing them to operate at high solid contents, typically on the order of 40–50% v/v, and high polymerization rates. Solid content will most likely be more limited by the viscosity of the slurry than by the requirements to remove heat of polymerization from the reactor. Higher solid contents also imply that downstream devolatilization steps are easier and less costly than with autoclaves. They are also run full of liquid, without a gas space, making it is easier to control hydrogen concentration in a loop reactor than in an autoclave. This permits better control over the polymer MWD.

The only agitation in the reactor is provided by a pump situated at the bottom of one of the legs. The pumps are crucial, as they must move very large volumes of slurry at velocities on the order of meters per second without becoming fouled or damaging the growing polymer particles. High fluid velocity (on the order of $10-30$ m s^{-1}) creates turbulence in the reactor, which helps prevent particle settling and improves the overall heat transfer coefficient of the reactor. However, it is necessary to avoid the formation of gas bubbles in the reactor whenever possible, as hydrogen and monomer will be favorably partitioned to the gas phase, and cavitation around the blades of the pump can be a problem.

As with FBRs, the overall conversion in a loop can be very high, with per pass conversions being less than 10%. In propylene bulk polymerization, the conversion per pass is limited by the amount of monomer needed to sufficiently suspend the slurry. As mentioned above, the RTD of the loop reactor is reasonably approximated by that of a CSTR. This means that the product made in these reactors is relatively homogeneous and that there are fewer levers with which to manipulate properties in these reactors than in certain types of gas-phase reactors such as the HSBR and the MZCR.

4.2.3
Solution Reactors

Solution reactors are commonly used for the production of EPDM rubbers with soluble Ziegler–Natta vanadium-based catalysts and for different grades of polyethylene with soluble Zielger–Natta or metallocene catalysts. There are no solution processes for polypropylene production, likely because it is difficult to find a soluble catalyst system that produces polypropylene with the required stereospecificity at the reaction conditions required for propylene solution polymerization. Because one can dissolve long-chain α-olefins, solution processes may have some advantages for the production of polyethylene with long-chain branches.

Solution processes use mostly autoclaves, but tubular reactors are also found either as prepolymerizers that feed into an autoclave or as "afterburners" or "trimmers" to attain high conversion at the outlet of the main autoclave(s). As compared to slurry- and gas-phase polymerization, solution processes are commonly operated at a much higher temperature ($\sim 140-250\,°C$) to maintain the polymer dissolved in the reaction medium and higher pressures. The higher temperatures, coupled with the fact that the active sites are not supported and therefore are not subject to mass transfer resistance (Chapter 7), mean that the polymerization rates in solution processes are much higher than those in gas and slurry reactors with supported catalyst, and therefore, the average residence times of solution autoclaves will be much shorter than those of the reactors discussed in the previous sections (typically 1 to 20 min as opposed to 1–4 h). Reactor volumes are typically an order of magnitude smaller than those commonly used in gas and slurry process, on the order of 3–15 m^3 for solution versus 50–150 m^3 for heterogeneous processes.

The short residence time possible in solution reactors is a distinct advantage during grade transition. The fact that the polymer is in solution is also beneficial for process control since solution viscosity can be used as a measure of polymer molecular weight, for instance. However, high solution viscosities are a limiting factor for these reactors, reducing the achievable polymer concentration in solution, especially for high molecular weight resins. Typical ethylene concentrations in LLDPE processes are on the order of 10–20 wt%, so even at very high conversions, the amount of polymer in solution remains relatively low. Solution reactors can also operate in a wider range of temperatures than slurry- and gas-phase reactors, for which the temperature should be high enough to permit high polymerization rates but not so high as to soften the polymer, causing particle agglomeration and reactor fouling. This wider temperature range allows for more flexibility in terms of catalyst types and polymer structural control. In addition, solution reactors can be used to produce polymer from very high to very low crystallinity, since there are no problems with reactor fouling caused by sticky low crystallinity polymers.

In terms of their RTDs, the stirred autoclaves used for polyethylene production can have a range of behaviors. Like their cousins the VSBR reactors used for gas-phase polypropylene and the autoclaves used to make polyolefins in slurry phase, these reactors (and their agitators) can be designed to behave more or less like CSTRs. This is desirable for certain types of products. For instance, if one wishes a unimodal MWD and relatively homogeneous product, ideal mixing will be important. Composition homogeneity is also a requirement of most EPDM rubbers since the formation of populations with higher crystallinity is generally not acceptable in the rubber industry.

However, depending on the reactor and agitator geometries, the viscosity of the solution, and the locations of the feed and product withdrawal streams, there might be significant deviations from the ideal CSTR model. Nonideal mixing can be controlled in a foreseeable and reproducible manner, for instance, by using stirred autoclaves with large aspect ratios (height-to-diameter ratio) with the appropriate mixer and feed points. In this case, it is possible to create zones of different temperature and different hydrogen and/or monomer concentration in the reactor. This means that one can produce polymers with broad or even bimodal MWD and copolymer composition distributions in a single reactor, even with standard Ziegler–Natta or single metallocene catalysts.

4.2.4
Summary of Reactor Types for Olefin Polymerization

Table 4.1 summarizes the major advantages and disadvantages of the different reactors discussed in Section 4.2. As mentioned above, some reactors are more complex in their operation than others but might offer the possibility of a wider product slate. Other reactors might have better heat transfer capacity, or be used for extremely high production rates, but be limited either in the range of products that can be made. The final choice will be down to the economics: which reactor will offer the best cost-benefit for a given market. In other words, there is no "best" overall reactor configuration.

Table 4.1 Summary of advantages and disadvantages for several reactor configurations.

Reactor type	Advantages	Disadvantages
Stirred autoclave	Low capital cost Simplicity of operation	Low heat transfer area Low space-time yield for gas and slurry
Loop	Low capital cost Simplicity of construction High heat removal High space-time yield	Can be prone to fouling
FBR	High heat removal for gas-phase reactor Wide product range possible High space-time yield	Delicate operation Large reactor, slow grade changes Formation of hot spots
HSBR	PFR characteristic allows for fast grade transfer, wider variety of polypropylene products	More complex reactor design Higher capital cost
MZCR	Different reacting zone allows production of product not possible with traditional reactors	Complex reactor design High capital cost Agglomeration can be an issue

4.3
Olefin Polymerization Processes

It can be seen from the block diagrams in Figure 4.1 that a polymerization process is far more than one or more reactors in series but rather comprises a collection of different stages, each containing one or more unit operations, intended for a separate task. For instance, we can discern the raw materials preparation (composed of monomer purification, catalyst preparation, etc.), the polymerization (containing one or more reactors), the postreactor purification (composed of deashing, degassing, etc.), and product treatment (pelletization and stabilization) and storage (product silos) stages. Nonetheless, a discussion of all the different parts of the polymerization process is beyond the scope of this book, so when we use the term "*process*," we are essentially referring to the polymerization stage. In the context of this book, a polymerization process is therefore a collection of one or more reactors, each playing a specific role. This simplification can be justified by the fact the reactors are the heart of any polymerization process and ultimately control the physical properties of the final material.

The simplest polymerization processes have only one polymerization reactor. Even though a series of polyethylene or polypropylene grades can be made with a single reactor, certain products require the presence of two or more reactors in series. Impact polypropylene resins, for instance, need two reactors, one to make

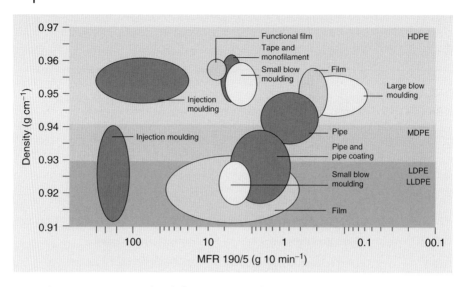

Figure 4.11 Density and melt flow ratio (inversely proportional to the molecular weight) requirements for different product applications. (*Source*: Used with permission from LyondellBasell Inc.)

the isotactic polypropylene phase and another to produce the ethylene/propylene rubber components, as discussed in Chapter 1. Likewise, HDPE resins with bimodal MWD and reverse composition also require the presence of two reactors in series, as also introduced in Chapter 1. The use of two reactors in series, despite increasing the plant capital cost, adds flexibility to the polyolefin production process and is becoming the standard in the industry nowadays.

As discussed in Chapter 1, polyethylene and polypropylene are in fact generic terms for polymers with a wide range of physical properties. For instance, the family of polyethylene products can be broken down into high-, medium-, and low-density polymers (HDPE, MDPE, and LDPE/LLDPE, respectively). Furthermore, as shown in Figure 4.11, the molecular weight is as important as the density in terms of determining the final polymer properties (in Figure 4.11, the molecular weight is reflected by the melt flow ratio[2]). Often, a combination of properties will be required; for instance, high molecular weight products (i.e., low MFR) are stiffer

---

2) The melt flow rate, or MFR, is an indication of the "flowability" of a molten polymer. Its value is the mass of molten polymer that flows through a standardized orifice at a given temperature under the pressure created by placing a standardized mass on the melted sample. Since the viscosity of a polymer melt is a function of the size of the molecules, the MFR is often used as a rough estimate of the molecular weight of the polymer. The MFR 190/5 indicated in Figure 4.11 refers to the mass of polymer that flows through the orifice at 190 °C with a 5 kg weight. The terms melt index (MI) and melt flow index (MFI) are also used for this concept. Some nomenclature complication arises from the fact that the term *melt flow ratio* can also be defined as the ratio of two MI measured using different weights and can therefore give an indication of the polydispersity index of the polymer. Melt flow rate should not, therefore, be confused with melt flow ratio.

and impart high tensile strength, whereas low molecular weight polymers are easier to process. Therefore, many products will require a mixture of both long and short chains; it is very common to find polyethylene products with broad or multimodal MWDs. In the case of polypropylene, the final products can also be mixtures of homopolymers and random copolymers or of semicrystalline homopolymers and amorphous elastomers.

Broadly speaking, there are two different approaches to make products with such mixtures of molecular weights and/or compositions: blending different products in an extruder or creating in-reactor product blends. The latter option is more effective than the former because it is almost impossible to achieve molecular-level mixing of two polymers (even if they have the same chemical structure) by mechanical mixing.

Figure 4.12 illustrates the concept of different routes to obtaining in-reactor blends of different chain lengths. One can produce a broad, unimodal MWD that contains appropriate amounts of long and short chains to produce polyethylene with a decent balance of mechanical strength and processability, or one can use a multimodal MWD and tailor the averages and relative quantities of each mode to obtain the desired balance of properties. A unimodal MWD can be made in one reactor with a conventional catalyst at a set temperature and pressure. A multimodal MWD requires either a mixture of single-site catalysts (supported or in solution) that are used in a single reactor with a unique set of operating conditions or, more commonly, one catalyst that it is exposed to different sets of conditions in a cascade

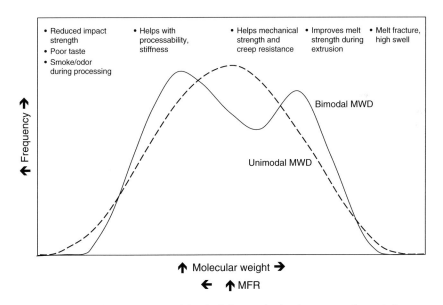

Figure 4.12 Bimodal versus unimodal polyethylene and related properties for each fraction. Using a bimodal process with two reactors results in finer control over the placement of the peaks and the relative fraction of each material. Unimodal products can be made with one reactor.

of reactor zones, for instance, two reactors in series. The multiple reactor process will clearly offer far more fine control over the relative quantities and sizes of the polymer molecules in the final product than will the single-reactor option because of the increased number of degrees of operating freedom. The question of which to choose will boil down to economics. Often, product from a single-reactor process will be less expensive because of the lower capital and operating costs. On the other hand, it might be preferable to pay for the well-defined molecular properties that can be obtained from a two-reactor process. Note that trimodal polyethylene is also coming into the market at the time this book is being written. While not a great deal of information is available on processes for trimodal polyethylene, it is not difficult to image that this will simply be an extension of the concepts used for bimodal products.

We conclude this chapter with a discussion of the selection of different processes for the production of polyolefins that will illustrate the ideas discussed above and show how the different reactors described in the previous section are used.

4.3.1
Polyethylene Manufacturing Processes

4.3.1.1 Slurry (Inert Diluent) Processes

Since operating with liquid ethylene is not pragmatic, all slurry processes for polyethylene production employ diluents to dissolve the monomer, comonomer, and hydrogen. These were the first commercial processes for the production of polyolefins. The basic process consists of a series of CSTRs, with the polymerization taking place in a heterogeneous catalyst suspended in an inert diluent. Many variations of the slurry process were developed in the early 1970s. Since the activity of these early Ziegler–Natta catalysts was relatively low, a series of CSTRs were needed to push the reaction to completion. Catalyst residue removal, called *deashing*, was also necessary, greatly increasing the capital costs of the process. With the advent of high activity catalysts, the polymerization can now be completed in one or two reactors without deashing. Despite being the first process for ethylene polymerization, the slurry process is still economically viable and competitive these days.

There are several competing hexane slurry processes, but LyondellBasell (Hostalen process), Equistar–Maruzen (also part of the LyondellBasell portfolio), and the Mitusi CX processes are the dominant ones. All three processes use similar reactor configurations composed of two to three stirred autoclaves in a cascade and similar operating conditions (Table 4.2) to make HDPE. All three processes use hexane as a diluent and are not adequate for the production of LLDPE since the amorphous fraction of this polymer would dissolve in the diluent, causing reactor fouling. The three-reactor configuration (Hostalen process) is shown in Figure 4.13. The autoclave reactors can be operated in series or in parallel, with the catalyst being fed only to the first reactor, along with ethylene and hydrogen. Typically, when a catalyst with a decay profile is used, low molecular weight homopolymer will be made in the first reactor. Since hydrogen is used to control the molecular weight and it can cause the polymerization rate to drop (see Figure 3.23a), it is added while

Table 4.2 Typical reactor conditions for slurry HDPE/MDPE processes.

Process	Reactor type	Diluent	Reactor temperature (°C)	Reactor pressure (bar)	Residence time
Mitsui	2 stirred autoclaves (series or parallel)	Hexane	80–85	<8	45 min per reactor
Basell (Hostalen process)	2–3 stirred autoclaves (series or parallel)	Hexane	75–85	5–10	1–5 h per reactor
Equistar-Maruzen-Nissan	1–2 stirred autoclaves (series or parallel)	Hexane	75–90	10–14	45 min to 2 h per reactor
Chevron Phillips	Single loop (multilegged)	Isobutane	85–100	30–40	1 h
Borealis–Borstar	2 loops[a]/FBR	Supercritical propane (loop)	85–100	60–65	–
Innovene S	1–2 loops	Isobutane	70–85	25–40	1 h

Reactor conditions differ for different product grades; the values shown here are only approximate averages found in these processes.
[a] First loop much smaller for prepolymerization.

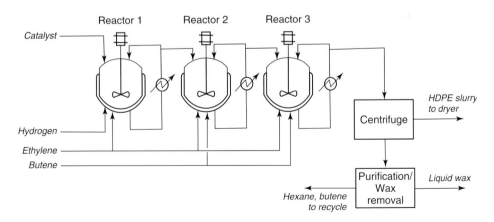

Figure 4.13 New Hostalen process from LyondellBasell.

the catalyst has its highest intrinsic activity. Higher molecular weight copolymers will be made in the subsequent reactors. In the second, and eventually third, reactor, it is usual to take advantage of the "comonomer-kick" (it is well known that small amounts of an α-olefin comonomer can increase the rate of polymerization of ethylene with respect to homopolymerization) to keep productivity up. This order of addition of catalyst, hydrogen, and comonomer is very common to most multireactor platforms, regardless of comonomer type or polymerization medium phase. Depending on the type of catalyst employed, the third reactor can be used to generate a very high or ultrahigh molecular weight component.

When ran in parallel, the reactors can operate at the same conditions to increase plant output or they can be operated at different conditions to produce different polymers.

Another important slurry process for polyethylene manufacture is the Phillips process. The Phillips process is based on the loop reactor, which can operate at higher monomer and slurry concentrations than stirred autoclaves and allows for higher space-time yields. As we mentioned above, the RTD in loop reactors is also that of a CSTR, but instead of a stirrer, an axial pump is used to circulate the reactants around a pipe loop. Isobutane is used as a diluent in the Phillips process. It has a lower boiling temperature than hexane, so can be more easily flashed off upon leaving the reactor. This eliminates the need for the centrifuge and drying steps used in the Mitsui CX process. In addition, the solubility of amorphous materials is lower in isobutane than in hexane, making it less challenging to produce lower density grades in this process. Phillips developed chromium-based catalysts that inherently make polyethylene with very broad MWD, thus lessening the need for a second reactor. This is considered to be one of the most efficient industrial versions of the slurry process. There are over 80 reactors of this type currently operating in the world, producing almost one-third of the world's HDPE. The Phillips process has been refined over the years, and the latest plants have impressive capacities of 320 kt per year using a single-loop reactor. Most of the Phillips reactors are for HDPE, although it is claimed that some LLDPE grades can be made with supported metallocene catalysts that can produce the comonomer (usually hexene) *in situ*. Since large amounts of 1-hexene in the continuous phase would lead to the increased solubility of LLDPE fractions, and thus reactor fouling, producing the comonomer *in situ* is advantageous because its concentration in the continuous phase remains relatively low. In addition, metallocene resins have much narrower chemical composition distribution, without the low crystallinity tail observed in Ziegler–Natta LLDPE resins.

Both INEOS and TOTAL use processes involving either a single loop for unimodal products or a cascade of two loops to make bimodal pipe grades. Their processes also employ isobutane as the diluent. In the cascade version of these processes, there will be a flash unit between the two reactors to remove any residual hydrogen, thereby improving the control over the high molecular weight product produced in the second reactor.

The other process that uses a loop reactor for polyethylene is the Borstar process from Borealis. The Borstar process has many unique features such as a loop reactor followed by a fluidized gas-phase reactor; it also uses supercritical propane as a diluent. Polyethylene has lower solubility in supercritical propane than in isobutane and, therefore, causes less fouling in the reactor. The higher operating pressure also allows the reactor to be operated at higher hydrogen concentration to produce polymers with higher melt indices (lower molecular weight averages). The catalyst fed to the loop is typically prepolymerized in a smaller loop under milder conditions than those found in the main reactor. Owing to the complexity of the process, however, it requires higher capital investment and only a few plants using this technology have been built to date.

Other processes have been patented that claim to take advantage of solubility differences to produce LLDPE in slurry processes. For instance, US Patent 6716936 to Equistar Chemicals (2004) describes a slurry process where the diluent is boiling propane. The authors claim that using the light solvent at subcritical conditions also decreases the amount of soluble material in solution, so this process can be used to produce lower-density material compared to the hexane-based processes. Whether or not this process is exploited commercially is not known, but the patent does serve to illustrate the importance that the type of diluent can have on the product range.

The main characteristics of slurry processes for ethylene polymerization are summarized in Table 4.2.

4.3.1.2 Gas-Phase Processes

Union Carbide, now part of Dow Chemical, was the first company to commercialize the technology for polyolefin production using fluidized bed gas-phase reactors. Since polymerization occurs in the gas phase, separation of the unreacted monomer from the polymer product is achieved simply by flashing off the monomer. Any low molecular weight polymer formed remains in the polymer particles, and no further separation is necessary. This means that gas-phase processes are true swing[3] processes capable of making all types of polyethylene resins, from linear low- to high-density products. For reasons discussed above, FBRs are the only type of gas-phase reactors used to make polyethylene. As shown in Figure 4.14, the process only requires an FBR, a product discharge system to get the polymer out of the reactor and flash off the monomer, and a purge column to remove any residual monomer and to deactivate the catalyst.

First-generation Unipol plants were often plagued by operational problems that caused runaway reactions and formed large lumps of polymer in the reactor. With improvements in catalyst development and in process control, this type of problem has largely been overcome. In addition, modern reactors now operate in condensed mode, which greatly increases their throughput. Univation now licenses the Unipol process for LLDPE.

Both HDPE and LLDPE can be made using the Unipol process, although this process has found broader acceptance for the production of LLDPE. Some Unipol

---

3) A "swing" process is one that can be used to make a full range of products with densities from 0.96 or more, down to $0.90\,\mathrm{g\,cm^{-3}}$.

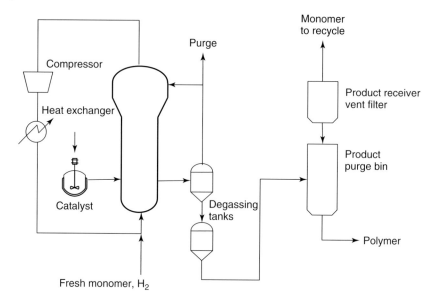

Figure 4.14 Unipol process for polyethylene manufacture.

plants were designed and operated in the *swing mode* between HDPE and LLDPE, but most plants are designed only for LLDPE production.

Several other companies have developed and are licensing gas-phase polyethylene technologies. They include the Innovene G process from INEOS and the Spherilene processes from Basell. All of them are based on the same principle of using a fluidized bed gas-phase reactor, although the operating mode and conditions differ among these processes. The Innovene G process (Figure 4.15) has a cyclone at the outlet to the reactor to eliminate fines from the recycle stream. Also shown in this figure is a separation loop below the reactor that recovers the condensable material from the recycle loop. In the Innovene process, the liquid coming out of this last stage is injected into the reactor at a point above the distributor plate, and dry gas is blown through the bottom of the reactor. Other condensed mode processes, such as the original Unipol process, see a mixture of gas and subcooled liquid blown in below the distributor plate.

The Spherilene line of processes from LyondellBasell comes in two main forms: the Spherilene S process that uses a single FBR and Ziegler or chromium catalysts and the Spherilene C process that is a cascade of two FBRs and mainly employs Ziegler catalysts. The chromium catalyst process was originally developed as Lupotech G. The Spherilene S and C processes are proposed with Ziegler catalysts, but do not offer condensed mode cooling. Rather, the reactors are cooled using propane instead of nitrogen as the process gas. Propane has a much higher heat capacity than nitrogen, and therefore, the continuous phase can absorb much more heat. LyondellBasell also claims that since the process streams contain no nitrogen, downstream gas purification operations are simpler and do not require membrane

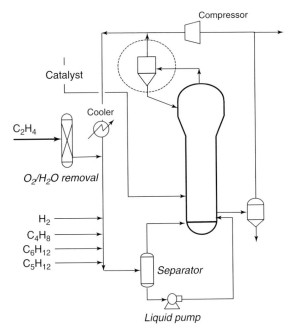

Figure 4.15 Innovene G flow diagram of the reactor block, including a cyclone at the outlet to the disengagement zone and a coolant section below the reactor for condensed mode operation.

separators. The older Lupotech process did include a condensed mode operation using hexane as the heat exchange inert.

Although it is not shown in the flow sheets of the gas-phase processes, a prepolymerization step is often used for Ziegler or metallocene catalysts in order to preserve the catalyst morphology and especially to prevent particle overheating. However, it should be noted that in circumstances where the catalysts activate slowly, with what is called a *buildup rate* (for instance, with chromium catalysts), this prepolymerization step is not as important. The buildup in catalyst activity means that they start slowly, and so polymer accumulates on the catalyst particles in a more controlled manner, simplifying the injection of virgin catalyst directly into the reactor (see Chapter 7 for a more detailed discussion of the importance of prepolymerization).

Finally, Mitsui also offers a gas-phase technology in the form of the Evolue platform. Like the Spherilene C process, it also employs two FBRs in series. On the other hand, it is also one of the rare processes where the catalyst (a metallocene) is fed to both reactors. In virtually all other cascade processes, the catalyst is almost always fed uniquely to the first reactor. Another major difference is that comonomer is fed to both reactors. Once again virtually all other processes will make a homopolymer in the first reactor. In fact, the second reactor in the Evolue process is bigger than the first (once again, this is usually not the case) because extra catalyst is fed into it. This process does not have the market penetration of the

Table 4.3 Typical reactor conditions for gas-phase polyethylene processes.

Process	Reactor type	Mode of operation[a]	Reactor temperature (°C)	Reactor pressure (bar)	Residence time (h per reactor)
Unipol (Univation)	1–2 FBR[b]	Condensed	90–110	20–25	~2
Spheriline Chrome (LyondellBasell)	FBR	Condensed	90–110	20–25	~2
Spherilene C/S (LyondellBasell)	1–2 FBR[c]	Dry	70–90	20–25	~1.5
Innovene (INEOS)	1 FBR	Condensed	90–110	20–25	~2
Evolue (Mitsui)	2 FBR	Dry	–	–	2–3

Reactor conditions differ for different product grades; the values shown here are only approximate averages found in these processes.
[a] No distinction here between condensed and supercondensed; the major difference is the quantity of liquid injected.
[b] Unipol I has 1 FBR, and the less common Unipol II has 2.
[c] Spherilene S has 1 FBR, and Spherilene C has 2.

Unipol and Innovene platforms, as it is probably used mainly for the production of specialty copolymers.

The main characteristics of fluidized bed processes for olefin polymerization are given in Table 4.3.

4.3.1.3 Mixed-Phase Processes

The Borstar process from Borealis uses a combination of slurry loops and FBRs to make a full range of polyethylene products. As shown in Figure 4.16, the process consists of two-loop reactors, where the polymerization occurs in supercritical propane, followed by a degassing step and then at least one (possibly two) FBRs. As mentioned above, polyethylene has lower solubility in supercritical propane than in isobutene, so the potential for reactor fouling is lower in supercritical propane loops. The price of using supercritical propane in the loops is the need for a higher pressure (65–80 bar) and temperatures of at least 85 °C. The higher operating pressure also allows the reactor to be operated at higher hydrogen concentration to produce polymers with higher melt indices (lower molecular weight averages). However, the lower solubility of the polymer and the wide range of hydrogen concentrations allow one to produce an extremely wide range of MFR in the main loop.

The purpose of the single-loop prepolymerizer is to improve the polymer particle morphology and give highly active catalysts time to reach an appropriate size for injection into the main loop (Chapter 7). This is necessary since the main loop polymerization conditions are such that the local reaction rates are quite high, and virgin catalyst particles, especially the highly active ones used in this process, would not maintain their integrity without prepolymerization.

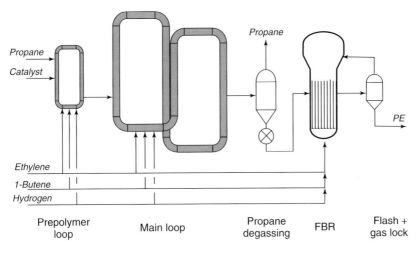

Figure 4.16 Borstar polyethylene process with prepolymerization loop, a main loop reactor (both under supercritical propane), and one FBR. The supercritical slurry reactors are separated from the gas phase by a degassing step and a gas lock. PE, polyethylene.

The inter-reactor flash removes all propane and hydrogen used in the loops and allows the slurry- and gas-phase reactors to be run under independent conditions. This is a feature that is commonly found in the mixed-phase processes used in polypropylene production discussed below.

Low molecular weight polymer is made in the loop, and higher molecular weight products in the FBR. The advantage of finishing in an FBR is that there one can add varying levels of butene comonomer without risking issues related to solubility (while the solubility of amorphous polyethylene is low in supercritical propane, it is not zero). This makes the Borstar polyethylene process a true swing process capable of producing LLDPE with densities ranging from over 0.96 down to below 0.92 g cm^{-3}. This FBR typically runs at 80 °C and approximately 20 bar. Some descriptions of this FBR indicate that there is a bottom-mounted agitator that can help with fluidization and mixing in the critical zone just above the distributor plate. A second FBR in this process would serve a function similar to that in the gas-phase processes mentioned above: allows one to make a copolymer phase with bimodal composition and/or MWDs.

The major limitation of this process is its complexity. It requires higher capital investment than other polyethylene processes, and only a few plants using this technology have been built to date.

4.3.1.4 Solution Processes

Like the HDPE slurry polymerization processes, current solution polymerization processes rely on one or two stirred autoclaves. The similarity stops there since solution polymerization takes place at much higher temperatures (up to 250 °C) and pressures (up to 100 bar). The average reactor residence times in solution processes are much shorter than in the previously discussed processes, in the order of a few

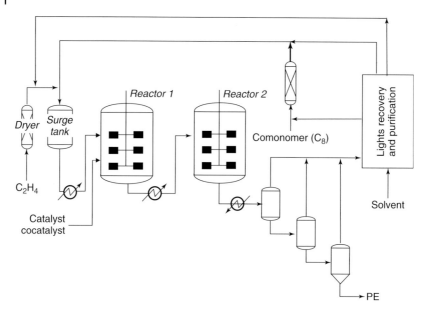

Figure 4.17 Dowlex solution process for polyethylene production.

minutes, and this allows for much faster grade transition. Solution polymerization also allows the use of higher α-olefin comonomers, such as 1-hexene, 1-octene (or even higher carbon numbers), which produces LLDPE with excellent properties. Solution processes are also most adaptable to metallocene catalyst technology because the catalyst does not need to be supported.

A process flow diagram for the Dowlex process by Dow Chemical is shown in Figure 4.17. Although it is the dominant process in solution polymerization in terms of production volumes, Dow does not license this technology to other companies. Other competing processes include the DSM Compact process, and the Sclairtech and Advanced Sclairtech (AST) processes by Nova Chemicals.

Standard configurations of the Dowlex and Nova processes use two autoclaves in series to produce bimodal resins with soluble Ziegler–Natta or metallocene catalyst. Even though it is possible to make bimodal resins using a mixture of soluble catalysts in a single reactor, two reactors in series tend to give better control over the MWD because the RTD of the cascade of reactors is narrower than that of a single autoclave. In the original Sclairtech process developed by DuPont, the reactor block included one autoclave followed by a tubular reactor to complete monomer conversion. The Compact process (as the name suggests) is simpler and uses only one autoclave reactor.

The Dowlex process uses a higher-boiling hydrocarbon solvent (Isopar, which is a mixture of aliphatic C_8 and C_9 cuts), whereas the Nova Chemicals and DSM processes use a lighter solvent such as cyclohexane. The use of the heavier solvent in Dowlex means that lower pressures and temperatures of reaction can provide

Table 4.4 Overview of solution polymerization processes.

Process	Reactor[a]	Solvent	Reactor temperature (°C)	Reactor pressure (bar)	Residence time (min)
Dowlex	2 CSTRs	ISOPAR E	150–200	25–30	∼30
Sclairtech	2 CSTRs or 1 CSTR + 1 PFR	Cyclohexane	∼300	∼138	∼30
Sclairtech AST	2 CSTRs	Light HC (proprietary)	<200	∼138	∼5–10
DSM Compact	1 CSTR	Lighter than octene	150–250	30–100	<10

[a] Can use CSTRs in cascade or in parallel.

the polymerization rates as in other processes. On the other hand, the use of lighter solvents facilitates the use of fewer postreactor separators to purify the product.

The stirred autoclaves used in solution processes are operated adiabatically, where the feed is heated to the reaction temperature using the heat of polymerization. While heat removal is not an issue in itself, it remains nonetheless critical to prevent overheating by keeping the rate and polymer contents at acceptable levels. The reactor effluent is usually reheated at the reactor outlet before the high-temperature separation steps and then fed molten into to pelletizer. This, combined with the short residence times and fast grade changes, is one of the principle advantages of solution processes. In certain conditions, these advantages can help to outweigh higher capital and operating costs associated with solution polymerization plants. A summary of the different solution processes discussed here is given in Table 4.4.

4.3.2
Polypropylene Manufacturing Processes

There are many parallels between polypropylene and polyethylene manufacturing processes, but owing to the different requirements of the polymer, there are significant differences between these processes as well. In a somewhat simplistic way, polyethylene processes aim at achieving the correct molecular weight (melt index) and copolymer composition (density), whereas polypropylene processes need to be able to provide an even wider range of products, including propylene homopolymer with right MWD, random propylene/ethylene copolymers (and sometimes propylene/ethylene/α-olefin terpolymers), and heterophasic impact copolymers where a rubber phase is created directly in the growing particles in order to get good in-reactor blends of thermodynamically incompatible materials. While propylene homopolymer can be produced in reactors of various configurations, for impact copolymer production, gas-phase reactors are the only valid choice because of the solubility of the propylene/ethylene rubber phase in the monomer and diluent.

4.3.2.1 Slurry (Inert Diluent) Processes

Like the slurry process for polyethylene manufacture, this is also the first commercial process for the production of polypropylene. The basic process includes a series of stirred autoclaves, and polymerization takes place with a heterogeneous catalyst suspended in an inert diluent. Many different configurations of the slurry process were developed in the early 1970s. Zielger–Natta catalysts available at that time for propylene polymerization had lower activities and produced polypropylene with a significant fraction of atactic material, which was reflected in the early process configuration. Since catalyst activity was low, a series of reactors were needed to push the reaction to completion. It is not unusual to see processes with five reactors in series, and some with even seven reactors in series, in processes designed at those earlier days. Deashing was required to remove the high level of catalyst residue from the product, and an atactic polypropylene removal step was also necessary. Different variations of the slurry process included the use of diluents ranging from C_6–C_{12} hydrocarbons.

Diluent slurry processes for polypropylene are expensive to build and to run because of the number of pieces of equipment involved. They have largely been replaced by the more efficient bulk and gas-phase processes. Most of the remaining diluent slurry plants in the world now focus on producing specialty polymers, as diluent slurry processes do offer some advantages over other bulk and gas-phase processes. An example is the production of high-crystallinity polypropylene (HCPP), where most of the atactic polymer is dissolved in the diluent and removed from the final product, increasing the crystallinity and stiffness of the resulting polymer.

4.3.2.2 Gas-Phase Processes

There are several different gas-phase processes for propylene polymerization, with distinct reactor configurations, and each one has its own unique attributes.

The Unipol process, initially developed for polyethylene production, has been extended to polypropylene manufacture. The process consists of a large fluidized bed gas-phase reactor for homopolymer and random copolymer production (propylene and ethylene or butene), and a second smaller reactor for impact copolymer production. The second reactor is smaller than the first one because only 20% of the production comes from the second reactor. This reactor typically has a lower pressure rating, as copolymerization is usually carried out at lower temperatures and pressures. Condensed mode operation is used in the homopolymer reactor but an inert diluent is not required because propylene (plus traces of inert propane) is partially fed as a liquid. The copolymerization reactor is operated purely in the gas phase. The Unipol process has a unique and complex product discharge system, which allows for very efficient recovery of unreacted monomer, but this does add complexity and capital cost to the process. The large reactor has a residence time of approximately 1 h, and that of the second reactor is shorter. This means that the grade transition times in this gas-phase process used for polypropylene are

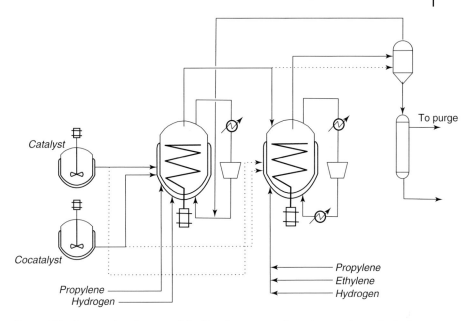

Figure 4.18 Process flow diagram of the Novolen process shown in cascade mode. Broken lines indicate the configuration for parallel mode operation.

competitive with those of loop reactors. The use of one homopolymerization reactor (with respect to two or three loops found in the Spheripol process described below) limits the control that can be exerted over the MWD, but this implies lower capital and operating costs.

Other processes using FBRs exist, including the Sumitomo process. This process employs two or three FBRs in series to produce different polypropylene grades. The reactors are smaller than those in the Unipol process, and lines typically operate at one-half to one-third of the productivity of the latter. The smaller capacities are compensated for by the ability to use three FBRs in series to narrow the RTD of the particles and thus make more uniform product with higher value added for certain types of automotive and appliance applications.

The Novolen process, shown in Figure 4.18, was originally developed by BASF and is now licensed by Lummus Technology. The process was one of the first gas-phase processes for polypropylene production and is still used to make over 7 million tonnes per year of polypropylene products. The process relies on two VSBRs that can operate either in cascade to make impact copolymers or in parallel to improve productivity for homopolymers and random copolymers. Pulses of powder are removed semicontinuously by a dipstick and transferred along a pressure gradient to the next reactor or to the degassing and monomer recovery section. Whether or not this process, like the Unipol process described above, has the full range of product possible with the four- to six-reactor processes described below is open for discussion. Nevertheless, the simplicity of the reactor design and operation makes this process attractive to a number of polypropylene producers.

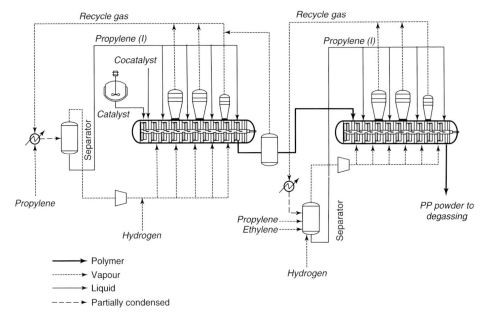

Figure 4.19 Flow sheet of Innovene and Horizone processes.

Both Japan Polypropylene Corporation (with the Horizon process) and INEOS (with the Innovene polypropylene process) also propose platforms built around stirred powder beds, but these reactors are the HSBR discussed above. Amoco and Chisso jointly developed this process starting from the mid-1970s to late 1970s, with the first Amoco plant coming on-line in 1979. A schema of this type of process is proposed in Figure 4.19. While similar, the Innovene and Horizone processes do have some differences. In particular, the Horizon process sees the two reactors arranged one above the other, so that the powder from the first reactor can flow under the influence of gravity to the second reactor; in the Innovene process, the reactors are on the same level, so a powder transfer system is required. Japan Polypropylene claims that this allows them to make higher ethylene content random copolymers, which are sticky and can block mechanical conveying lines. As with the Novolen process, both of these processes can probably also be run in cascade or parallel modes, with all or part of the powder from the first reactor being sent to a purification step rather than to the second reactor.

In cascade mode, used for impact copolymers, the catalyst and cocatalyst are injected at separate points in the initial section of the first reactor. This is a critical step, and the two components must be properly mixed before being conveyed to the rest of the bed. If they are injected too closely, then rapid activation can lead to lump formation and loss of activity; if the injection points are too far apart, then the catalyst will not be properly activated. The first reactor is cooled by spraying liquid propylene (and eventually butane, although it is not known how widely spread this practice is) over the powder in the bed. While it is possible to cool the second reactor

4.3 Olefin Polymerization Processes

Table 4.5 Comparison of different gas-phase polypropylene manufacture processes.

Process	Reactor type		Temperature of PP reactor (°C)	Reactor pressure (bar)	Residence time[a] (h per reactor)
	PP	EPR			
Unipol	FBR (condensed)	FBR	60–70	25–30	~1
Novolen	VSBR (condensed)	VSBR	80–85	30–35	~1
Innovene/Horizone	HSBR (condensed)	HSBR	65–85	25–30	~1

[a] EPR, ethylene–propylene rubber.

in this manner, the amount of liquid that can safely be sprayed on the powder in this bed will be inversely proportional to the quantity of rubber being made (recall that the amorphous elastomeric phase will dissolve in liquid hydrocarbons). In the cascade mode, a separator between the two reactors removes most of the residual process gases from the homopolymer powder before it is sent into the second reactor.

If we compare Figures 4.18 and 4.19, it can be seen that the residence time and mixing in the two powder beds will be quite different. Thus, the big advantage of the Innovene and Horizon processes is the enhanced polymer properties possible with the plug-flow-like reactor characteristics and fast grade transition possible because of this RTD. However, the complexity of the reactor (particularly its horizontal impeller) increases capital cost and can make cleaning and other maintenance operations time consuming and expensive.

Table 4.5 summarizes the most important characteristics of these gas-phase process for the production of polypropylene.

4.3.2.3 Mixed-Phase Processes

In this type of processes, polymerization to make homopolymer and random copolymers with low ethylene content takes place in liquid propylene without the use of an inert diluent. This is a significant simplification over the traditional diluent slurry process, as propylene can be separated from the polymer by flashing and there is no need for the extensive diluent recovery system. Running in liquid monomer significantly enhances the reaction rate and productivity, but one typically uses at least two well back-mixed reactors in series in order to control the MWD.

The dominant process in this market segment is the Spheripol process by Basell, shown in Figure 4.20 for a two-loop and a single-FBR configuration. Similar to the dominance achieved by the Phillips process in HDPE, roughly one-third of the world's polypropylene is produced using the Spheripol process. The Spheripol process uses loop reactors to make propylene homopolymers. In this process, a small loop reactor is used to prepolymerize the catalyst; the main polymerization, for homopolymer or random copolymer, takes place in one- or two-loop reactors,

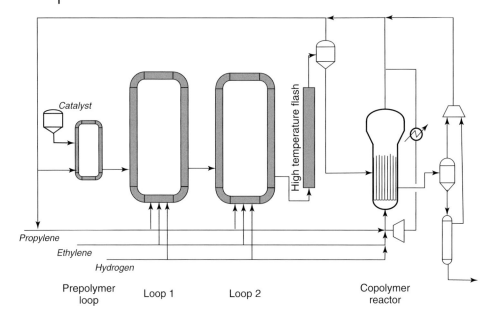

Figure 4.20 Spheripol process for polypropylene manufacture.

each with two to eight legs depending on production requirements. Typically, two main loops are required for homopolymer production.

For impact copolymer production, a gas-phase reactor is required after the loop reactor because of the limited solubility of ethylene in liquid propylene. A high-temperature flash separation is included between the liquid and gas sections of the reactor in order to remove as much hydrogen as possible (gives better molecular weight control in the copolymer reactors) and to be sure that the powder injected in the FBR is as dry as possible. A second (or even third) FBR can be included in the process. Increasing the number of copolymer reactors allows one to obtain more homogeneous (in terms of rubber content) copolymers because of the narrowed RTD. Typically, residence times in the copolymer reactors are very small – about 10–20% – compared to the loops because of the high specific reaction rates of the copolymerization step.

Occasionally, ethylene-rich polymers will be made in the final reactor to produce CATALLOY resins. Also, although not shown here, other process variants include feeding 1-butene to the copolymer reactor train to obtain a variety of different properties.

Basell has recently developed the Spherizone process, which replaces the FBR shown in Figure 4.20 with a circulating bed reactor. Other than the different reactor configuration, the Spherizone process is essentially the same as the Spheripol process.

Another competing bulk propylene processes is the Mitsui Hypol process. The original Mitsui Hypol process is based on two stirred autoclave reactors in place of the loops shown for Spheripol in Figure 4.20, followed by two stirred fluidized

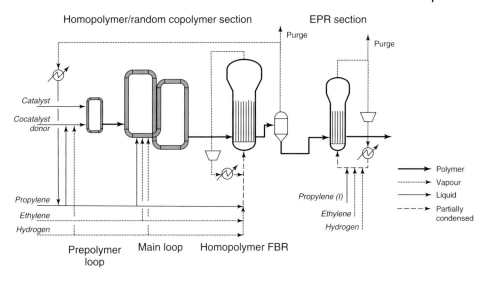

Figure 4.21 Borstar PP process with a single rubber reactor.

bed gas-phase reactors. The stirred autoclaves are replaced by loop reactors in the Hypol II process, giving higher throughput and further reducing capital costs. The gas-phase reactors are also used in homopolymerization, enabling them to make polymer with a wider range of properties. The gas-phase reactors in the Hypol processes are proprietary stirred fluidized beds, with wall scrapers in order to eliminate fouling and sheeting problems. It is claimed that this helps achieve higher rubber contents when making impact copolymers than is possible in a conventional FBR.

The third major player in the area of slurry/gas-phase processes is Borealis with the Borstar polypropylene process shown in Figure 4.21. The platform for making homopolymer and random copolymer is similar in conception to the process shown in Figure 4.16, with a prepolymerizer loop followed by a main homopolymerization loop and an FBR. While the Borstar polyethylene loop reactors run with supercritical propane as the diluent, the Borstar polypropylene loop reactors use propylene as the continuous medium, and reaction conditions in the loop reactors can be altered so that the propylene is either sub- or supercritical. It is said that the advantage of running in supercritical propylene is to avoid the formation of gas bubbles in the continuous phase. The use of an FBR in the homopolymer/random copolymer section allows for the production of stickier random copolymers than might be possible in a loop reactor. The next section of the process, which is cut off from the first by a gas lock and a separator, is used to make ethylene/propylene rubber copolymers in one, or eventually two, smaller FBRs. The separator allows the composition in the two parts of the plant to be controlled independently. It is possible to add an additional FBR if high rubber contents are required. The FBR(s) in the second section is smaller than the FBR used for the homopolymerization because the reaction rates are fairly high and the residence times are typically shorter.

In summary, mixed-phase processes are the workhorses of the polypropylene industry. As noted above, stirred autoclaves are giving way to loop reactors to take advantage of the better heat transfer capacity and thus higher production rates. These processes contain upward of six reactors divided between a block dedicated to homopolymers and random copolymers and a second block used for sticky copolymers and that contains only FBRs. The two blocks are almost always separated by some type of gas lock/flash in order to offer better control over the properties of the different phases (of course, this is not an issue with homopolymers and random copolymers). This gives producers a huge range of product grades that they can make. However, capital investment for these plants will be high, and grade changes and process stability can be an issue.

4.4
Conclusion

This chapter was intended to the give the reader an overview of the different olefin polymerization reactor configurations and how one or more reactors can be assembled to create the heart of an industrial process.

There is at least one issue that we have not touched upon here that merits mention, the fact that very specific types of catalysts are strongly associated with different process platforms. In fact, most of the companies mentioned in this chapter will propose their own catalysts for license with the reactor platforms. This is an issue that is outside the scope of this book, as it involves not only just the "chemical" nature of the catalyst but also technological issues such as how the catalyst precursor, activators, or external donors are mixed and physically injected into the reactor. The activation of the catalyst *in situ* can be associated with the heat removal capacity of the reactor, the design of the agitators, and so on. However, a great deal of this type of information remains proprietary for obvious reasons, so we cannot speculate on the details. Nonetheless, it is important to point out that this is something that, in the world of industrial polymer production, is particular to the polyolefin industry.

References

1. http://encyclopedia.airliquide.com/ encyclopedia.asp. (accessed 23 March 2012).
2. Shepard, J.W., Jezl, J.L., Peters, E.F., and Hall, R. D. (1976) Divided horizontal reactor for the vapor phase polymerization of monomers at different hydrogen levels. US Patent 3,957,488.
3. Dittrich, C.J. and Mutsers, S.M.P. (2007) On the residence time distribution in reactors with non-uniform velocity profiles: the horizontal stirred bed reactor for polypropylene production. *Chem. Eng. Sci.*, **62**, 5777–5793.
4. Covezzi, M. and Mei, G. (2001) The Multizone Circulating Reactor Technology, pp. 4059-4067, Copyright.

Further Reading

Because of the proprietary nature of the technology and its operation, information on industrial processes is not very abundant in the open literature. Nonetheless, some interesting articles on reactors can be found. Here, we limit ourselves to proposing a select few articles that talk about aspects such as mixing or RTDs.

Some articles on technology overviews include the following by engineers from Borealis that look at general aspects related to polyethylene processes:

Knuuttila, H., Lehtinen, A., and Nummila-Pakarinen, A. (2004) Advanced polyethylene technologies – controlled material properties. *Adv. Polym. Sci.*, **169**, 13–27.

For an article describing the mechanics of mixing in a VSBR,

Cooker, B. and Neddermann, R.M. (1987) A theory of the mechanics of the helical ribbon powder agitator. *Powder Technol.*, **50**, 1–13.

The basic operating principles of the MZCR are described in the following:

Covezzi, M. and Mei, G. (2001) The multizone circulating reactor technology. *Chem. Eng. Sci.*, **56**, 4059–4067.

An article from Univation outlines some of the developments in the Unipol fluidized bed:

Burdett, I. (2008) New innovations drive gas phase PE technology. *Hydrocarbon Eng.*, **13** (13), 67–76.

5
Polymerization Kinetics

> When you cannot express it in numbers, your knowledge is of a meager and unsatisfactory kind.
>
> *William Thomson, Lord Kelvin (1824–1907)*
>
> An ounce of algebra is worth of a ton of verbal argument.
>
> *J. B. S. Haldane (1892–1964)*

5.1
Introduction

In this chapter, we use mathematical models to describe olefin polymerization kinetics with coordination catalysts. For a given catalyst/cocatalyst system, polymerization kinetics depend only on the concentration of reagents and temperature at the catalytic active sites. We have already discussed in the preceding chapters how these concentrations and temperatures may be affected by mass and heat transfer resistances at the mesoscale and macroscale levels. For solution processes, these conditions may be equal to the bulk reactor conditions, greatly simplifying the modeling problem; however, for slurry- and gas-phase processes using supported catalysts, these conditions may differ from the bulk reactor values and become a function of the radial position in the polymer particle.

We will assume that the polymerization conditions at the active sites are either known or estimated from bulk reactor values. In Chapter 7, we develop mesoscale models to show how the concentrations of reagents, as well as temperature, may vary inside the polymer particle. If these gradients are significant, the models developed in this chapter are only valid locally, at a given radial position in the polymer particle. Finally, in Chapter 8, we discuss ways to connect microscale, mesoscale, and macroscale in one unified approach for modeling olefin polymerization reactors.

Our first mathematical model starts with a simple subset of the fundamental model for polymerization kinetics. This model is gradually modified throughout this chapter to include additional elementary steps, or alternative polymerization kinetic schemes. This approach has the advantage of keeping the mathematical complexity to a minimum when we introduce the basic modeling concepts, without sacrificing the main conclusions about the phenomena under study.

When modeling polymerization kinetics, we must keep in mind two main objectives. The first is akin to that of any chemical reaction engineering model: to describe how fast reactants are consumed, products are formed, and energy is exchanged between the system and its surroundings. This piece of information is essential for chemical reactor design, operation, optimization, and control. The second objective, the description of the polymer microstructure, is unique to polymers and distinguishes polymerization reaction engineers from their colleagues working with small molecules. Most of our modeling effort, in fact, is dedicated to this second objective. Polymer microstructure models are covered in Chapter 6, but their foundations are laid out in the present chapter.

Most polymerization kinetic studies with coordination catalysts are done in semibatch reactors operated at constant pressure and temperature. Typically, the cocatalyst is added with the solvent to help scavenge any impurity that may be present in the system. For ethylene or propylene homopolymerization in slurry- or gas-phase reactors, the monomer is fed on demand to keep the reactor pressure constant. An in-line mass flowmeter measures the flow rate of gaseous monomer to the reactor, and the monomer uptake curve is reported as the polymerization rate. The presence of liquid comonomers, such as 1-hexene and 1-octene, complicates this simple approach. Two modes of operation are commonly used in this case: the liquid comonomer is fed either in batch mode at the beginning of the polymerization or continuously to the reactor at the rate it is consumed by the polymerization to avoid composition drift. When the liquid comonomer is fed in batch mode (a more popular, but not the best, approach), low conversions are required to avoid significant composition drift that may affect the results. When the comonomer is fed continuously, a gas composition analyzer (commonly, a fast gas chromatograph or infrared detector) is required to send a feedback signal to the pump controlling the comonomer feed rate. Continuous feeding of the liquid comonomer eliminates comonomer drift effects but is more complex to implement experimentally. The polymerization is started with the injection of a pulse of catalyst after all other conditions (reactant concentrations and temperature) have reached their set points and is terminated after a given time by rapidly venting the reactor contents and adding a catalyst poison such as methanol.

A typical semibatch reactor setup for ethylene/α-olefin copolymerization is illustrated in Figure 5.1.

In the case of bulk propylene polymerization, a reactor calorimeter is required to measure the rate of polymerization. The reactor is operated in batch mode with liquid propylene, and the measured heat of polymerization is translated into the polymerization rate curve through an energy balance around the reactor system.

Independently of the system under study, kinetic curves for olefin polymerization are classified into buildup and decay types, as illustrated in Figure 5.2. Buildup curves are common for older heterogeneous Ziegler–Natta catalysts and Phillips catalysts, while decay curves are usually observed with newer heterogeneous Ziegler–Natta catalysts, metallocenes, and late transition metal catalysts. The

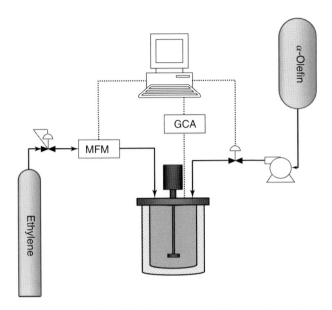

Figure 5.1 Typical semibatch reactor system for measuring polymerization kinetics. (MFM, mass flow meter; GCA, gas composition analyzer.

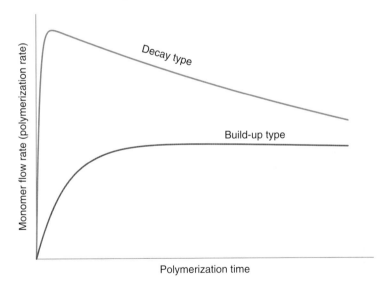

Figure 5.2 Representative types of olefin polymerization kinetic curves.

curves in Figure 5.2 are just representative of the general polymerization kinetics behavior with coordination catalysts; there are, of course, other intermediate types between these two extreme cases.

In the following sections, we discuss mathematical models that can be used to describe these polymerization kinetic curves.

Table 5.1 Homopolymerization mechanism for the determination of the main polymerization kinetic constants.

Description	Chemical equation	Rate constant
Activation	$C + Al \to P_0$	k_a
Propagation	$P_r + M \to P_{r+1}$	k_p
First-order deactivation	$P_r \to C_d + D_r$	k_d
Second-order deactivation	$2P_r \to 2C_d + 2D_r$	k_d^*

5.2
Fundamental Model for Polymerization Kinetics

The fundamental model for polymerization kinetics was introduced in Chapter 3. It is still the most widely used model to describe olefin polymerization kinetics with single- and multiple-site catalysts. The fundamental model can describe most of the phenomena observed in industrial and laboratory reactors, especially if semiempirical corrections are added to its parameters when needed. On the other hand, important phenomena such as the comonomer and hydrogen effects on the polymerization rate, and change of polymerization reaction order can only be described with extensions of the fundamental model. Unfortunately, there is no general agreement on these alternative models; the fundamental model is still the most accepted way to describe the main polymerization steps with coordination catalysts.

5.2.1
Single-Site Catalysts

5.2.1.1 Homopolymerization

We can model most olefin polymerization kinetic curves with relatively simple expressions that are derived from the funadamental model for polymerization kinetics. The most important reactions are listed in Table 5.1: site activation, propagation, and catalyst deactivation. Since, according to the funadamental model, chain transfer reactions are assumed to have no effect on the polymerization rate, we do not need to include them in the present treatment. They are introduced later in this chapter when we present some alternative models for polymerization kinetics and are described in more detail in Chapter 6, when we start developing mathematical models to describe molecular weight distributions.

The polymerization rate, R_p (mol·l^{-1}·s^{-1}), is given by the equation

$$R_p = -\frac{d[M]}{dt} = k_p[M] \sum_{r=0}^{\infty} [P_r] = k_p[M][Y_0] \qquad (5.1)$$

where [M] is the monomer concentration at the active sites, $[P_r]$ is the molar concentration of living chains with length r, $[Y_0]$ is the total molar concentration of living chains in the reactor, and k_p is the propagation rate constant (l·mol^{-1}·s^{-1}).

In a semibatch reactor such as the one shown in Figure 5.1, the polymerization rate can be obtained by dividing the monomer feed flow rate to the reactor, F (mol·s^{-1}), by the reactor volume, V_R

$$R_p = \frac{F}{V_R} \quad (5.2)$$

For all practical purposes, $[Y_0]$ is equal to the molar concentration of active sites in the reactor, since the fraction of polymer-free active sites is negligible due to the high activity of coordination catalysts; $[Y_0]$ can be calculated using the chemical equations in Table 5.1, as shown below.

The molar balance for the catalyst precursor added at the beginning of polymerization in a semibatch reactor is

$$\frac{d[C]}{dt} = -k_a[Al][C] \quad (5.3)$$

where $[C]$ and $[Al]$ are the concentrations of the catalyst and cocatalyst, respectively.

Since the cocatalyst is generally present in large excess, we may assume the product $k_a[Al] = K_a$ to be nearly constant, which permits the straightforward solution of Eq. (5.3) with the initial condition $[C](0) = [C_0]$

$$[C] = [C_0] \exp(-K_a t) \quad (5.4)$$

Developing a molar balance for Y_0, for the case of first-order deactivation, we find

$$\frac{d[Y_0]}{dt} = K_a[C] - k_d[Y_0] = K_a[C_0]\exp(-K_a t) - k_d[Y_0] \quad (5.5)$$

Equation (5.5) is a first-order linear differential equation that can be solved with the initial condition $[Y_0](0) = 0$ to obtain

$$[Y_0] = \frac{1 - e^{-K_a\left(1 - \frac{k_d}{K_a}\right)t}}{1 - \frac{k_d}{K_a}} [C_0] e^{-k_d t} \quad (5.6)$$

Finally, substituting Eq. (5.6) into Eq. (5.1), we obtain the expression for olefin polymerization rate with a single-site catalyst that follows first-order deactivation kinetics

$$R_p = \frac{F}{V_R} = k_p[M] \frac{1 - e^{-K_a\left(1 - \frac{k_d}{K_a}\right)t}}{1 - \frac{k_d}{K_a}} [C_0] e^{-k_d t} \quad (5.7)$$

Equation (5.7) describes the two typical polymerization kinetic profiles shown in Figure 5.2. The parameters K_a (or k_a), k_p, and k_d are estimated by fitting experimental monomer uptake curves with Eq. (5.7). Figure 5.3a–c illustrates how changing the values of each one of these parameters affects the shape of the polymerization kinetic curves.

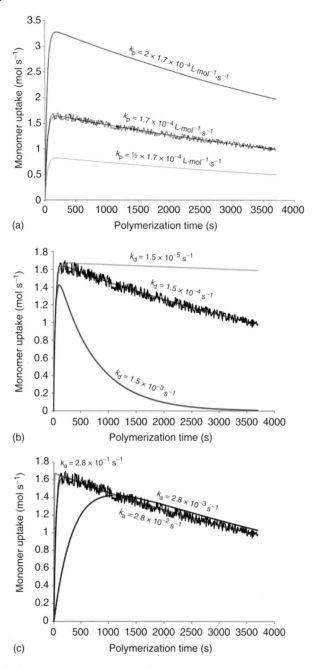

Figure 5.3 Sensitivity analysis for the parameters (a) k_p, (b) k_d, and (c) K_a in Eq. (5.7). The best fit for the experimental data (ethylene polymerization with a metallocene) is given by $k_p = 1.7 \times 10^4 \, \text{l} \cdot \text{mol}^{-1} \cdot \text{s}^{-1}$, $k_d = 1.4 \times 10^{-4} \, \text{s}^{-1}$, and $K_a = 2.8 \times 10^{-2} \, \text{s}^{-1}$. Other simulation parameters are $[C_0] = 1.0 \times 10^{-4} \, \text{mol} \cdot \text{l}^{-1}$ and $[M] = 1.0 \, \text{mol} \cdot \text{l}^{-1}$.

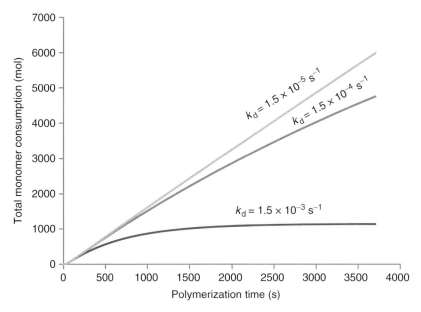

Figure 5.4 Sensitivity analysis for the parameters k_d in Eq. (5.8). Simulation parameters: $K_a = 2.8 \times 10^{-2}\,s^{-1}$, $k_p = 1.7 \times 10^4\,l\cdot mol^{-1}\cdot s^{-1}$, $[C_0] = 1.0 \times 10^{-4}\,mol\cdot l^{-1}$, and $[M] = 1.0\,mol\,l^{-1}$. Compare with Figure 5.3b.

The total number of moles of monomer consumed during polymerization in a semibatch reactor, n_M, is calculated from the integration of Eq. (5.7) assuming [M] is constant

$$n_M = \frac{k_p[M][C_0]V_R}{1 - \dfrac{k_d}{K_a}} \int_0^t \left[1 - e^{-K_a\left(1 - \frac{k_d}{K_a}\right)t}\right] e^{-k_d t} dt$$

$$= \frac{k_p[M][C_0]V_R}{1 - \dfrac{k_d}{K_a}} \left(\frac{1 - e^{-k_d t}}{k_d} - \frac{1 - e^{-K_a t}}{K_a}\right) \quad (5.8)$$

Figure 5.4 shows how the total monomer consumption varies as a function of time for several values of the catalyst deactivation rate constant k_d.

Second-order catalyst deactivation is also commonly reported for metallocene catalysts and may result from the bimolecular deactivation mechanism mentioned in Chapter 3. In this case, the equation for the number of moles of living chains is

$$\frac{d[Y_0]}{dt} = K_a[C_0]\exp(-K_a t) - k_d^*[Y_0]^2 \quad (5.9)$$

The solution of Eq. (5.9) is given in Example 5.3, but since the activation of most metallocenes is practically instantaneous ($K_a \to \infty$), we may reduce Eq. (5.9) to the simpler form

$$\frac{d[Y_0]}{dt} = -k_d^*[Y_0]^2 \tag{5.10}$$

which can be solved with the initial condition $[Y_0](0) = [C_0]$ to obtain

$$[Y_0] = \frac{[C_0]}{1 + [C_0]k_d^*t} \tag{5.11}$$

Substituting Eq. (5.11) into Eq. (5.1), we obtain the following solution for the rate of polymerization under second-order deactivation kinetics and instantaneous site activation

$$R_p = \frac{k_p[M][C_0]}{1 + [C_0]k_d^*t} \tag{5.12}$$

Therefore, for catalysts that decay following second-order kinetics, the reciprocal of the polymerization rate is a linear function of time, that is,

$$\frac{1}{R_p} = \frac{1}{k_p[M][C_0]} + \frac{k_d^*}{k_p[M]}t \tag{5.13}$$

which is a very simple relationship to test for second-order decay kinetics of single-site catalysts.

Instantaneous site activation may also be assumed for first-order catalyst decay, simplifying Eq. (5.5) to the following differential equation

$$\frac{d[Y_0]}{dt} = -k_d[Y_0] \tag{5.14}$$

with the solution for $[Y_0](0) = [C_0]$ given by

$$[Y_0] = [C_0]\exp(-k_d t) \tag{5.15}$$

and polymerization rate by

$$R_p = k_p[M][C_0]\exp(-k_d t) \tag{5.16}$$

Consequently, for catalysts that follow first-order decay kinetics and instantaneous site activation, the logarithm of the rate of polymerization is a linear function of time

$$\ln R_p = \ln\left(k_p[M][C_0]\right) - k_d t \tag{5.17}$$

It is illustrative to compare the shapes of the monomer uptake curves expected from first- and second-order catalyst decay mechanisms. To allow for a meaningful comparison, we set the half-life, $t_{1/2}$, of both catalysts to the same value. For first-order decay, $t_{1/2}$ is found by setting $[Y_0] = [C_0]/2$ in Eq. (5.15)

$$t_{1/2} = \frac{\ln 2}{k_d} \tag{5.18}$$

Similarly, from Eq. (5.11), for second-order decay kinetics,

$$t_{1/2}^* = \frac{1}{[C_0]k_d^*} \tag{5.19}$$

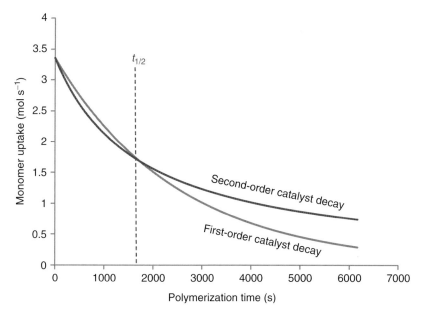

Figure 5.5 Influence of catalyst decay order on polymerization kinetics. Simulation parameters: $K_a \to \infty$ (instantaneous site activation), $k_p = 3.3 \times 10^4 \, l \cdot mol^{-1} \cdot s^{-1}$, $k_d = 4.0 \times 10^{-4} \, s^{-1}$ (first order), $k_d^* = 5.77 \, l^2 \cdot mol^{-2} \cdot s^{-1}$ (second order), $[C_0] = 1.0 \times 10^{-4} \, mol \cdot l^{-1}$, and $[M] = 1.0 \, mol \cdot l^{-1}$. For this simulation, $t_{1/2} = t_{1/2}^* = 1733 \, s$ (Eqs. (5.18) and (5.19)).

Therefore, by setting $t_{1/2} = t_{1/2}^*$, we obtain a relation between the two deactivation rate constants

$$\frac{k_d}{k_d^*} = [C_0] \ln 2 \tag{5.20}$$

Figure 5.5 shows that catalysts that deactivate according to a second-order decay law lose their activity faster than those that deactivate following first-order kinetics for times below $t_{1/2}$, but this behavior is reversed for polymerization times higher than the half-life time of the catalyst.

Examples 5.1 and 5.2 investigate the effect of catalyst concentration and pressure on the ethylene polymerization kinetics with a metallocene catalyst.

Example 5.1: Polymerization Kinetic Parameter Estimation for a Single-Site Catalyst: Effect of Catalyst Concentration

Figure E.5.1.1 shows the ethylene uptake curves in a semibatch solution polymerization reactor using *rac*-Et(Ind)$_2$ZrCl$_2$ at an ethylene pressure of 120 psi and polymerization temperature of 120 °C. All variables were kept constant during the polymerization, except for the catalyst concentration that was allowed to vary from 6.5×10^{-9} to $4.6 \times 10^{-8} \, mol \, l^{-1}$.

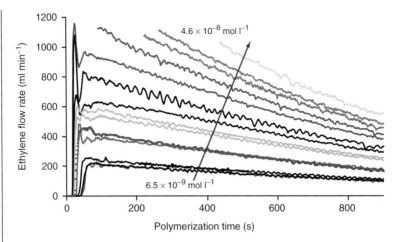

Figure E.5.1.1 Ethylene uptake rates at different catalyst concentrations (indicated in the figure) in a semibatch solution reactor operated at an ethylene partial pressure of 120 psi and a temperature of 120 °C. Lines of the same shades of gray indicate replicate runs.

As expected, the polymerization rate increases steadily with increasing catalyst concentration in the reactor. Catalyst activation seems to be instantaneous, and the shape of the ethylene uptake curves appears to follow the first-order model described with Eq. (5.16). Therefore, it is expected that a plot of $\ln R_p \times t$ should be linear, with slope k_d and intercept $\ln(k_p[M]C_0)$, as demonstrated in Eq. (5.17). Figure E.5.1.2 shows that the experimental data follows the expected trend for this first-order model. Notice that

$$R_p = \frac{F}{V_R}$$

as stated in Eq. (5.2).

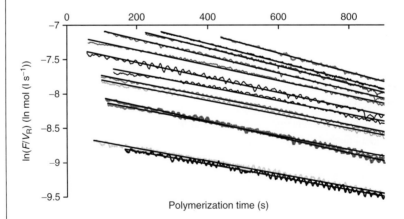

Figure E.5.1.2 Plot of $\ln R_p \times$ polymerization time for the data shown in Figure E.5.1.1.

From the slope and intercept of the curves shown in Figure E.5.1.2, it is possible to estimate the values for the propagation frequency, $k_p[M] = 1.13 \times 10^5$ s^{-1}, and deactivation rate constant, $k_d = 1.11 \times 10^{-3}$ s^{-1}. The value of k_p can be obtained from the product $k_p[M]$ if the solubility of ethylene in toluene at the polymerization conditions is known.

Polymer yield, m_P, can be obtained by the integration of Eq. (5.16)

$$m_P = 28\,\text{g mol}^{-1} \times k_p[M][C_0]V_R \int_0^t \exp(-k_d t)\,dt$$

$$= 28\,\text{g mol}^{-1} \times \frac{k_p[M][C_0]V_R}{k_d}\left[1 - \exp(-k_d t)\right]$$

where 28 g mol^{-1} is the molar mass of ethylene. For a given polymerization time, the polymer yield depends linearly on the initial catalyst concentration, as illustrated in Figure E.5.1.3.

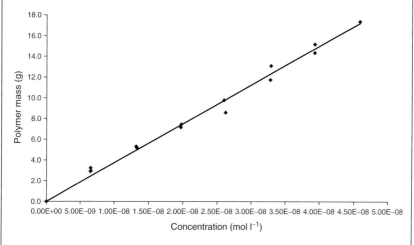

Figure E.5.1.3 Final polymer yield as a function of catalyst concentration for the data shown in Figure E.5.1.1.

Example 5.2: Polymerization Kinetic Parameter Estimation for a Single-Site Catalyst: Effect of Ethylene Pressure

The same system described in Example 5.1 is used now to study the effect of ethylene pressure on the polymerization rate. Figure E.5.2.1 shows the ethylene uptake curves during polymerization in a semibatch solution reactor using *rac*-Et(Ind)$_2$ZrCl$_2$ at a catalyst concentration of 1.32×10^{-8} mol l^{-1} and polymerization temperature of 120 °C. All variables were kept constant during polymerization, except for ethylene pressure that was allowed to vary from 40 to 200 psi.

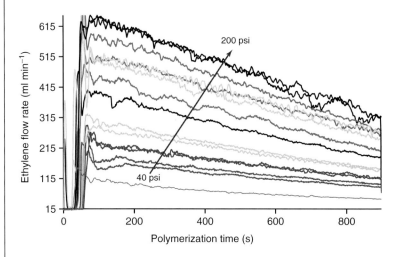

Figure E.5.2.1 Ethylene uptake rates at different ethylene pressures (indicated in the figure) in a semibatch solution reactor operated at an ethylene partial pressure of 120 psi and a temperature of 120 °C. Lines of the same shades of gray indicate replicate runs.

The polymerization rate appears to be directly proportional to the ethylene pressure. Catalyst activation is again observed to be instantaneous. Figure E.5.2.2 shows that the plot of ln $R_p \times t$ is linear, and therefore, the polymerization rate is first order with respect to ethylene concentration.

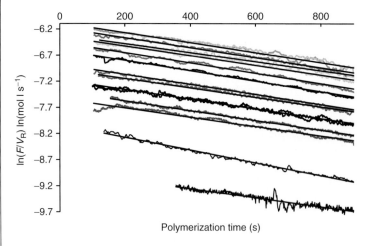

Figure E.5.2.2 Plot of ln $R_p \times$ polymerization time for the data shown in Figure E.5.2.1.

From the slope and intercept of the curves shown in Figure E.5.2.2, it is possible to estimate the values for the propagation frequency, $k_p[M] = 1.15 \times 10^5$ s^{-1}, and deactivation rate constant, $k_d = 1.11 \times 10^{-3}$ s^{-1}, which agree with the ones previously obtained during the catalyst concentration study in Example 5.1.

Finally, the plot of polymer yield as a function of ethylene concentration is linear (Figure E.5.2.3) as expected for first-order behavior with respect to monomer concentration.

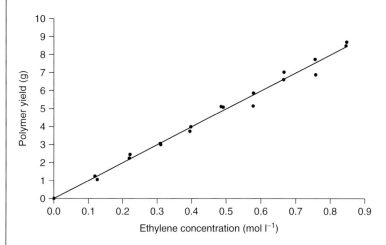

Figure E.5.2.3 Polymer yield as a function of ethylene concentration for the data shown in Figure E.5.2.1.

Therefore, we can conclude that polyethylene made with rac-Et(Ind)$_2$ZrCl$_2$ in a solution polymerization reactor under these polymerization conditions follows the fundamental model for olefin polymerization with coordination catalysts, with first-order dependency on ethylene and catalyst concentration and first-order catalyst deactivation kinetics.

Although we defer the discussion on polymer microstructure to Chapter 6, the polydispersity indices of all polymers made under these conditions is near to the theoretical value of 2.0, which indicates single-site behavior. Polydispersity indices and molecular weight averages as a function of ethylene pressure are shown in Figure E.5.2.4.

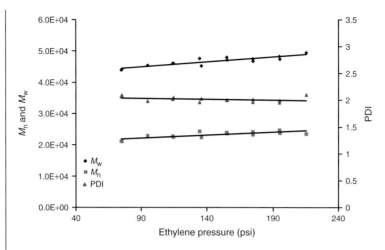

Figure E.5.2.4 Polydispersity index (PDI) number and weight average molecular weights (M_n and M_w) as a function of ethylene pressure.

Example 5.3: Polymerization Kinetics for Second-Order Deactivation and Noninstantaneous Activation

Equation (5.9), for noninstantaneous site activation, was not solved in the text because its solution requires the use of Bessel functions. This example shows that the instantaneous activation assumption works very well for values of K_a that are not too small, as generally observed for most coordination catalysts, since the reaction between the catalyst precursor and the cocatalyst is generally fast and often the catalyst precursor is precontacted with the cocatalyst even before being fed to the reactor.

The solution of Eq. (5.9) is given by[1]

$$R_p = k_p[M] \sqrt{\frac{K_a[C_0]}{k_d^*}} \frac{\beta_t \left[I_1(\alpha) K_1(\alpha \beta_t) - I_1(\alpha \beta_t) K_1(\alpha) \right]}{I_1(\alpha) K_0(\alpha \beta_t) + I_0(\alpha \beta_t) K_1(\alpha)}$$

where the parameters α and β are defined as

$$\alpha = 2 \sqrt{\frac{k_d^*[C_0]}{K_a}}$$

---

[1] MAPLE™ is an excellent software package for solving these types of ordinary differential equations.

$$\beta_t = e^{-\frac{K_a}{2}t}$$

and I_i and K_i are the modified Bessel functions of first and second kind and order i, respectively.

Figure E.5.3.1 compares R_p for three different catalyst activation frequencies with that for instantaneous activation, given in Eq. (5.12). We notice that even for the relatively slow activation frequencies of 0.1 and 0.5 s^{-1}, there is practically no difference from the instantaneous activation curve. Therefore, unless the experimental results indicate slow catalyst activation, Eq. (5.12) can describe the rate of polymerization under second-order catalyst deactivation kinetics.

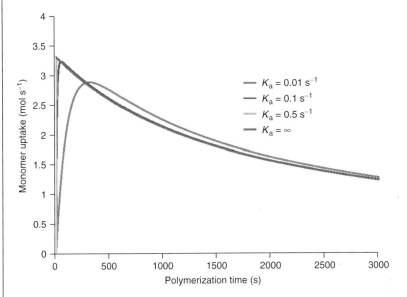

Figure E.5.3.1 Comparison of polymerization rates for instantaneous ($K_a = \infty$) and noninstantaneous catalyst activation. Other simulation data is the same as for Figure 5.5.

5.2.1.2 Copolymerization

Two models are commonly used to describe copolymerization with transition metal catalysts: the Bernoullian model and the terminal model. Table 5.2 shows the propagation steps for these two models. The Bernoullian model assumes that only the type of comonomer coordinating to the active site influences the value of k_p, while in the terminal model, the propagation rate constant depends also on the type of comonomer[TM] last inserted into the polymer chain. The latter approach is closer to reality since, due to steric and electronic effect, the type of monomer added to

Table 5.2 Bernoullian and terminal copolymerization models.

Model type	Propagation steps[a]	Rate constants
Bernoullian	$P_r + A \rightarrow P_{r+1}$	k_{pA}
	$P_r + B \rightarrow P_{r+1}$	k_{pB}
Terminal	$P_r^A + A \rightarrow P_{r+1}^A$	k_{pAA}
	$P_r^A + B \rightarrow P_{r+1}^B$	k_{pAB}
	$P_r^B + A \rightarrow P_{r+1}^A$	k_{pBA}
	$P_r^B + B \rightarrow P_{r+1}^B$	k_{pBB}

[a] A and B represent monomer type; the superscripts A and B indicate the type of monomer at the end of the polymer chain.

the chain end is likely to influence the value of k_p for the next propagating step, but it requires the estimation of two additional rate constants. Higher-order models, such as the penultimate model, where the last two comonomer units added to the polymer chain affect the value of k_p of the coordinating monomer are rarely used in coordination polymerization because there is little experimental data to justify the use of a more complex copolymerization model for olefins.[2]

For a Bernoullian model, the polymerization rate is given by the expression,

$$R_p = (k_{pA}[A] + k_{pB}[B])[Y_0] \tag{5.21}$$

where [A] and [B] are the concentrations of comonomers A and B (for instance, ethylene and 1-hexene), respectively, at the active site, and k_{pA} and k_{pB} are their respective propagation rate constants. Equation (5.21) is more conveniently expressed as

$$R_p = (k_{pA}f_A + k_{pB}f_B)[M][Y_0] \tag{5.22}$$

where f_A and f_B are the molar fractions of the comonomers A and B, and $[M] = [A] + [B]$ is the total comonomer concentration. It is possible, therefore, to define a *pseudo-propagation rate constant*

$$\hat{k}_p = k_{pA}f_A + k_{pB}f_B \tag{5.23}$$

so that the copolymerization rate reduces to a form similar to that for homopolymerization

$$R_p = \hat{k}_p[M][Y_0] \tag{5.24}$$

2) The Bernoullian, terminal, and penultimate models are also called the zeroth-, first-, and second-order Markovian models, respectively, indicating how many monomers attached to the chain end affect the polymerization rate constant of the coordinating monomer.

The Bernoullian model, albeit appealing because of its simplicity, is often not adequate to describe the vast majority of copolymerizations with single- and multiple-site coordination catalysts. The terminal model, on the other hand, has much wider applicability, but it also requires the estimation of four polymerization rate constants. The equation for the polymerization rate with the terminal model is

$$R_p = (k_{pAA}[A] + k_{pAB}[B])[Y_{0A}] + (k_{pBA}[A] + k_{pBB}[B])[Y_{0B}] \tag{5.25}$$

where Y_{0A} and Y_{0B} are the number of moles of the living polymer chains terminated in monomers A and B, respectively, and k_{pij} are the four possible combinations for propagation rate constants.

Equation (5.25) is more conveniently written as

$$R_p = (k_{pAA}\phi_A f_A + k_{pAB}\phi_A f_B + k_{pBA}\phi_B f_A + k_{pBB}\phi_B f_B)[M][Y_0] \tag{5.26}$$

where ϕ_A and ϕ_B are the molar fractions of the living polymer chains terminated in monomers A and B, respectively.

Consequently, we can define the pseudo-propagation rate constant as

$$\hat{k}_p = k_{pAA}\phi_A f_A + k_{pAB}\phi_A f_B + k_{pBA}\phi_B f_A + k_{pBB}\phi_B f_B \tag{5.27}$$

and use this definition in Eq. (5.24) to calculate the polymerization rate for the terminal model.

The parameters ϕ_A and ϕ_B are easily calculated by realizing that for high polymers, the rate of AB insertions must be equal to the rate of BA insertions, that is,

$$k_{pAB}[Y_{0A}][B] = k_{pBA}[Y_{0B}][A] \tag{5.28}$$

or equivalently,

$$k_{pAB}\phi_A f_B = k_{pBA}\phi_B f_A \tag{5.29}$$

This hypothesis is called the long chain approximation (LCA), and it is very easy to justify. Consider the following hypothetical polymer chain

AABBABABABABBBBABAAAABBBAAABBBBAABAAABAB

The number of times monomer A is followed by monomer B is $n_{AB} = 11$, and the number of times monomer B is followed by monomer A is $n_{BA} = 10$. Had we added one more unit of monomer A to the right end of the chain,

AABBABABABABBBBABAAAABBBAAABBBBAABAAABABA

the number of AB and BA changes would be exactly the same, that is $n_{AB} = n_{BA} = 11$, which is an equivalent way to postulate Eq. (5.28) or (5.29). It is easy to conclude that for any polymer chain, $n_{AB} = n_{BA} \pm 1$. For long chains, this ± 1 difference is negligible and we can safely use Eqs. (5.28) and (5.29).

Consequently, since $f_B = 1 - f_A$ and $\phi_B = 1 - \phi_A$ for a binary copolymer,

$$\phi_A = \frac{k_{pBA} f_A}{k_{pAB}(1-f_A) + k_{pBA} f_A} \tag{5.30}$$

Therefore, if the molar fractions f_A and f_B are kept constant during the polymerization, the value of $\hat{k}_p$ will remain unaltered and all the equations we derived above for homopolymerization can also be used for copolymerization.

A similar approach can be used with terpolymers and higher copolymers. Using the terminal model, we can define the pseudo-propagation constant as

$$\hat{k}_p = \sum_{i=1}^{m} \sum_{j=1}^{m} k_{pij} \phi_i f_j \tag{5.31}$$

where m is the number of monomer types ($m = 1, 2, 3, \ldots$) in the system.

Similarly, the LCA approximation for terpolymers and higher copolymers is given by

$$\sum_{i=1,\neq j}^{m} k_{pij} \phi_i f_j - \sum_{i=1,\neq j}^{m} k_{pij} \phi_j f_i = 0, \; j = 1, m-1 \tag{5.32}$$

These $m-1$ equations can be solved for the m unknown ϕ_i using the following additional relation

$$\sum_{i=1}^{m} \phi_i = 1 \tag{5.33}$$

The site activation and deactivation steps are generally considered to be the same for homo- and copolymerization. This is a very important simplifying assumption because it enables us to describe the more complex copolymerization process as a homopolymerization process, *provided that the comonomer composition is kept constant during polymerization*. Figure 5.6 illustrates such an example, where Eq. (5.7) was used to calculate the rate of polymerization by substituting k_p with $\hat{k}_p$ estimated with Eq. (5.27). In Figure 5.6, A is the fast comonomer (for instance, ethylene), whereas B is the slow comonomer (propylene or an α-olefin).

Therefore, if enough experiments with varying comonomer composition are done, it is in principle possible to estimate all four propagation constants k_{pAA}, k_{pAB}, k_{pBA}, and k_{pBB} using a combination of Eqs. (5.7), (5.27), and (5.30). Unfortunately, the polymerization kinetics of several coordination catalysts is affected by the *comonomer effect*, which is discussed in Section 5.2, making the estimation of these parameters more challenging than it might appear now.

If composition drift takes place during polymerization, however, the value of $\hat{k}_p$ will also vary and the approach outlined above cannot be used directly. This is why copolymerization experiments used for estimating polymerization kinetic parameters should be done in the absence of comonomer composition variation.

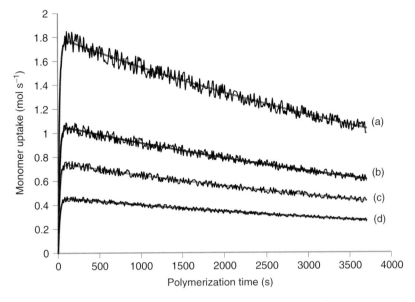

Figure 5.6 Simulated copolymerization curves for several comonomer molar fractions: (a) $f_A = 0.9$, (b) $f_A = 0.85$, (c) $f_A = 0.75$, and (d) $f_A = 0.6$. Simulation parameters: $K_a = 5 \times 10^{-2}\,\mathrm{s}^{-1}$, $k_d = 1.5 \times 10^{-4}\,\mathrm{s}^{-1}$, $k_{pAA} = 2.0 \times 10^4\,\mathrm{l\cdot mol^{-1}\cdot s^{-1}}$, $k_{pAB} = 2.0 \times 10^3\,\mathrm{l\cdot mol^{-1}\cdot s^{-1}}$, $k_{pBA} = 1.0 \times 10^4\,\mathrm{l\cdot mol^{-1}\cdot s^{-1}}$, $k_{pBB} = 10\,\mathrm{l\cdot mol^{-1}\cdot s^{-1}}$, $[C_0] = 1.0 \times 10^{-4}\,\mathrm{mol\cdot l^{-1}}$, and $[M] = 1.0\,\mathrm{mol\cdot l^{-1}}$.

It is also possible to obtain an independent estimation of the reactivity ratios $r_A = k_{pAA}/k_{pAB}$ and $r_B = k_{pBB}/k_{pBA}$ by analyzing the copolymer chemical composition with the Mayo–Lewis equation, as described in Chapter 6.

5.2.2
Multiple-Site Catalysts

Multiple-site catalysts, such as heterogeneous Ziegler–Natta and Phillips catalysts, pose several challenges to parameter estimation but are not essentially harder to model than the single-site catalysts described in the previous section. In general, we assume that a catalyst has two or more site types if it produces polymers with broad microstructural distributions, as discussed in Chapter 1. Each active site type, however, is governed by a set of distinct polymerization rate constants and behaves essentially as a single-site catalyst. For instance, the polymerization rate of a catalyst that has two site types is given by

$$R_p = \left(k_{p1}[Y_{01}] + k_{p2}[Y_{02}]\right)[M] \tag{5.34}$$

where the subscripts 1 and 2 identify the site types.

Since each site type is assumed to follow the same polymerization mechanism of a single-site catalyst, the equations developed in Section 5.2.1 for homo- and copolymerization are also valid for multiple site catalyts. Consequently, we can apply Eq. (5.7) to model the behavior of a two-site catalyst as follows:

$$R_p = k_{p1}[M] \frac{\left\{1 - \exp\left[-K_{a1}\left(1 - \frac{k_{d1}}{K_{a1}}\right)t\right]\right\}\exp(-k_{d1}t)}{1 - \frac{k_{d1}}{K_{a1}}}[C_{01}]$$

$$+ k_{p2}[M] \frac{\left\{1 - \exp\left[-K_{a2}\left(1 - \frac{k_{d2}}{K_{a2}}\right)t\right]\right\}\exp(-k_{d2}t)}{1 - \frac{k_{d2}}{K_{a2}}}[C_{02}] \qquad (5.35)$$

or more concisely,

$$R_p = \left(\frac{x_1 k_{p1}\left\{1 - \exp\left[-K_{a1}\left(1 - \frac{k_{d1}}{K_{a1}}\right)t\right]\right\}\exp(-k_{d1}t)}{1 - \frac{k_{d1}}{K_{a1}}} + \frac{(1-x_1)k_{p2}\left\{1 - \exp\left[-K_{a2}\left(1 - \frac{k_{d2}}{K_{a2}}\right)t\right]\right\}\exp(-k_{d2}t)}{1 - \frac{k_{d2}}{K_{a2}}} \right) \times [M][C_0]$$

(5.36)

where x_1 is the molar fraction of site type 1 at the beginning of polymerization and $[C_0]$ is the total concentration of active sites at $t = 0$. A similar approach could have been followed for the other rate laws we developed in the previous section.

Figure 5.7 shows that a single-site model may not be able to model the monomer uptake data for ethylene polymerization with a heterogeneous Ziegler–Natta catalyst, but the fit is much improved, especially for higher polymerization times, when a two-site model is used. The model prediction shows that both site types are activated very rapidly, but site type 1 is significantly more stable than site type 2, which quickly deactivates after approximately 2000 s of polymerization.

It is also interesting to see that neither the plot of $\ln R_p \times t$ nor the plot of $R_p^{-1} \times t$ are linear in this case (Figure 5.8), confirming that the polymerization is not governed by single-site first- or second-order propagation rate kinetics.

In general, for a catalyst with n distinct site types, we can write

$$R_p = \left(\sum_{i=1}^{n} k_{pi}[Y_{0i}]\right)[M] \qquad (5.37)$$

or

$$R_p = \left(\sum_{i=1}^{n} k_{pi} x_i\right)[M][Y_0] \qquad (5.38)$$

where x_i is the molar fraction of site type i in the reactor at a given polymerization time.

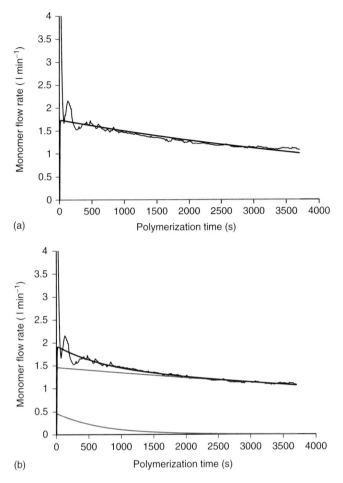

Figure 5.7 Ethylene polymerization rate with a heterogeneous Ziegler–Natta catalyst: (a) single-site model: $K_a = 0.2779\ \text{s}^{-1}$, $k_{d1} = 1.476 \times 10^{-4}\ \text{s}^{-1}$, and $k_{p1} = 1.743 \times 10^4\ \text{l}\cdot\text{mol}^{-1}\cdot\text{s}^{-1}$ and (b) two-site model: $x_1 = 0.87$, $K_{a1} = 0.2778\ \text{s}^{-1}$, $k_{d1} = 8.46 \times 10^{-5}\ \text{s}^{-1}$, $k_{p1} = 1.678 \times 10^4\ \text{l}\cdot\text{mol}^{-1}\cdot\text{s}^{-1}$, $x_2 = 0.13$, $K_{a2} = 0.2817\ \text{s}^{-1}$, $k_{d2} = 1.470 \times 10^{-3}\ \text{s}^{-1}$, $k_{p2} = 3.475 \times 10^4\ \text{l}\cdot\text{mol}^{-1}\cdot\text{s}^{-1}$.

The parameter estimation problem gets even more complex when the number of active site types is increased, possibly leading to several local minima. As a general rule of thumb, two active site types can describe most polymerization kinetics curves, and unless additional information is available about the polymer, it is difficult to justify the use of additional site types to model polymerization kinetics alone. In Chapter 6 we see, however, that two site types are usually not sufficient to describe the molecular weight distribution of polyolefins made with heterogeneous Ziegler–Natta and Phillips catalysts. In this case, the molecular weight distribution information may be used to help us decide how many more site types are required to describe the system under consideration.

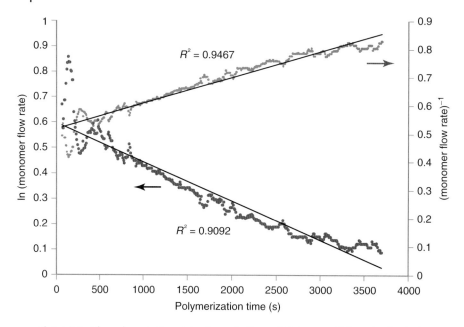

Figure 5.8 The polymerization data shown in Figure 5.7 does not follow simple single-site kinetics of first or second order.

5.2.3
Temperature Dependence of Kinetic Constants

Polymerization constants depend on the temperature according to Arrhenius law. For the propagation step, this dependency is given by the equation

$$k_p = A_p \exp\left(-\frac{E_p}{RT}\right) \tag{5.39}$$

where A_p and E_p are the preexponential constant and activation energy for propagation, respectively, R is the gas constant, and T is absolute temperature. Similar equations can be written for k_a, k_d, and the chain transfer constants.

Since the activation energy is always positive, all rate constants increase when the polymerization temperature is raised. On the other hand, observed polymerization rates with coordination catalysts generally pass through a maximum as the temperature increases and then decrease for higher temperatures, because the activation energy for the catalyst deactivation step, E_d

$$k_d = A_d \exp\left(-\frac{E_d}{RT}\right) \tag{5.40}$$

is always higher than the activation energy for propagation. Such phenomenon is illustrated in Figure 5.9, and it is a common feature of polymerization with coordination catalysts.

Figure 5.9 also illustrates a very important point when comparing the activity of different catalysts or of the same catalyst at different temperatures: to report

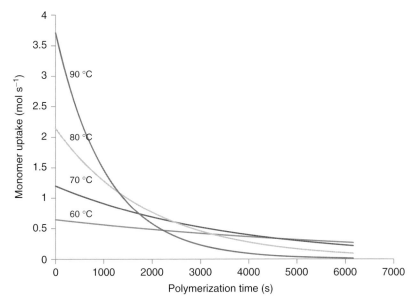

Figure 5.9 Effect of temperature on polymerization rate when $E_d > E_p$. Model parameters $K_a \to \infty$ (instantaneous activation), $A_d = 1.0 \times 10^6 \, s^{-1}$, $E_d = 1.5 \times 10^4 \, cal \cdot mol^{-1}$, $A_p = 1.0 \times 10^{13} \, l \cdot mol^{-1} \cdot s^{-1}$, $E_p = 1.4 \times 10^4 \, cal \cdot mol^{-1}$, $[M] = 1.0 \, mol \cdot l^{-1}$, and $C_0 = 1.0 \times 10^{-4} \, mol \cdot l^{-1}$.

catalyst activities as the polymer yield at the end of the polymerization is a rather crude way to compare catalysts. In Figure 5.9, the yield after 1000 s for the polymerization performed at 90 °C is the highest, but this is not the case after 6000 s have elapsed because of the different temperature responses of the polymerization and deactivation rates. Therefore, it is always preferable to report the complete monomer uptake curve (or its equivalent kinetic parameters) when comparing different catalysts or polymerization conditions.

This general trend is also easily visualized by calculating the total monomer consumption using one of the rate equations we derived above. We use the case of instantaneous site activation, Eq. (5.16), for the sake of simplicity

$$n_M = k_p[M][C_0]V_R \int_0^t \exp(-k_d t) dt$$

$$= \frac{k_p}{k_d}[M][C_0]V_R \left[1 - \exp(-k_d t)\right] \quad (5.41)$$

Substituting Arrhenius law, Eqs. (5.39) and (5.40), and simplifying the resulting expression, we obtain

$$n_M = \frac{A_p}{A_d} \exp\left(-\frac{E_p - E_d}{RT}\right)[M][C_0]V_R \left\{1 - \exp\left[-A_d \exp\left(-\frac{E_d}{RT}\right)t\right]\right\} \quad (5.42)$$

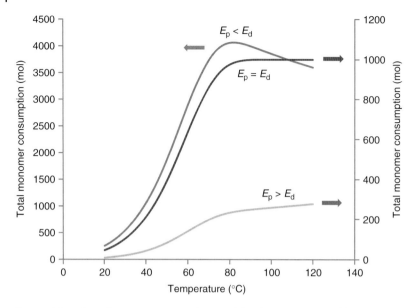

Figure 5.10 Effect of temperature on polymer yield with varying E_p. Model parameters are the same as in Figure 5.9, but E_p decreases in the order 1.6×10^4 cal·mol^{-1} ($E_p > E_d$), 1.5×10^4 cal·mol^{-1} ($E_p = E_d$), and 1.4×10^4 cal·mol^{-1} ($E_p < E_d$).

Therefore, if $E_d > E_p$ then $-(E_p - E_d)/RT > 0$, consequently increasing the polymerization temperature will eventually lead to a decrease in polymer yield, as illustrated in Figure 5.10. Interestingly, if $E_d = E_p$, the polymer yield reaches a plateau, and if $E_d < E_p$, the yield always increases when the temperature is raised; these two cases, however, are not observed experimentally with coordination polymerization catalysts.

As usual, the case of multiple-site catalyst is more complicated because different active site types may have distinct activation energies for activation, propagation, and deactivation, leading to intermediate temperature responses, but the general treatment shown here for single-site catalysts should also be applicable to these more complex systems.

5.2.4
Number of Moles of Active Sites

All the polymerization kinetic models we have described in this chapter depend on the initial concentration or number of moles of active sites in the reactor, n_{C_0}. Unfortunately, it is very difficult to estimate this parameter.

It is easy to measure the weight, w_C, of catalyst added to the reactor. If the catalyst is a soluble metallocene or late transition metal complex, the number of moles is calculated directly from its molar mass, M_C

$$n_{\bar{C}_0} = \frac{w_C}{M_C} \qquad (5.43)$$

(The reason why we are using $n_{\tilde{C}_0}$ instead of n_{C_0} to represent the number of catalyst sites in the reactor will become clear in a moment.)

If a heterogeneous catalyst is used (supported single-site, Ziegler–Natta or Phillips), we can find out the mass fraction of catalyst molecules, x_C, on the support by elemental analysis and then calculate $n_{\tilde{C}_0}$.

$$n_{\tilde{C}_0} = x_C \frac{w_C}{M_C} \tag{5.44}$$

Therefore, measuring the total number of transition metal sites on the catalyst is not particularly difficult. Unfortunately, we cannot be certain that all catalyst precursor molecules become active sites because of incomplete activation by the cocatalyst, by poisoning with reactor impurities and by-products, due to deactivation reactions with functional groups on the support, or simply because of steric hindrance effects caused by the support surface. Consequently, $n_{\tilde{C}_0}$ is not the number we should use in our polymerization rate equations.

Some experimental techniques, notably active site tagging with radioactive carbon monoxide or alcohols, have been attempted to determine the exact number of sites present on heterogeneous Ziegler–Natta catalysts, but the results from this studies have been inconclusive. It is believed that only 1–10% of the Ti atoms of a $TiCl_4/MgCl_2$ is active for polymerization, but currently, it is not possible to arrive at a more precise number.

Consequently, when we use n_{C_0} as the initial condition in the models we developed above, we mean only the number of moles of active sites, not the total number of moles of catalyst introduced in the reactor, that is,

$$n_{C_0} = \eta n_{\tilde{C}_0} \tag{5.45}$$

where η, an efficiency factor varying between 0 and 1, accounts for the incomplete activation of the catalyst precursor sites.

Since the parameter η is hard to estimate, we generally use $[\tilde{C}_0]$ in our polymerization rate equations and estimate an apparent propagation rate constant instead of the intrinsic propagation rate. For instance, Eq. (5.7) could be expressed as

$$R_p = k_p[M] \frac{\left\{1 - \exp\left[-K_a\left(1 - \frac{k_d}{K_a}\right)t\right]\right\}\exp(-k_d t)}{1 - \frac{k_d}{K_a}} \eta[\tilde{C}_0]$$

$$= \tilde{k}_p[M] \frac{\left\{1 - \exp\left[-K_a\left(1 - \frac{k_d}{K_a}\right)t\right]\right\}\exp(-k_d t)}{1 - \frac{k_d}{K_a}} [\tilde{C}_0] \tag{5.46}$$

where $\tilde{k}_p = \eta k_p$ is an apparent polymerization rate constant.

Evidently, if η is affected by the polymerization conditions, the parameter estimation, even for homopolymerization with a single-site catalyst, becomes more complex. Unfortunately, there is no technique to determine reliable values for $[C_0]$ or η, and olefin coordination polymerization models must live with this shortcoming. On the other hand, if we take a pragmatic approach, the parameter η

can be used as a semiempirical correction factor for polymerization behavior that cannot be properly described with fundamental models. This approach may not be very elegant, but it gives the modeler an additional degree of freedom when trying to describe these complex catalytic systems.

5.3
Nonstandard Polymerization Kinetics Models

As mentioned in Chapter 3, the polymerization kinetics scheme we call the fundamental model does not explain all aspects of olefin polymerization kinetics with coordination catalysts. Notably, there are three main phenomena that the fundamental model fails to describe.

1) Polymerization orders with respect to monomer different from 1.
2) Hydrogen concentration effects on the polymerization rate.
3) Polymerization rate enhancement effects resulting from the copolymerization of ethylene and α-olefins.

Several polymerization mechanisms have been proposed for these phenomena, but no consensus exists on a single, unifying explanation for all of them. The complexity is compounded by the fact that these phenomena are observed with metallocene, Ziegler–Natta, and Phillips catalysts. It is not surprising that a single model may fail to describe all these catalyst systems. Nonetheless, some trends are prevalent enough to permit a generic treatment, even if we run the risk of oversimplifying a rather complex problem, along Haldane's precept that "an ounce of algebra is worth a ton of verbal argument". In this section, we will work out the mathematical details for some of these models.

5.3.1
Polymerization Orders Greater than One

Several models have been proposed to explain polymerization orders higher than unity. Among them, Ystenes's trigger mechanism assumes that a monomer molecule can only be inserted into the carbon–metal bond of the growing polymer chain when a second molecule "triggers" its insertion. Table 5.3 lists the main elementary steps of the trigger mechanism. We will assume that catalyst activation is instantaneous and that catalyst deactivation is negligible to simplify the mathematical treatment.

From Table 5.3, the rate of polymerization is given as

$$R_p = -\frac{d[M]}{dt} = k_p[M] \sum_{r=0}^{\infty} [P_r \cdot M] = k_p[M][Y_0 \cdot M] \tag{5.47}$$

As usual, for transition states, we can make the steady-state approximation for $[Y_0 \cdot M]$

$$\frac{d[Y_0 \cdot M]}{dt} = k_f[M][Y_0] - k_r[Y_0 \cdot M] - k_p[M][Y_0 \cdot M] = 0 \tag{5.48}$$

5.3 Nonstandard Polymerization Kinetics Models

Table 5.3 Trigger mechanism for homopolymerization.

Description	Chemical equations	Rate constants
Activation	$C + Al \rightarrow P_0$	$k_a \rightarrow \infty$
Reversible coordination	$P_r + M \rightleftarrows P_r \cdot M$	k_f, k_r
Propagation	$M + P_r \cdot M \rightarrow P_{r+1} + M$	k_p

to obtain the following expression

$$[Y_0 \cdot M] = \frac{k_f[M][Y_0]}{k_r + k_p[M]} \quad (5.49)$$

Finally, substituting Eq. (5.49) into Eq. (5.50) we get

$$R_p = \frac{k_p k_f [M]^2 [Y_0]}{k_r + k_p[M]} \quad (5.50)$$

Since $[Y_0]$ in Eq. (5.47) is a function of time, it may be substituted with expressions such as those given in Eq. (5.6), (5.11), or (5.15), or any other catalyst deactivation law, to describe how the polymerization rate varies as a function of time.

Inspecting Eq. (5.50), it is easy to see that the polymerization order with respect to monomer concentration may vary from 1 when $k_p[M] \gg k_r$

$$R_p \cong k_f[M][Y_0] \quad (5.51)$$

to 2 when $k_p[M] \ll k_r$

$$R_p \cong \frac{k_p k_f}{k_r}[M]^2[Y_0] = k_p^*[M]^2[Y_0] \quad (5.52)$$

Orders of reaction in the interval [1,2] result when the polymerization conditions are between these two limiting cases.

Example 5.4 further explores this interesting phenomenon.

Example 5.4: Polymerization Kinetic Parameter Estimation for a Single-Site Catalyst: Effect of Catalyst Concentration

Figure E.5.4.1 shows the ethylene uptake curves during polymerization in a semibatch solution reactor using CGC-Ti (a monocyclopentadienyl catalyst belonging to the constrained geometry catalyst family) at a polymerization temperature of 120 °C. All variables were kept constant during polymerization, except for the ethylene pressure that was allowed to vary from 25 to 220 psi.

The shape of the ethylene uptake curves looks very different from those shown in Figure E.5.2.1, when $Et(Ind)_2ZrCl_2$ was used as a catalyst

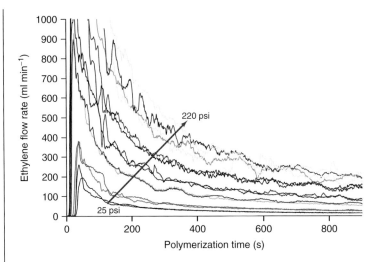

Figure E.5.4.1 Ethylene uptake rates at different ethylene pressures (indicated in the figure) in a semibatch solution reactor operated at a temperature of 120 °C.

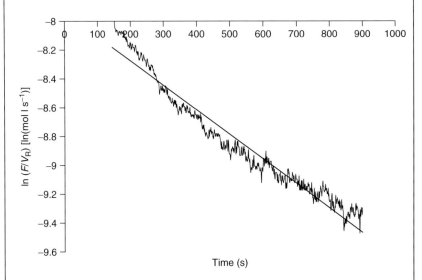

Figure E.5.4.2 Plot of ln R_p × polymerization time for the data shown in Figure E.5.4.1.

under similar ethylene polymerization condition. Indeed, these profiles do not fit first-order catalyst deactivation kinetics because the plot of ln $R_p \times t$ is not linear, as illustrated in Figure E.5.4.2 for one representative polymerization run. On the other hand, when the reciprocal of the monomer uptake curves are plotted versus polymerization time (Figure E.5.4.3), a linear relation is observed, as expected from the bimolecular deactivation mechanism described with Eq. (5.13).

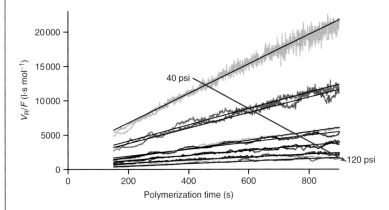

Figure E.5.4.3 Plot of $1/R_p \times$ polymerization time for the data shown in Figure E.5.4.1.

In addition, the polymerization order with respect to ethylene concentration (or partial pressure) is not linear throughout the whole range of experimental conditions. Figure E.5.4.4 shows that the polymer yield seems to increase linearly for higher ethylene pressures but deviates from linearity for pressures below approximately 80 psi.

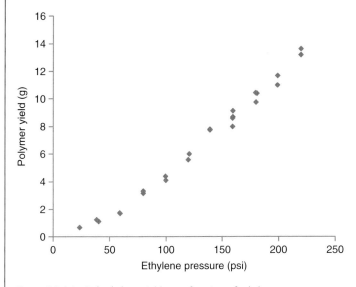

Figure E.5.4.4 Polyethylene yield as a function of ethylene pressure.

This behavior is consistent with the trigger mechanism described in this section and a bimolecular deactivation mechanism. Therefore, we may combine Eqs. (5.11) and (5.50) to describe the polymerization kinetics for this system as follows:

$$R_p = \frac{k_p k_f [M]^2}{k_r + k_p[M]} \times \frac{[C_0]}{1 + [C_0]k_d^* t}$$

The polymer yield for this system can be calculated as

$$m_P = \left(28 \text{gmol}^{-1}\right) \times \frac{k_p k_f [M]^2 [C_0] V_R}{k_r + k_p[M]} \int \frac{dt}{1 + [C_0]k_d^* t}$$

$$= \left(28 \text{gmol}^{-1}\right) \times \frac{k_p k_f [M]^2 V_R}{k_d^* (k_r + k_p[M])} \ln\left(1 + [C_0]k_d^* t\right)$$

or more concisely,

$$m_P = \left(28 \text{gmol}^{-1}\right) \times \frac{K_1 [M]^2 V_R}{1 + K_2[M]} \ln\left(1 + [C_0]k_d^* t\right)$$

where

$$K_1 = \frac{k_p k_f}{k_d^* k_r}$$

$$K_2 = \frac{k_p}{k_r}$$

We will assume that Henry's law describes the relation between ethylene pressure and concentration in the solvent (toluene)

$$[M] = K_H P_M$$

where K_H is the Henry's law constant for ethylene dissolved in toluene at $120\,^\circ\text{C}$ and P_M is the partial pressure of ethylene in the reactor. Finally, the expression for polymer yield becomes

$$m_P = \left(28 \text{gmol}^{-1}\right) \times \frac{K_1' P_M^2 V_R}{1 + K_2' P_M} \ln\left(1 + [C_0]k_d^* t\right)$$

where $K_1' = K_1 K_H$ and $K_2' = K_2 K_H$.

Figure E.5.4.5 shows that the equation we developed for polymer yield fits the experimental data well with the following model parameters $K_1' = 5.31\ \text{mol} \cdot \text{l}^{-1} \cdot \text{psi}^{-2}$, $K_2' = 0.011\ \text{psi}^{-1}$, and $k_d^* = 5.8 \times 10^{-2}\ \text{s}^{-1}$. Therefore, for this system, the trigger mechanism may be used to explain the change from second to first order, as the ethylene pressure is increased.

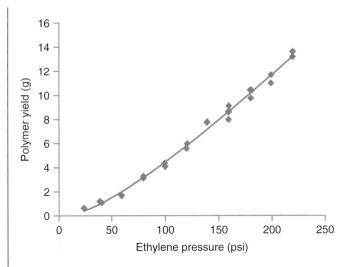

Figure E.5.4.5 Polyethylene yield as a function of ethylene pressure: experimental × model fit ($[C_0] = 5.47 \times 10^{-7}$ mol·l^{-1} and $V_R = 222.8$ ml).

Notice, however, that other models described below can also predict the order change depicted in Figure E.5.4.5. Consequently, the good experimental data description shown above is not a definite proof that the trigger mechanism is the explanation for this observed behavior.

5.3.2
Hydrogen Effect on the Polymerization Rate

Besides working as a chain transfer agent, hydrogen affects the olefin polymerization rate with most coordination catalysts in two opposite ways: it decreases the polymerization rate of ethylene but increases the polymerization rate of propylene. Although the possibility of hydrogen affecting the nature of the active sites (for instance, through reduction reactions) cannot be completely discarded, it does not seem to be the main explanation behind this phenomenon because its effect is reversible: when hydrogen is removed from the reactor, the polymerization rate is restored to its original values, as already shown in Chapter 3 (Figure 3.23). A representative example, in the case of propylene polymerization with a heterogeneous Ziegler–Natta catalyst, is shown in Figure 5.11: the introduction of hydrogen after 30 min of polymerization causes the rate to increase to the same level it had when hydrogen was present since the beginning of the polymerization. On the other hand, when hydrogen is vented off the reactor after 30 min of polymerization, the rate decreases to the value it would have been if the polymerization proceeded without hydrogen since its beginning (the fast decreasing propylene flow rates

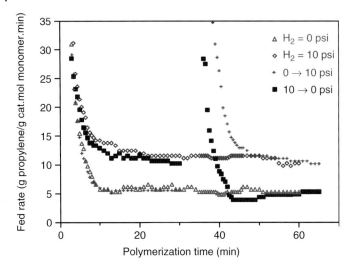

Figure 5.11 Effect of hydrogen on the polymerization rate of propylene with a heterogeneous Ziegler–Natta catalyst in a semibatch slurry reactor.

observed in the figure at the beginning of each transition are related to the reactor pressurization with propylene and are not caused by fast catalyst decay).

In addition, when the same catalyst is used for the homopolymerization of ethylene or propylene, hydrogen increases the polymerization rate of propylene, but decreases the rate of ethylene. These effects have been observed for several heterogeneous Ziegler–Natta catalysts, as well as for homogeneous and supported metallocene catalysts. Therefore, it does not seem that a permanent change in catalyst nature can be attributed to hydrogen use during polymerization.

Two main hypotheses have been developed to explain the effect of hydrogen on the polymerization rate of ethylene. The simplest one proposes that the metal hydride site that results after transfer to hydrogen is not very reactive toward ethylene insertion. Consequently, increasing the hydrogen concentration in the reactor leads to a higher fraction of lower-reactivity metal hydride sites, resulting in lower polymerization rates. Since β-hydride elimination produces metal hydride sites that are undistinguishable from those made by transfer to hydrogen, as explained in Chapter 3, a portion of these lower-reactivity sites will be always present during polymerization.

A set of reactions that exemplifies this mechanism is listed in Table 5.4. The four most common transfer reaction steps (β-hydride elimination and transfer to monomer, hydrogen, and cocatalyst) are included in the table. Transfer to hydrogen, as well as β-hydride elimination, generate low-reactivity metal hydride sites represented by P_H. To simplify the mathematical treatment, we assumed that transfer to cocatalyst creates a site type similar to that generated by transfer to ethylene, P_1, which is a relatively good assumption if triethyl aluminum (TEA) is used as activator. In addition, we will also assume that the propagation rate for sites P_1 is the same as for longer chains, which may not be completely accurate, but

Table 5.4 Mechanism for hydrogen effect on the polymerization rate of ethylene.

Description	Chemical equations	Rate constants
Propagation	$P_r + M \rightarrow P_{r+1}$	k_p
Transfer to hydrogen	$P_r + H_2 \rightarrow P_H + D_r$	k_{tH}
β-Hydride elimination	$P_r \rightarrow P_H + D_r$	$k_{t\beta}$
Transfer to ethylene	$P_r + M \rightarrow P_1 + D_r$	k_{tM}
Transfer to cocatalyst	$P_r + Al \rightarrow P_1 + D_r$	k_{tAl}
Initiation of P_H	$P_H + M \rightarrow P_1$	$k_{iH} \ll k_p$

is necessary to simplify the mathematical treatment below. The negative hydrogen effect on the polymerization rate is thus modeled by setting $k_{iH} < k_p$.

Let us define the instantaneous polymerization rate as

$$R_p = k_p[M][Y_0] = k_p[M]\phi[C^*] \tag{5.53}$$

where we excluded $[P_H]$ from the definition of $[Y_0]$

$$[Y_0] = \sum_{r=1}^{\infty} [P_r] \tag{5.54}$$

The parameter ϕ is the fraction of active sites that have one or more monomers growing on them

$$\phi = \frac{[Y_0]}{[Y_0] + [P_H]} = \frac{1}{1 + \frac{[P_H]}{[Y_0]}} \tag{5.55}$$

and $[C^*]$ is the total concentration of active sites in the reactor at a given time

$$[C^*] = [P_H] + \sum_{r=1}^{\infty} [P_r] = [P_H] + [Y_0] \tag{5.56}$$

Assuming that the catalyst precursor initiation is instantaneous, the molar balance for $[Y_0]$ becomes

$$\frac{d[Y_0]}{dt} = k_{iH}[M][P_H] - (k_{t\beta} + k_{tH}[H_2])[Y_0] \tag{5.57}$$

Making the steady-state approximation, we can calculate the ratio $[Y_0]/[P_H]$

$$\frac{[P_H]}{[Y_0]} = \frac{k_{tH}[H_2] + k_{t\beta}}{k_{iH}[M]} \tag{5.58}$$

and we finally get

$$\phi = \frac{1}{1 + \frac{k_{tH}[H_2] + k_{t\beta}}{k_{iH}[M]}} \tag{5.59}$$

Therefore, the expression for the instantaneous polymerization rate becomes

$$R_p = \frac{k_p[M]}{1 + \dfrac{k_{tH}[H_2] + k_{t\beta}}{k_{iH}[M]}}[C^*] \qquad (5.60)$$

or

$$R_p = \frac{k_{iH}[M]}{\dfrac{k_{iH}}{k_p} + \dfrac{k_{tH}[H_2] + k_{t\beta}}{k_p[M]}}[C^*] \qquad (5.61)$$

Some interesting inferences can be drawn from Eqs. (5.60) and (5.61). For high polymers, the propagation frequency is much higher than the chain transfer frequency; therefore, the second term in the denominator of Eq. (5.61) must be much smaller than one.[3] If the constants for propagation and P_H initiation are comparable, then

$$\frac{k_{iH}}{k_p} \gg \frac{k_{tH}[H_2] + k_{t\beta}}{k_p[M]} \qquad (5.62)$$

and Eq. (5.60) is reduced to the usual form for the propagation rate equation predicted by the fundamental model for olefin polymerization kinetics, Eq. (5.1). Contrarily, if k_{iH} is much smaller than k_p, a polymerization rate reduction effect will be observed and apparent polymerization orders higher than one may also result. For this to occur, however, the initiation frequency for the metal hydride sites must be comparable to the termination frequency that generates these sites, that is, $k_{iH}[M] \approx k_{tH}[H_2] + k_{t\beta}$.

An alternative mechanism was proposed by Kissin to explain the hydrogen effect on the polymerization rate of ethylene. It resembles the slow metal hydride model we have just described, but it is based on a different hypothesis. Instead of postulating that metal hydride sites have slow initiation rates, it hypothesizes that active sites having only one ethylene unit form a stable state via β-agostic interaction between the metal center and the hydrogen attached to the β-carbon atom in the chain, slowing down the second ethylene insertion and reducing the overall polymerization rate. Since these "dormant" sites may be formed after ethylene is inserted onto a metal hydride site, increasing the concentration of hydrogen will increase the fraction of dormant sites in the active site population. Figure 5.12

---

3) If transfer to monomer and β-hydride elimination are the dominant chain-transfer mechanisms in this case, this ratio is equal to the reciprocal of the number average chain length, r_n.

$$\frac{1}{r_n} = \frac{k_{tH}[H_2] + k_{t\beta} + k_{tM}[M] + k_{tAl}[Al]}{k_p[M]} \ll 1.0$$

As explained in more detail in Chapter 6, if all transfer reactions in Table 5.4 contribute to chain length control, a more generic expression results.

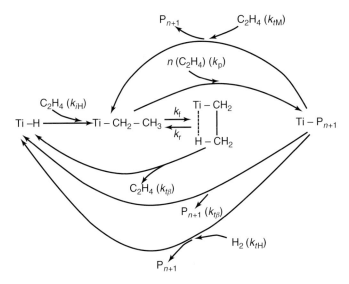

Figure 5.12 Mechanism for the formation of a "dormant" β-agostic interaction site.

illustrates this mechanism. As shown in the diagram, transfer to ethylene and β-hydride elimination also produce sites that may lead to the formation of dormant sites according to this mechanism. It is not clear whether sites formed after transfer to cocatalyst would also undergo the same process, but we will assume that they may also form dormant sites.

Table 5.5 lists the kinetic steps for the mechanism described in Figure 5.12. We assumed that all sites with one or more monomer units propagate with the same rate constant k_p, although small deviations may be observed for very short chains such as P_1. We have also assumed that the site formed after transfer to cocatalyst may form a dormant site, which may not be true, since the aluminum atom may interfere in the proposed β-agostic interaction. Finally, the dormant sites $\tilde{P}_1$ are produced through a reversible first-order reaction, as proposed in Figure 5.12.

Table 5.5 Mechanism for hydrogen effect on the polymerization rate of ethylene, assuming a dormant β-agostic interaction site.

Description	Chemical equations	Rate constants
Propagation	$P_r + M \rightarrow P_{r+1}$	k_p
Transfer to hydrogen	$P_r + H_2 \rightarrow P_H + D_r$	k_{tH}
β-Hydride elimination	$P_r \rightarrow P_H + D_r$	$k_{t\beta}$
Transfer to ethylene	$P_r + M \rightarrow P_1 + D_r$	k_{tM}
Transfer to cocatalyst	$P_r + Al \rightarrow P_1 + D_r$	k_{tAl}
Initiation of P_H	$P_H + M \rightarrow P_1$	k_{iH}
β-Agostic interaction	$P_1 \rightleftarrows \tilde{P}_1$	k_f, k_r

The polymerization rate equation for this mechanism is also given by Eq. (5.53)

$$R_p = k_p[M][Y_0] = k_p[M]\phi[C^*] \tag{5.53}$$

with a small modification in the definition of $[C^*]$

$$[C^*] = [P_H] + [\tilde{P}_1] + \sum_{r=1}^{\infty}[P_r] = [P_H] + [\tilde{P}_1] + [Y_0] \tag{5.63}$$

Consequently, the parameter ϕ is redefined as

$$\phi = \frac{[Y_0]}{[Y_0] + [P_H] + [\tilde{P}_1]} = \frac{1}{1 + \frac{[P_H]}{[Y_0]} + \frac{[\tilde{P}_1]}{[Y_0]}} \tag{5.64}$$

The molar balance for the dormant sites is given by

$$\frac{d[\tilde{P}_1]}{dt} = k_f[P_1] - (k_r + k_{t\beta})[\tilde{P}_1] \tag{5.65}$$

Assuming fast equilibrium

$$[\tilde{P}_1] = \frac{k_f}{k_r + k_{t\beta}}[P_1] = K_{eq}[P_1] \tag{5.66}$$

where

$$K_{eq} = \frac{k_f}{k_r + k_{t\beta}} \tag{5.67}$$

Similarly, the balances for $[P_1]$ and $[P_H]$ are expressed as

$$\frac{d[P_1]}{dt} = k_{iH}[P_H][M] + k_r[\tilde{P}_1] + K_T^1[Y_0] - k_f[P_1] - k_p[P_1][M] \tag{5.68}$$

$$\frac{d[P_H]}{dt} = k_{t\beta}[\tilde{P}_1] + K_T^H[Y_0] - k_{iH}[P_H][M] \cong K_T^H[Y_0] - k_{iH}[P_H][M] \tag{5.69}$$

In Eq. (5.69) we assumed that the β-hydride elimination frequency for the dormant sites was negligible in comparison with the chain transfer frequency resulting from transfer to hydrogen of the higher chains in the system, $K_T^H[Y_0] \gg k_{t\beta}[\tilde{P}_1]$, which is a reasonable assumption in the presence of hydrogen.

The lumped constants K_T^1 and K_T^H are the frequencies of the transfer reactions that produce ethylene-terminated and metal hydride sites, respectively, which according to our mechanism are given by

$$K_T^1 = k_{tM}[M] + k_{tAl}[Al] \tag{5.70}$$

$$K_T^H = k_{t\beta} + k_{tH}[H_2] \tag{5.71}$$

Solving for $[P_H]$ and $[P_1]$ instantaneously, we get

$$\frac{[P_H]}{[Y_0]} = \frac{K_T^H}{k_{iH}[M]} \tag{5.72}$$

$$\frac{[P_1]}{[Y_0]} = \frac{K_T}{k_p[M]} \tag{5.73}$$

where

$$K_T = K_T^1 + K_T^H \tag{5.74}$$

Substituting Eqs. (5.66), (5.72), and (5.73) into Eq. (5.64)

$$\phi = \frac{1}{1 + \dfrac{K_T^H}{k_{iH}[M]} + \dfrac{K_{eq} K_T}{k_p[M]}} \tag{5.75}$$

Therefore, if K_{eq} is set to zero (no formation of the β-agostic dormant site), the model becomes the same as the previous one that assumed a slow initiation of the metal hydride site given by Equation (5.59). All other conditions being the same, as the value of K_{eq} increases, so does the fraction of sites that remain in the dormant state, thus decreasing ϕ and the overall polymerization rate.

Example 5.5 compares these two mechanisms for the negative effect of hydrogen concentration on the polymerization rate of ethylene with coordination catalysts.

Example 5.5: Comparison of Slow Metal Hydride and Dormant β-Agostic Interaction Site Models

Figure E.5.5.1 compares the values of the parameter ϕ predicted using the slow metal hydride site ($K_{eq} = 0$) with the dormant β-agostic interaction model for two values of K_{eq} for a simulated single-site catalyst. We also assumed that $k_{iH} < k_p$ for this catalyst. As expected, the fraction of sites that is active for polymerization decreases as the value of K_{eq} is increased. Similarly, ϕ (and the observed polymerization rate) decreases as the transfer to hydrogen frequency, as represented by the parameter K_T^H, increases. A significant deviation from the slow metal hydride site model is only apparent for large values of the parameter K_{eq}, indicating that the β-agostic interaction state proposed in Figure 5.12 must be highly favored to explain a sizeable polymerization rate reduction. Figure E.5.5.2 compares similar simulations for the case when $k_p = k_{iH}$. Evidently, in this case, the metal hydride model cannot predict a decrease in the polymerization rate on introduction of hydrogen, but this effect can still be effectively described with the β-agostic interaction model.

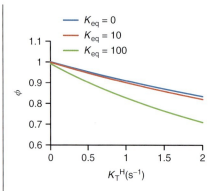

Figure E.5.5.1 Value of the parameter ϕ in Eq. (5.53) as a function of K_{eq} (defined in Eq. (5.67)) and the frequency of transfer to hydrogen, K_T^H. Other model parameters: $K_{iH} = 10\,s^{-1}$, $K_p = 1000\,s^{-1}$, and $K_T^1 = 0.1\,s^{-1}$.

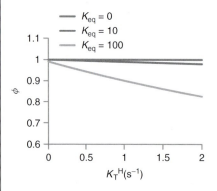

Figure E.5.5.2 Value of the parameter ϕ in Eq. (5.53) as a function of K_{eq} (defined in Eq. (5.67)) and the frequency of transfer to hydrogen K_T^H. Other model parameters: $K_{iH} = K_p = 1000\,s^{-1}$, $K_T^1 = 0.1\,s^{-1}$.

It is also interesting to notice that an apparent second-order behavior may result because of either mechanism. For instance, substitute Eq. (5.75) into Eq. (5.53) and rearrange the equation to obtain the expression

$$R_p = \frac{k_{iH} k_p^2 [M]^2 [C^*]}{k_p \left(k_{tH}[H_2] + k_{t\beta} + k_{iH}[M]\right) + K_{eq} k_{iH} \left(k_{tH}[H_2] + k_{t\beta} + k_{tM}[M] + k_{tAl}[Al]\right)}$$

which will predict a second-order dependency of the polymerization rate on monomer concentration for low monomer concentration values. This behavior is clear in Figure E.5.5.3 and resembles the results we explained above using the trigger mechanism. As often happens with kinetic models, several mechanisms may be able to describe the same behavior, and it is only through detailed and well-designed experiments that we may be able to discriminate between them.

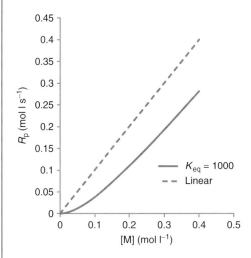

Figure E.5.5.3 Simulated polymerization rate using a linear ($\phi = 1$) and the β-agostic interaction model $\phi = f([M])$. Model parameters: $k_p = 10000 \, l \cdot mol^{-1} \cdot s^{-1}$, $k_{iH} = 100 \, l \cdot mol^{-1} \cdot s^{-1}$, $k_{tH} = 1 \, l \cdot mol^{-1} \cdot s^{-1}$, $k_{t\beta} = 0.5 \, s^{-1}$, $k_{tM} = 0.1 \, l \cdot mol^{-1} \cdot s^{-1}$, $k_{tAl} = 1 \, l \cdot mol^{-1} \cdot s^{-1}$, $K_{eq} = 1000$, $[H_2] = 1 \, mol \cdot l^{-1}$, $[C^*] = 0.0001 \, mol \cdot l^{-1}$, $[Al] = 0.001 \, mol \cdot l^{-1}$.

Finally, if we are able to ignore the effect of hydrogen on catalyst deactivation, we can combine kinetic equations such as the one introduced in Eq. (5.7) with the definition for ϕ to calculate, in a relatively straightforward manner, the effect of hydrogen on the polymerization rate curves as a function of time

$$R_p = k_p[M] \frac{1 - e^{-K_a\left(1 - \frac{k_d}{K_a}\right)t}}{1 - \frac{k_d}{K_a}} \phi[C_0]e^{-k_d t} \tag{5.76}$$

Contrary to ethylene polymerization, the rate of propylene polymerization increases with the introduction of hydrogen for the majority of Ziegler–Natta and metallocene catalysts. In addition, this effect is reversible; removal of hydrogen from the reactor will restore the rate to its original value. Some hypotheses have been proposed to explain this phenomenon; the most widely accepted one postulates that the propagation rate of 2-1 terminated chains is much slower than the propagation rate of 1-2 terminated sites. In the absence of hydrogen, these 2-1 terminated living chains form dormant sites that reduce the overall polymerization rate. When hydrogen is added to the reactor, chain transfer reactions free up those

Table 5.6 Mechanism for hydrogen rate enhancement for propylene polymerization.

Description	Propagation steps	Rate constants
Propagation	$P_r^{12} + M \rightarrow P_{r+1}^{12}$	$k_{p12,12}$
	$P_r^{12} + M \rightarrow P_{r+1}^{21}$	$k_{p12,21}$
	$P_r^{21} + M \rightarrow P_{r+1}^{12}$	$k_{p21,12}$
	$P_r^{21} + M \rightarrow P_{r+1}^{21}$	$k_{p21,21}$
Transfer to H_2	$P_r^{12} + H_2 \rightarrow P_H + D_r$	k_{tH12}
	$P_r^{21} + H_2 \rightarrow P_H + D_r$	k_{tH21}
Initiation	$P_H + M \rightarrow P_1^{12}$	k_{i12}
	$P_H + M \rightarrow P_1^{21}$	k_{i21}

dormant sites for further polymerization. This mechanism has been supported by several polymerization kinetic and chain-end analysis investigations.

From a modeling point of view, this mechanism is similar to the binary copolymerization model shown in Table 5.2, where the insertion of a propylene molecule can generate either a 1-2 (fast) or a 2-1 (slow) terminated active site, as depicted in Table 5.6. For simplicity, we assumed that transfer to hydrogen was the only chain transfer mechanism, but additional chain transfer steps may be added to the kinetic scheme without changing the general conclusions from the model.

The overall polymerization rate for this system is given by the equation

$$R_p = (k_{p12,12} + k_{p12,21})[Y_0^{12}][M] + (k_{p21,12} + k_{p21,21})[Y_0^{21}][M] \quad (5.77)$$

where $[Y_0^{12}]$ and $[Y_0^{21}]$ are the concentrations of 1-2 and 2-1 terminated living chains, respectively.

Equation (5.77) can be rendered in the more convenient form shown below

$$R_p = \{(k_{p12,12} + k_{p12,21})\phi^{12} + (k_{p21,12} + k_{p21,21})(1 - \phi^{12})\}[Y_0][M] \quad (5.78)$$

where ϕ^{12} is the molar fraction of 1-2 terminated living chains, calculated making the LCA

$$\phi^{12} = \frac{[Y_0^{12}]}{[Y_0^{12}] + [Y_0^{21}]} \quad (5.79)$$

The molar balance for the 1-2 terminated living chains is given by

$$\frac{d[Y_0^{12}]}{dt} = k_{i12}[P_H][M] + k_{p21,12}[Y_0^{21}][M] - (k_{tH12}[H_2] + k_{p12,21}[M])[Y_0^{12}] \quad (5.80)$$

and the molar balance for the metal hydride sites, P_H, is

$$\frac{d[P_H]}{dt} = \left(k_{tH12}[Y_0^{12}] + k_{tH21}[Y_0^{21}]\right)[H_2] - (k_{i12} + k_{i21})[P_H][M] \quad (5.81)$$

Solving Eqs. (5.80) and (5.81) simultaneously at steady state, we obtain after certain manipulations the following expression for the fraction of living chains terminated in 1-2 propylene insertions

$$\phi^{12} = \frac{k_{i12}k_{tH21}[H_2] + k_{p21,12}(k_{i12} + k_{i21})[M]}{(k_{i12}k_{tH21} + k_{i21}k_{tH12})[H_2] + (k_{p21,12} + k_{p12,21})(k_{i12} + k_{i21})[M]} \quad (5.82)$$

It is possible to calculate the hydrogen rate enhancement effect on propylene polymerization with single-site catalysts by combining Eqs. (5.82) and (5.78). Since the main premise is that $k_{p12,12}$ far exceeds all other propagation rate constants, the overall polymerization rate will be nearly proportional to ϕ^{12}.

Example 5.6 illustrates the use of this model.

Example 5.6: Hydrogen Effect on Propylene Polymerization Rate

Equation (5.82) can be modified to the more convenient dimensionless form

$$\phi^{12} = \frac{1 + x_{12}\tau_{21}\frac{[H_2]}{[M]}}{1 + \pi + \left[x_{12}\tau_{21} + (1 - x_{12})\frac{\tau_{12}}{\pi}\right]\frac{[H_2]}{[M]}}$$

where

$$x_{12} = \frac{k_{i12}}{k_{i12} + k_{i21}}$$

$$\tau_{12} = \frac{k_{tH12}}{k_{p12,21}}$$

$$\tau_{21} = \frac{k_{tH21}}{k_{p21,12}}$$

$$\pi = \frac{k_{p12,21}}{k_{p21,12}}$$

The parameter x_{12} is the fraction of chains that are initiated with 1-2 propylene insertions; τ_{12} and τ_{21} are the ratios of the rate constants for chain transfer to hydrogen to cross-propagation for 1-2 terminated chains and 2-1 terminated chains, respectively; and π is the ratio between the two cross-propagation constants (head-to-head and tail-to-tail). For regiospecific catalysts, we expect

x_{12} to be relatively high. If we make the assumption that $k_{tH12} \cong k_{tH21}$, then the 2-1 dormant site model requires that $\tau_{12} < \tau_{21}$ because the 2-1 terminated site is supposed to have very low propagation rate, that is, $k_{p21,12} \ll k_{p12,21}$. For the same reason, $\pi > 1$.

Figure E.5.6.1 shows how the fraction of chains terminated in 1-2 insertions varies as a function of the $[H_2]/[M]$ ratio for several values of the parameter π. As the relative hydrogen concentration in the reactor increases, so does the value of ϕ^{12}, since the sluggish 2-1 sites are terminated by transfer to hydrogen and free to polymerize again. The fraction of 1-2 terminated sites for a given $[H_2]/[M]$ ratio is higher for smaller π values since this implies that 1-2 to 2-1 insertions will be less frequent for a given constant $k_{p21,12}$ or, equivalently, 2-1 to 1-2 insertions will occur more often for a constant $k_{p12,21}$ value. The former option is generally assumed for this case, since it is assumed that $k_{p12,21} \gg k_{p21,12}$.

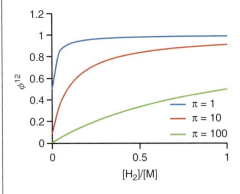

Figure E.5.6.1 Simulated fraction of 1-2 terminated living polymer chains. Model parameters: $x_{12} = 1$, $\tau_{12} = 0.1$, and $\tau_{21} = 100$.

It is also instructive to follow how ϕ^{12} varies with the $[H_2]/[M]$ ratio for several values of the parameter τ_{21}, as illustrated in Figure E.5.6.2. As expected, the value of ϕ^{12} increases as τ_{21} increases because of the higher transfer to hydrogen rate for 2-1 terminated chains. The results shown in Figures E.5.6.1 and E.5.6.2 show that hydrogen can influence the fraction of chains in the dormant 2-1 insertion mode and, therefore, affect the polymerization rate of propylene. Since we assume that propagation rate constant for 1-2 to 1-2 insertion is much higher than the other ones in highly regioselective catalysts, Eq. (5.78) can be approximated as

$$R_p \cong k_{p12,12} \phi^{12} [Y_0][M]$$

and the polymerization rate is approximately proportional to the fraction of 1-2 terminated sites.

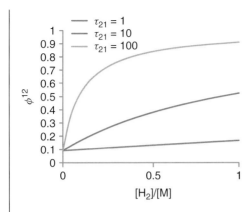

Figure E.5.6.2 Simulated fraction of 1-2 terminated living polymer chains. Model parameters: $x_{12} = 1$, $\tau_{12} = 0.1$, and $\pi = 10$.

A similar model can be developed for multiple-site catalysts, with each site type characterized by a different set of kinetic parameters. Heterogeneous Ziegler–Natta catalysts are generally assumed to have some site types that are very regioselective and therefore are less affected by hydrogen, while having other less regioselective site types that would be more sensitive to reactor hydrogen addition.

5.3.3
Comonomer Effect on the Polymerization Rate

A somewhat puzzling behavior is observed when α-olefins are copolymerized with ethylene. When a small fraction of α-olefin is introduced with ethylene in the reactor, the polymerization rate increases to values higher than for ethylene alone. If the fraction of α-olefin in the reactor keeps increasing, the polymerization rate reaches a maximum, and then starts decreasing. Since ethylene has a much higher homopolymerization rate than α-olefins, this observation implies that α-olefins change the polymerization kinetics of the existing sites, create new types that are inactive in their absence, or facilitate the diffusion of monomer molecules to the active sites. Although mass transfer effects may indeed be a factor for polymerization with heterogeneous catalysts, the fact that α-olefin rate enhancement effects are observed even for solution polymerization clearly precludes it as the sole explanation for this phenomenon.

A few mechanisms have been proposed to explain the comonomer rate enhancement effect. Karol proposed that the α-olefin comonomer may act as a ligand, coordinating with vacancies on the active sites. Different olefins would, therefore, create active sites with distinct propagation rate constants or activate sites that were inactive in their absence. The trigger mechanism seen above has also been used to explain this effect: different olefins would have different abilities to "trigger"

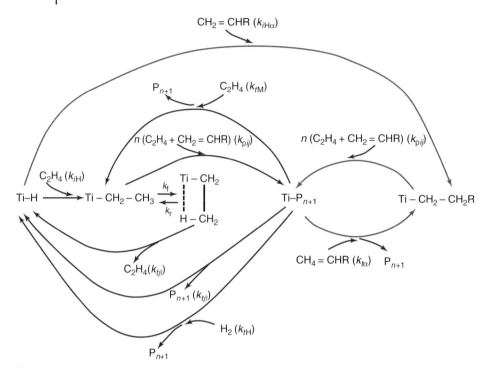

Figure 5.13 Ethylene/α-olefin copolymerization mechanism with the formation of a "dormant" β-agostic interaction site after transfer to ethylene. R represents alkyl group.

monomer insertion. According to this model, α-olefins would be able to increase the propagation rate of ethylene by forming a more favorable transition state complex before insertion.

The model proposed by Kissin, used above to analyze the hydrogen effect on ethylene polymerization rate, can be extended to include the comonomer effect. According to its proponents, chain transfer to α-olefin would lead to an active site with weaker β-agostic interaction, therefore bypassing the reversible dormant state shown in Figure 5.12. Even though this mechanism is not considered the definite explanation for this phenomenon, it is supported by the fact that polymer molecular weight does decrease on the introduction of α-olefins for ethylene polymerization with most catalysts, an observation associated with increased frequency of transfer to comonomer. Figure 5.13 shows the catalyst cycle previously depicted in Figure 5.12 extended for copolymerization. Notice how transfer to comonomer and the initiation of metal hydride sites with an α-olefin avoid the formation of the Ti–CH$_2$–CH$_3$ species that may lead to the formation of sites in the dormant state.

Table 5.7 summarizes the most important kinetic steps using the Bernoullian model for copolymerization. The terminal model could also have been used but would result in a more complex and, for our purposes, less useful, set of equations. As usual, A represents ethylene and B the α-olefin comonomer.

Table 5.7 Copolymerization mechanism (Bernoullian model) including a β-agostic interaction site.

Description	Chemical equations	Rate constants
Propagation	$P_r + A \rightarrow P_{r+1}$	k_{pA}
	$P_r + B \rightarrow P_{r+1}$	k_{pB}
Transfer to hydrogen	$P_r + H_2 \rightarrow P_H + D_r$	k_{tH}
β-Hydride elimination	$P_r \rightarrow P_H + D_r$	$k_{t\beta}$
Transfer to monomer	$P_r + A \rightarrow P_1^A + D_r$	k_{tA}
	$P_r + B \rightarrow P_1^B + D_r$	k_{tB}
Transfer to cocatalyst	$P_r + Al \rightarrow P_1 + D_r$	k_{tAl}
Initiation of P_H	$P_H + A \rightarrow P_1^A$	k_{iHA}
	$P_H + B \rightarrow P_1^B$	k_{iHB}
β-Agostic interaction	$P_1^A \rightleftarrows \tilde{P}_1^A$	k_f, k_r

The polymerization rate equation for this mechanism is given by

$$R_p = (k_{pA}[A] + k_{pB}[B])[Y_0] = [k_{pA}f_A + k_{pB}(1-f_A)][Y_0][M] \quad (5.83)$$

where

$$f_A = \frac{[A]}{[A]+[B]} \quad (5.84)$$

$$[M] = [A] + [B] \quad (5.85)$$

Recalling the concept of pseudokinetic constants seen above, we can rewrite Eq. (5.83) in a way similar to Eq. (5.53)

$$R_p = \hat{k}_p[M][Y_0] = \hat{k}_p[M]\phi[C^*] \quad (5.86)$$

where

$$\hat{k}_p = k_{pA}f_A + k_{pB}(1-f_A) \quad (5.87)$$

and $[C^*]$ and ϕ are given by expressions very similar to Eqs. (5.63) and (5.64)

$$[C^*] = [P_H] + [\tilde{P}_1^A] + \sum_{r=1}^{\infty}[P_r] = [P_H] + [\tilde{P}_1^A] + [Y_0] \quad (5.88)$$

$$\phi = \frac{[Y_0]}{[Y_0] + [P_H] + [\tilde{P}_1^A]} = \frac{1}{1 + \frac{[P_H]}{[Y_0]} + \frac{[\tilde{P}_1^A]}{[Y_0]}} \quad (5.89)$$

The only difference between the copolymerization and homopolymerization models is that transfer to comonomer and initiation of P_H sites by comonomer bypass the pathway that leads to the formation of the dormant site $\tilde{P}_1^A$.

The molar balance for the dormant sites is given by

$$\frac{d[\tilde{P}_1^A]}{dt} = k_f[P_1^A] - (k_r + k_{t\beta})[\tilde{P}_1^A] \tag{5.90}$$

Assuming fast equilibrium

$$[\tilde{P}_1^A] = \frac{k_f}{k_r + k_{t\beta}}[P_1^A] = K_{eq}[P_1^A] \tag{5.91}$$

Similarly, the balances for $[P_1^A]$ and $[P_H]$ are expressed respectively as

$$\frac{d[P_1^A]}{dt} = k_{iHA}[P_H]f_A[M] + k_r[\tilde{P}_1^A] + K_T^{1,A}[Y_0] - k_f[P_1^A] - \hat{k}_p[P_1^A][M] \tag{5.92}$$

$$\frac{d[P_H]}{dt} = k_{t\beta}[\tilde{P}_1^A] + K_T^H[Y_0] - \hat{k}_{iH}[P_H][M] \cong K_T^H[Y_0] - \hat{k}_{iH}[P_H][M] \tag{5.93}$$

where the pseudo-initiation rate constant was defined as

$$\hat{k}_{iH} = k_{iHA}f_A + k_{iHB}(1 - f_A) \tag{5.94}$$

The lumped constant $K_T^{1,A}$, the frequency of the transfer reaction that produces ethylene-terminated sites, results from a slight modification of Eq. (5.70)

$$K_T^{1,A} = k_{tA}f_A[M] + k_{tAl}[Al] \tag{5.95}$$

and the expression for K_T^H, the frequency of metal hydride site formation, is still given by Eq. (5.71).

Solving for $[P_H]$ and $[P_1^A]$ instantaneously, we get

$$\frac{[P_H]}{[Y_0]} = \frac{K_T^H}{\hat{k}_{iH}[M]} \tag{5.96}$$

$$\frac{[P_1^A]}{[Y_0]} = \frac{f_A k_{iHA} K_T + (1 - f_A) k_{iHB} K_T^{1,A}}{\hat{k}_{iH} \hat{k}_p[M]} \tag{5.97}$$

where

$$K_T = K_T^{1,A} + K_T^H \tag{5.98}$$

Notice that these equations reduce to their homopolymerization forms if $f_A = 1$.

Finally, substituting Eqs. (5.91), (5.96), and (5.97) into Eq. (5.89), we obtain the final expression for the fraction of "nondormant" sites in the reactor

$$\phi = \frac{1}{1 + \dfrac{K_T^H}{\hat{k}_{iH}[M]} + K_{eq} \times \dfrac{f_A k_{iHA} K_T + (1 - f_A) k_{iHB} K_T^{1,A}}{\hat{k}_{iH} \hat{k}_p[M]}} \tag{5.99}$$

Example 5.7 uses this model to investigate the effect of hydrogen and 1-olefin concentrations on the rate of polymerization with coordination catalysts.

Example 5.7: Effect of Hydrogen and α-Olefin Concentration on the Copolymerization Rate of Ethylene Using the Dormant β-Agostic Interaction Site Model

If α-olefin comonomers had no influence on the nature and/or number of active sites in the catalyst, we should expect that the rate of polymerization would decrease steadily when the concentration of α-olefin increased in the reactor. For the β-agostic interaction model, this condition is observed when the value of $K_{eq} = 0$, that is, where there is no formation of the dormant site shown in Figure 5.13. Figure E.5.7.1 shows how the product $\phi \times \hat{k}_p$ would behave in this case. A monotonic decrease is observed when the fraction of comonomer increases, since $k_{pA} > k_{pB}$. This, however, is not the behavior observed with single- or multiple-site coordination catalysts.

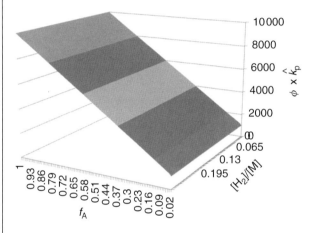

Figure E.5.7.1 Product $\phi \times \hat{k}_p$ as function of $[H_2]/[M]$ ratio and fraction of ethylene in the reactor (f_A) when $K_{eq} = 0$ (no dormant site). Model parameters: $k_{pA} = k_{iHA} = 10000 \text{ l} \cdot \text{mol}^{-1} \cdot \text{s}^{-1}$, $k_{pB} = k_{iHB} = 1000 \text{ l} \cdot \text{mol}^{-1} \cdot \text{s}^{-1}$, $k_{tH} = 1 \text{ l} \cdot \text{mol}^{-1} \cdot \text{s}^{-1}$, $k_{t\beta} = 0.1 \text{ s}^{-1}$, $k_{tA} = k_{tB} = 1 \text{ l} \cdot \text{mol}^{-1} \cdot \text{s}^{-1}$, $k_{tAl} = 0 \text{ l} \cdot \text{mol}^{-1} \cdot \text{s}^{-1}$, $[M] = 4 \text{ mol} \cdot \text{l}^{-1}$, $[C^*] = 0.0001 \text{ mol} \cdot \text{l}^{-1}$, and $[Al] = 0.001 \text{ mol} \cdot \text{l}^{-1}$.

When K_{eq} is increased to a value of 10, the kinetic profile is altered drastically, as shown in Figure E.5.7.2. In this case, addition of a small fraction of the α-olefin will substantially increase the product $\phi \times \hat{k}_p$ and therefore the polymerization rate, as described by Eq. (5.86). Finally, Figure E.5.7.3 shows simulation results when $K_{eq} = 100$, that is, when the formation of the dormant site is strongly favored. In this case, the maximum rate will take place when a higher fraction of comonomer is present in the reactor, but the maximum polymerization rate possible would not be as high as in the case with $K_{eq} = 10$, all other factors being the same.

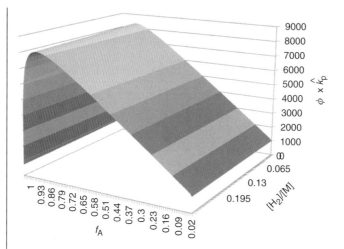

Figure E.5.7.2 Product $\phi \times \hat{k}_p$ as function of $[H_2]/[M]$ ratio and fraction of ethylene in the reactor (f_A) when $K_{eq} = 10$ (dormant sites are slightly favored). Model parameters are the same as in Figure E.5.7.1.

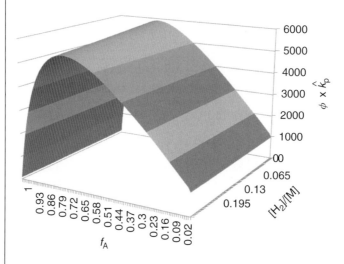

Figure E.5.7.3 Product $\phi \times \hat{k}_p$ as function of $[H_2]/[M]$ ratio and fraction of ethylene in the reactor (f_A) when $K_{eq} = 100$ (dormant sites are strongly favored). Model parameters are the same as in Figure E.5.7.1.

It is also interesting to see that in the presence of comonomer, the hydrogen rate reduction effect is much less noticeable, a factor also observed experimentally with many catalysts.

The comonomer effect is a complex mechanism that certainly cannot be described in its entirety with the model we presented in this section. Besides, we kept the model as simple as possible (including the Bernoullian kinetics assumption) to reduce the mathematical complexity to a point where analytical solutions were possible. We believe this approach illustrates the main features of the phenomenon under investigation and gives it a reasonable quantitative description. More detailed models can be used to describe this behavior, but they would not be amenable to the simple mathematical treatment shown above.

Regardless of the mechanism adopted to describe this phenomenon, one thing is clear: the propagation rate constant measured for ethylene alone cannot be used when ethylene is polymerized with α-olefins, since the value of k_{pAA} for copolymerization is always higher than the value of k_p for homopolymerization with the same catalyst. This is an important characteristic of coordination catalysts and a clear departure from simpler polymerization kinetic schemes, such as the ones for free radical polymerization.

Finally, if a heterogeneous catalyst is used, we cannot discard the possibility that mass transfer limitations may also affect the apparent copolymerization rate of ethylene and α-olefins. As discussed in more detail in Chapter 7, monomer can only diffuse through the region of amorphous polymer surrounding the active sites. As the fraction of α-olefin in the copolymer increases, the size of the polymer crystallite domains decreases and the amount of amorphous polymer phase that is permeable to monomer diffusion increases, thus facilitating the access of monomer to the catalyst sites.

5.3.4
Negative Polymerization Orders with Late Transition Metal Catalysts

Another "anomalous" polymerization kinetic behavior is observed when olefins are polymerized with some late transition metal catalysts, such as the Ni-diimine catalysts discussed in Chapter 3. The polymerization rate initially increases with monomer concentration, reaches a maximum, and then decreases on further monomer concentration increase. This behavior is illustrated in Figure 5.14 for ethylene polymerization in a slurry reactor with a Ni-diimine catalyst at three different temperatures, where the turnover frequency (TOF) is plotted as a function of ethylene concentration in the reactor.

The *TOF* is defined as the average rate of polymerization divided by the concentration of catalyst in the reactor

$$\text{TOF} = \frac{R_p}{[C^*]} \quad (5.100)$$

Since these polymerization experiments were relatively short (less than 10 min), and the catalyst deactivation rate was negligible during this period, it can be assumed that the concentration of active sites in the reactor remains practically the same throughout the polymerization.

Table 5.8 Ethylene polymerization mechanism with Ni-diimine catalysts.

Description	Chemical equations	Rate constants
Propagation	$P_r + M \rightarrow P_{r+1}$	k_p
Formation of dormant site	$P_r + mM \rightleftharpoons \tilde{P}_r$	k_c, k_c^-

A mechanism has been proposed to explain this behavior by assuming that a latent state is formed when two or more (m) monomers coordinate to the active site, as listed in Table 5.8.

Therefore, the concentration of active and latent sites at a given time in the reactor is given by

$$[Y_0] = \sum_{r=1}^{\infty} [P_r] \tag{5.101}$$

$$[\tilde{Y}_0] = \sum_{r=1}^{\infty} [\tilde{P}_r] \tag{5.102}$$

and the overall concentration of sites is expressed as

$$[C^*] = [Y_0] + [\tilde{Y}_0] \tag{5.103}$$

Therefore, as we did for all the models covered in Section 5.3, we express the polymerization rate with Eq. (5.53) and derive an expression for ϕ

$$\phi = \frac{[Y_0]}{[Y_0] + [\tilde{Y}_0]} = \frac{1}{1 + \dfrac{[\tilde{Y}_0]}{[Y_0]}} \tag{5.104}$$

Inspection of the simple mechanism in Table 5.8 tells us that the molar balance for $[Y_0]$ is given by

$$\frac{d[Y_0]}{dt} = k_c^-[\tilde{Y}_0] - k_c[Y_0][M]^m \tag{5.105}$$

which can be solved at steady state to give

$$\frac{[\tilde{Y}_0]}{[Y_0]} = \frac{k_c[M]^m}{k_c^-} = K_c[M]^m \tag{5.106}$$

where

$$K_c = \frac{k_c}{k_c^-} \tag{5.107}$$

Substituting Eqs. (5.106) and (5.104) in Eq. (5.53) we obtain the polymerization rate expression for the mechanism proposed in Table 5.8

$$R_p = k_p[M]\phi[C^*] = \frac{k_p[M][C^*]}{1 + K_c[M]^m} \tag{5.108}$$

This equation has two limiting solutions: when the monomer concentration is small, $K_c[M]^m \ll 1$ and the rate of polymerization increases linearly with [M]; however, for large [M], $K_c[M]^m \gg 1$ and the polymerization rate assumes a negative order with respect to monomer concentration

$$R_p \cong \frac{k_p}{K_c}[C^*][M]^{(1-m)} \qquad (5.109)$$

which explain the decrease in TOF as ethylene concentration increases after a certain limiting value, as shown in Figure 5.14.

Finally, Eq. (5.100) becomes

$$\text{TOF} = \frac{k_p[M]}{1 + K_c[M]^m} \qquad (5.110)$$

Figure 5.14 shows that this equation can fit the experimental data relatively well, either when $m = 2$ or when it is allowed to vary.

The examples discussed in this section show that the variety of polymerization kinetics behaviour associated with coordination catalysts can be difficult to describe with a single model. It is absolutely true that no single model is capable of describing the polymerization kinetics with all types of coordination catalysts, even single-site ones. The models reviewed in this section give only an overview of the several different schemes presented in the literature to cover these catalytic systems. Each system must be approached with care and analyzed based on reliable experimental polymerization kinetics data, as no single model is capable of fitting the variety of situations encountered when olefins are polymerized with coordination catalysts.

5.4
Vapor-Liquid-Solid Equilibrium Considerations

All the equations derived in this chapter presuppose that the concentrations of monomer, comonomer, hydrogen, and any other reactive species are known at the active site. As discussed in Chapter 7, in the case of heterogeneous catalysts, these concentrations will depend on mass and heat transfer resistances taking place inside (intraparticle) and outside (interparticle) the polymer particles, as well as on the solubility of these reagents in the polymer phase as it is formed during polymerization. Even for solution polymerization, the monomer has to dissolve in the polymerization medium before reaching the active sites (unless the reactor is completely filled with liquid, a condition common in industrial solution polymerization reactors, but not in laboratory-scale systems).

For gas-phase polymerization, the monomer has to be absorbed on the polymer phase before it can diffuse to the active site. Monomer solubility in polymer is not a trivial problem because besides depending on monomer type and concentration, polymer type, and temperature, it also depends on the crystallinity of the nascent polymer. Some equations of state that can be used to describe these systems are listed at the end of this chapter under Further Readings, but the simplest approach

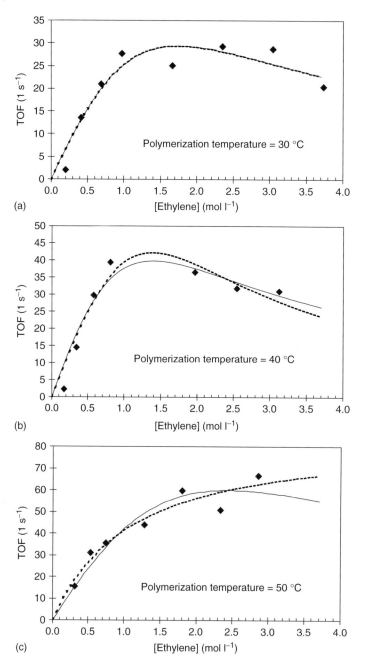

Figure 5.14 Ethylene polymerization rate with a Ni-diimine catalyst at (a) 30 °C, (b) 40 °C, and (c) 50 °C. Experimental points shown are compared with predictions from a second-order model (solid lines) and a variable-order model (dashed lines) (orders $m = 2.00$, 2.32, and 1.04, for (a), (b), and (c), respectively).

is to use a partition coefficient of the form

$$[M]_s = K_{g-s}[M]_s = K_{g-s}\frac{ZP_M}{RT} = K^*_{g-s}P_M \tag{5.111}$$

where the subscripts g and s indicate the gas and solid (polymer) phases, respectively; P_M is the partial pressure of monomer (or any other reagent in the gas phase); Z is the compressibility factor, and K_{g-s} and K^*_{g-s} are the experimental partition coefficients based on gas concentration and partial pressure, respectively. The partition coefficients for a given monomer/polymer system are a function of temperature, but normally, the simple linear relationship predicted by Eq. (5.111) holds true for a relatively broad range of gas pressures. The main difficulty is that the partition coefficient needs to be measured for the nascent polymer (as opposed to a film made by melting the polymer particles) to replicate the conditions existing in the reactor. In addition, $[M]_s$ in Eq. (5.109) is the concentration at the polymer surface, which may not be the same as the concentration at the active sites, depending on how severe are the intraparticle mass and heat transfer resistances. We move this discussion to Chapter 7, where these phenomena are analyzed in detail.

An analogous equation is needed for liquid propylene (or other liquid α-olefins) polymerization to account for propylene absorption on polypropylene

$$[M]_s = K_{l-s}[M]_s \tag{5.112}$$

where the subscript l represents the liquid phase.

For ethylene (hydrogen or other gaseous α-olefins) polymerization in slurry reactors, we must consider both gas–liquid and liquid–solid equilibria

$$[M]_l = K_{g-l}[M]_g = K_{g-l}\frac{P_M}{RT} = K^*_{g-l}P_M \tag{5.113}$$

$$[M]_s = K_{l-s}[M]_l = K_{l-s}K^*_{g-l}P_M = K'_{g-s}P_M \tag{5.114}$$

In these equations, we assumed that a linear relationship exists between the concentrations in the two phases. This assumption is accurate for the relatively low pressures encountered during the polymerization of olefins with coordination catalysts in most industrial processes.

Another assumption made in these equations is that the concentrations in the different phases are in equilibrium. This may not be the case for polymerizations with catalysts having extremely high activities during the initial seconds or minutes of polymerization. Since the average residence times in most olefin polymerization reactors is in the order of several minutes (for solution processes) to hours (for all other processes), the phase equilibrium assumption is well justified for most practical situations.

We can combine these simple phase equilibrium relations with the rate equations developed in this chapter to relate the polymerization rate to the monomer pressure or concentration in the reactor. For instance, in a typical semibatch slurry reactor, such as the one shown in Figure 5.1, the polymerization rate expression given in

Eq. (5.7) can be combined with Eq. (5.114) to give

$$R_p = k_p K_{g-s}^* P_M \frac{1 - e^{-K_a\left(1 - \frac{k_d}{K_a}\right)t}}{1 - \frac{k_d}{K_a}} [C_0] e^{-k_d t}$$

$$= k_p^* P_M \frac{1 - e^{-K_a\left(1 - \frac{k_d}{K_a}\right)t}}{1 - \frac{k_d}{K_a}} [C_0] e^{-k_d t} \tag{5.115}$$

If K_{g-s}^* is known, one can proceed to solve for k_p; if not, the apparent kinetic rate constant k_p^*, combining polymerization kinetics and gas–liquid/liquid–solid equilibria, can be estimated instead.

Further Reading

Several books and book chapters deal with olefin polymerization kinetics with Ziegler–Natta and metallocene catalysts.

Odian, G. (2004) *Principles of Polymerization*, 4th edn, John Wiley & Sons, Hoboken.

Kissin, Y.V. (2008) *Alkene Polymerization Reactions with Transition Metal Catalysts*, Elsevier, London.

Keii, T. (2004) *Heterogeneous Kinetics: Theory of Ziegler-Natta-Kaminsky Polymerization*, Kodansha, Tokyo.

Soares, J.B.P. and Simon, L.C. (2005) in *Handbook of Polymer Reaction Engineering* (eds T. Meyer and J. Keurentjes), Wiley-VCH Verlag GmbH, Weinheim, p. 365.

Soares, J.B.P., McKenna, T.F., and Cheng, C.P. (2007) in *Polymer Reaction Engineering* (ed. J.M. Asua), Blackwell Publishing, Oxford, p. 29.

Polymerization kinetics with late transition metal catalysts has not been studied so systematically, but some review articles are also available in this area.

Ittel, S.D. and Johnson, L.K. (2000) *Chem. Rev.*, **100**, 1169.

The use of reactor calorimeters to monitor the polymerization kinetics of liquid propylene (Samson, J.J.C., Weickert, G., Heerze, A.E., and Westerterp, K.R. (1998) *AIChE J.*, **44**, 1424; Samson, J.J.C., Bosman, P.J., Weickert, G., and Westerterp, K.R. (1999) *J. Polym. Sci., Part A: Polym. Chem.*, **37**, 219) and other slurry polymerizations (Lavanchy, F., Fortini, S., and Meyer, T. (2004) *Org. Process Res. Dev.*, **8**, 504; Mantelis, C.A., Barbey, R., Fortini, S., and Meyer, T. (2007) *Macromol. React. Eng.*, **1**, 78) has also been covered in the recent literature.

The pseudokinetic constant approach for copolymerization was first suggested by Hamielec, A.E., MacGregor, J.F., and Penlidis, A. (1987) *Makromol. Chem. Macromol. Symp.*, **10/11**, 521; Hamielec, A.E., MacGregor, J.F., and Penlidis, A. (1989) Copolymerization, in *Comprehensive Polymer Science* (ed. G. Allan), Pergamon Press, Oxford, pp. 17–31 and further validated by Xie, T. and Hamielec, A.E. (1993) *Makromol. Chem. Theory Simul.*, **2**, 421; Xie, T. and Hamielec, A.E. (1993) *Makromol. Chem. Theory Simul.*, **2**, 455.

This approach is not restricted to polymers made by coordination polymers since it is applicable to any polymer for which the LCA is valid.

Absolute measurement of active site concentration on Ziegler–Natta catalysts has been tried by Bukatov, G.D., Goncharov, V.S., and Zakharov, V.A. (1995) *Macromol. Chem. Phys.*, **196**, 1751; Bukatov, G.D. and Zakharov, V.A. (2001) *Macromol. Chem. Phys.*, **202**, 2003; Tait, P.J., Zohuri, G.H., Kells, A.M., and McKenzie, I.D. (1995) *Ziegler Catalysts* (eds G. Fink, R. Müllhaupt, and H.H. Brintzinger), Springer-Verlag, Berlin, Heidelberg, p. 343.

Stopped-flow reactors, modeled in Chapter 6, have also been used to measure this quantity.

Keii, T., Terano, M., Kimura, K., and Ishi, K. (1987) *Makromol. Chem. Rapid Commun.*, **8**, 583.

Mori, H. and Terano, M. (1977) *Trends Polym. Sci.*, **5**, 314.

Mori, H., Iguchi, H., Hasebe, K., and Terano, M. (1997) *Macromol. Chem. Phys.*, **198**, 1249.

The hydrogen effect on polymerization rate has also been the subject of several investigations. The experimental evidence of several groups argues in favor of the "2-1 insertion dormant site" hypothesis to explain the hydrogen rate enhancement effect on propylene polymerization.

Busico, V., Cipullo, R., and Corradini, P. (1992) *Makromol. Chem. Rapid Commun.*, **13**, 15.

Tsutsui, T., Kashiwa, N., and Mizuno, A. (1990) *Makromol. Chem. Rapid Commun.*, **11**, 565.

Chadwick, J.C., Miedema, A., and Sudmeijer, O. (1994) *Macromol. Chem. Phys.*, **195**, 167.

For ethylene polymerization, when the opposite effect is observed, no such consensus exists, but the most common models involve the slow M-H site or the β-agostic interaction state described in this chapter.

Kissin, Y.V. and Rishina, L.A. (2002) *J. Polym. Sci., Part A: Polym. Chem.*, **40**, 1353.

Kissin, Y.V., Rishina, L.A., and Vizen, E.I. (2002) *J. Polym. Sci., Part A: Polym. Chem.*, **40**, 1899.

Kissin, Y.V., Mink, R.I., Nowlin, T.E., and Brandolini, A.J. *Top. Catal.*, **7**, 69.

Kissin, Y.V. (1989) *J. Mol. Catal.*, **56**, 220.

Garoff, T., Johansson, S., Pesonen, K., Waldvogel, P., and Lindgren, D. (2002) *Eur. Polym. J.*, **38**, 121.

Polymerization rate enhancement by α-olefin is still a controversial subject. As discussed above, both the trigger mechanism (Ystenes, M. (1991) *J. Catal.*, **129**, 383) and the approach followed by Kissin can model this behavior, but other feasible alternative explanations have also been proposed, including mass transfer limitations discussed in Chapter 7. Harmon Ray and coworkers provide a useful review of these nonstandard polymerization kinetics mechanisms.

Shaffer, W.K.A. and Ray, W.H. (1997) *J. Appl. Polym. Sci.*, **65**, 1053.

The polymerization kinetic results shown in Examples 5.1, 5.2, and 5.4 were published by Mehdiabadi, S. and Soares, J.B.P. (2009) *Macromol. Symp.*, **285**, 101.

The hydrogen effect on propylene polymerization discussed in Figure 5.11 was reported by Soares, J.B.P. and Hamielec, A.E. (1996) *Polymer*, **37**, 4606.

The negative rate effect on the rate of ethylene polymerization with a Ni-diimine catalyst explained in Figure 5.14 was first published by Simon, L.C., Williams, C.P., Soares, J.B.P., and de Souza, R.F. (2001) *J. Mol. Catal. A: Chem.*, **165**, 55.

Finally, our treatment of vapor-liquid-solid equilibria in this chapter was very superficial, but much more detailed studies exist in the literature.

Sanchez, I.C. and Lacombe, R.H (1978) *Macromolecules*, **11**, 1145.

Orbey, H., Bokis, C.P., and Chen, C.C. (1998) *Ind. Eng. Chem.*, **37**, 4481.

Jog, P.K., Chapman, W.G., Gupta, S.K., and Swindol, R.D. (2002) *Ind. Eng. Chem. Res.*, **41**, 887.

Gauter, K. and Heidemann, R.A. (2001) *Fluid Phase Equilib.*, **183**, 87.

6
Polyolefin Microstructural Modeling

> When I am working on a problem, I never think about beauty. I think only of how to solve the problem. But when I have finished, if the solution is not beautiful, I know it is wrong.
>
> *Richard Buckminster Fuller (1895–1983)*
>
> A great deal of my work is just playing with equations and seeing what they give.
>
> *Paul Dirac (1902–1984)*

6.1
Introduction

Polyolefin microstructure can be defined by its distributions of molecular weight (MWD), chemical composition (CCD), and long-chain branching (LCB) and, in the case of polypropylene, by stereo- and regioregularity.

Often, polyolefin molecular properties are characterized by one or more averages instead of their complete distributions, but this practice leads to an incomplete description of the resin since polyolefins that have similar molecular property averages may have distinct microstructural distributions and physical properties. Industrially, the main measurements of molecular weight averages are still the melt flow index and the melt flow index ratio, while density is used to gauge the fraction of α-olefin in copolymers. Even though these indices are widely applied by manufacturing and processing companies to characterize polyolefin resins, we must keep in mind that they are only crude measurements that originated when more sophisticated characterization techniques were not as widely available as today.

This chapter is dedicated to mathematical models that describe polyolefin microstructural distributions. Equations with the highest formal beauty are associated with these distributions. They give us a deep understanding of the polymerization mechanism with single-site and multiple-site coordination catalysts and enable us to quantify the influence of mesoscale and macroscale phenomena on the microstructure of polyolefins.

We discuss two microstructural modeling approaches: the method of instantaneous distributions and Monte Carlo simulation. The former approach relies on analytical solutions of population balances; when it can be used, it is the best way to quantify the microstructure of polyolefins. The latter is a more powerful technique because it can handle cases without analytical solution.

6.2
Instantaneous Distributions

The method of instantaneous distributions involves three main modeling steps: (i) selection of a polymerization mechanism; (ii) establishment of the respective population balances for living and dead chains; and (iii) analytical solution (at steady state or instantaneously) of the population balances.

Instantaneous distributions have been derived for the MWD (or, equivalently, the chain length distribution (CLD)), CCD, and LCB of polyolefins made with single-site catalysts. The equivalent distributions for polymer made with multiple-site catalysts are generally represented as weighted summations of two or more single-site distributions, which are associated to different active site types present on the catalyst. This modeling approach assumes that each active site type on a multiple-site catalyst acts as a separate single-site catalyst, with its own chain propagation and chain transfer constants. There is still debate whether this is the "true" depiction of the multiple-site nature of heterogeneous Ziegler–Natta and Phillips catalysts; it is, however, very useful as a mathematical modeling approach and provides a convenient theoretical framework for understanding these catalytic systems.

6.2.1
Molecular Weight Distribution

6.2.1.1 Single-Site Catalysts

The MWD (or CLD) of polyolefins made with single-site coordination catalysts is described by Flory–Schulz most probable distribution.[1] It is instructive to derive Flory distribution; we use the simplest possible polymerization mechanism, one where a chain can either grow by propagation or terminate through a chain transfer reaction, as shown in Table 6.1. The inclusion of additional elementary steps, such as catalyst activation and deactivation, and other transfer steps, does not affect the final mathematical expression of Flory distribution.

We derive population balances[2] for the living and dead chains in the reactor using the mechanism depicted in Table 6.1, assuming that the polymerization happens in a CSTR operated at steady state with average residence time t_R. It is possible to

---

1) For simplicity, we call it Flory distribution for the rest of this chapter.
2) For our purposes, population balances are molar balances for chains of lengths 1, 2, 3, ..., n.

Table 6.1 A very simple coordination polymerization mechanism.

Description	Chemical equation	Rate constant
Propagation	$P_r + M \rightarrow P_{r+1}$	k_p
Transfer to H_2	$P_r + H_2 \rightarrow P_0 + D_r$	k_{tH}

derive the same equations without assuming steady-state CSTR operation, but the mathematical procedure is more involved.

For monomer-free active sites, that is, metal hydride sites obtained after transfer to hydrogen (P_0), the following population balance applies at steady state

$$\frac{dP_0}{dt} = k_{tH}[H_2]Y_0 - k_p[M]P_0 - \frac{P_0}{t_R} = 0 \tag{6.1}$$

where, as defined in Chapter 5, $Y_0 = \sum_{r=1}^{\infty} P_r$.

Solving for P_0, we find

$$P_0 = \frac{k_{tH}[H_2]}{k_p[M] + \frac{1}{t_R}} Y_0 = \frac{\frac{k_{tH}[H_2]}{k_p[M]}}{1 + \frac{1}{k_p[M]t_R}} Y_0 \tag{6.2}$$

An important simplification is applied to Eq. (6.2). In typical olefin polymerization reactors, the product $k_p[M]t_R$ is large because residence times are long and polymerization rates are high; consequently, $1 \gg \frac{1}{k_p[M]t_R}$. Therefore, Eq. (6.2) is simplified to

$$P_0 \cong \frac{k_{tH}[H_2]}{k_p[M]} Y_0 = \tau Y_0 \tag{6.3}$$

where the parameter τ is the ratio between the chain transfer frequency, $k_{tH}[H_2]$ (s^{-1}), and the monomer propagation frequency, $k_p[M]$ (s^{-1}).[3] This parameter appears frequently in this chapter.

A similar molar balance can be written for chains with length 1

$$\frac{dP_1}{dt} = k_p[M](P_0 - P_1) - k_t[H_2]P_1 - \frac{P_1}{t_R} = 0 \tag{6.4}$$

Solving for P_1 at steady state, under the same simplification that $1 \gg \frac{1}{k_p[M]t_R}$

$$P_1 = \frac{k_p[M]}{k_p[M] + k_t[H_2] + \frac{1}{t_R}} P_0 \cong \frac{1}{1+\tau} P_0 \tag{6.5}$$

3) Note that $\frac{1}{\tau} = \frac{k_p[M]}{k_{tH}[H_2]}$ is equal to the kinetic chain length, or number average chain length (r_n), of the polymer, as defined in introductory polymer science textbooks.

Analogously, for chains of length 2,

$$\frac{dP_2}{dt} = k_p[M](P_1 - P_2) - k_t[H_2]P_2 - \frac{P_2}{t_R} = 0 \tag{6.6}$$

Solving for P_2 under the same assumptions

$$P_2 = \frac{k_p[M]}{k_p[M] + k_t[H_2] + \frac{1}{t_R}} P_1 \cong \frac{1}{1+\tau} P_1 \tag{6.7}$$

We can substitute Eq. (6.5) into Eq. (6.7) to express P_2 as a function of P_0

$$P_2 = \left(\frac{1}{1+\tau}\right)^2 P_0 \tag{6.8}$$

By inspection of Eq. (6.8), a recursive pattern becomes apparent for chains with length r

$$P_r = \left(\frac{1}{1+\tau}\right)^r P_0 \tag{6.9}$$

Substituting Eq. (6.3) into Eq. (6.9), we obtain an expression for P_r as a function of the total number of moles of living chains and the parameter τ

$$P_r = \tau \left(\frac{1}{1+\tau}\right)^r Y_0 \tag{6.10}$$

Finally, Flory distribution is obtained as

$$f_r = \frac{P_r}{Y_0} = \tau \left(\frac{1}{1+\tau}\right)^r \tag{6.11}$$

Because the parameter τ is much smaller than 1 ($k_p[M] \gg k_{tH}[H_2]$ for high molecular weight polymers), the following approximation is valid[4]

$$\left(\frac{1}{1+\tau}\right)^r \cong e^{-r\tau} \tag{6.12}$$

expressing Flory distribution in its usual form

$$f_r = \tau e^{-r\tau} \tag{6.13}$$

Equation (6.13) is a probability density function describing the number of polymer chain moles with a given length r, also called the *number CLD*. The usual properties

---

4) This identity can be demonstrated by comparing the MacLaurin series of the left- and right-hand sides of Eq. (6.12) $\left(\frac{1}{1+\tau}\right)^r = 1 - r\tau + \frac{r(r+1)}{2}\tau^2 - \frac{r(r+1)(r+2)}{3!}\tau^3 + \cdots \cong 1 - r\tau + \frac{r^2}{2}\tau^2 - \frac{r^3}{3!}\tau^3 + \cdots$; $e^{-r\tau} = 1 - r\tau + \frac{r^2}{2}\tau^2 - \frac{r^3}{3!}\tau^3 + \cdots$ where, r was assumed to be a large integer in the first expression.

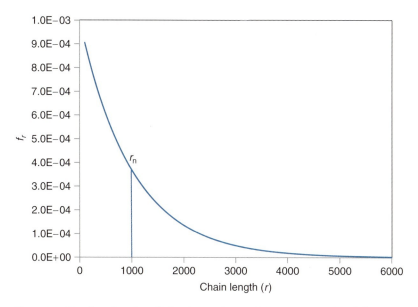

Figure 6.1 Number chain length distribution for a polyolefin with $r_n = 1 \times 10^3$ ($\tau = 1 \times 10^{-3}$).

of probability density functions apply to Flory distribution. For instance, the distribution is normalized

$$\int_0^\infty f_r dr = 1 \tag{6.14}$$

and the molar fraction of chains with lengths in the interval $[r, r + \Delta r]$ is calculated by

$$\Delta f_{r,r+\Delta r} = \int_r^{r+\Delta r} f_r dr = \left(1 - e^{-\Delta r \tau}\right) e^{-r\tau} \tag{6.15}$$

As usual, the mean (the number average chain length, r_n) is calculated as (compare with footnote 3)

$$r_n = \int_0^\infty r f_r dr = \frac{1}{\tau} \tag{6.16}$$

Figure 6.1 depicts the number CLD for a polymer with $r_n = 1 \times 10^3$ (in the case of polyethylene, the polymer would have $M_n = 2.8 \times 10^4$, since the molar mass of ethylene is 28 g mol^{-1}). It is interesting to see that the number of moles of the shorter chains is higher than the number of moles of the longer chains. We are not used to this representation because MWDs and CLDs are commonly shown on mass, not molar, basis.

To obtain the *mass* or *weight* CLD from the number CLD, the following transformation is needed

$$w_r = \frac{r f_r}{\int_0^\infty r f_r dr} = \frac{r \tau e^{-r\tau}}{\int_0^\infty r \tau e^{-r\tau} dr} = \frac{r \tau e^{-r\tau}}{\frac{1}{\tau}} = r \tau^2 e^{-r\tau} \tag{6.17}$$

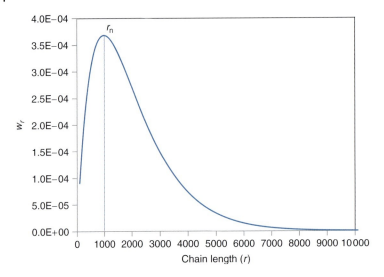

Figure 6.2 Weight chain length distribution for a polyolefin with $r_n = 1 \times 10^3$ ($\tau = 1 \times 10^{-3}$).

where the term in the denominator is included to normalize the resulting weight CLD.

Figure 6.2 depicts the weight CLD for the same polymer presented in Figure 6.1. The weight CLD has a more familiar aspect, showing that even though the shorter chains are present in higher numbers, they account for lower mass fractions. Curiously, the peak molecular weight in w_r is equal to r_n, as can be proven by solving the equation $\frac{dw_r}{dr} = 0$ for τ.

It is also easy to calculate the number average chain length, r_w, by integrating Eq. (6.17)

$$r_w = \int_0^\infty r w_r dr = \frac{2}{\tau} \tag{6.18}$$

Therefore, the theoretically expected polydispersity (PDI) of polymers made with single-site coordination catalysts is equal to 2

$$\text{PDI} = \frac{r_w}{r_n} = \frac{2}{\tau} \times \tau = 2 \tag{6.19}$$

The weight CLD is also a probability density function; therefore, the following properties apply

$$\int_0^\infty w_r dr = 1 \tag{6.20}$$

$$\Delta w_{r,r+\Delta r} = \int_r^{r+\Delta r} w_r dr = \left\{ 1 + r\tau - [1 + \tau(r + \Delta r)] e^{-\Delta r \tau} \right\} e^{-r\tau} \tag{6.21}$$

Another useful Flory distribution form is the weight CLD in logarithmic scale, the usual way to report MWDs measured by gel permeation chromatography (GPC).

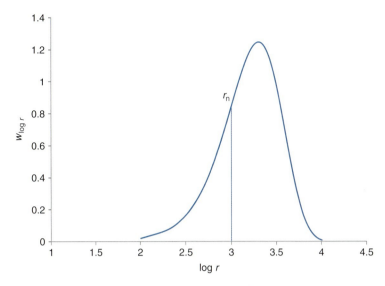

Figure 6.3 Log-scale mass chain length distribution for a polyolefin with $r_n = 1 \times 10^3$ ($\tau = 1 \times 10^{-3}$).

In this case, the following transformation

$$w_r dr = w_{\log r} d \log r \tag{6.22}$$

is used to indicate that the areas under both distributions are equivalent for a given chain length increment. Consequently,

$$w_r dr = w_{\log r} \frac{\log e}{r} dr \tag{6.23}$$

and

$$w_{\log r} = \frac{1}{\log e} r^2 \tau^2 e^{-r\tau}$$
$$= 2.3026 \times r^2 \tau^2 e^{-r\tau} \tag{6.24}$$

that is

$$w_{\log r} = 2.3026 \times r w_r \tag{6.25}$$

Figure 6.3 illustrates the weight CLD in log scale, with the familiar appearance of MWDs depicted in GPC reports.

The three forms of Flory distribution described above were derived for the living polymer chains. Is the CLD of the dead polymer accumulated in the reactor described by the same equations? The answer to this question is yes, provided that the reactor is operated under steady-state conditions. If this condition is not met, Flory distribution is still valid for both living and dead chains, but only at a given instant in time. This point is elaborated below.

First, let us assume that the reactor is operated at steady state. The population balance for dead polymer chains of length r is given by the expression

$$\frac{dD_r}{dt} = k_{tH}[H_2]P_r - \frac{D_r}{t_R} = 0 \tag{6.26}$$

Solving for D_r at steady state, we find

$$D_r = k_{tH}[H_2]t_R P_r \tag{6.27}$$

Equation (6.27) tells us that the number of moles of dead chains with length r is simply a multiple, by the factor $k_{tH}[H_2]t_R$, of the number of moles of living chains with the same length. Since this factor is the same for every dead chain in the reactor (assuming that the chain transfer frequency does not depend on chain length), the CLD of the dead chains is the same as the CLD of the living chains; we just accumulate more dead chains in the reactor as the average residence time increases. This conclusion also applies to semibatch reactors operated at steady-state conditions.

What happens when the reactor is not at steady state? In this case, Flory distribution is only valid instantaneously. An instantaneous distribution predicts the microstructure that is formed at the polymerization conditions existing at a given instant in time in the reactor. To use an analogy, instantaneous distributions are snapshots of the polymer microstructure. Let us extend this analogy a little further: if the subject in the picture is not moving (steady state), snapshots taken at different times look exactly the same; however, if the subject is in movement, a snapshot taken at a given time will be different from another taken at a later or earlier time (nonsteady state) – seen in sequence, the snapshots will resemble a movie strip. Figure 6.4 illustrates this behavior when the concentration of hydrogen decreases in the reactor, producing polymer with increasing molecular weight averages. This situation may happen, for instance, during a low-to-high molecular weight grade transition.

If we know the instantaneous CLDs at several times during non-steady-state operation, we can calculate the cumulative CLD made during the time interval $[t, t + \Delta t]$, provided the timescale for the dynamic phenomena taking place in the reactor, such as changes in temperature and reactant concentrations, is larger than the timescale for growing a polymer chain. Mathematically, this is equivalent to assuming that $\frac{dP_r}{dt} = 0$ for a given differential time interval dt, which is also the assumption we made when deriving Flory distribution. If the conditions in the reactor change at rates that are similar to the lifetime of a polymer chain, however, instantaneous distributions are no longer valid. Instantaneous distributions are solved assuming that all conditions are constant during one "snapshot" – using our analogy again, the subject may move but not faster than the shutter speed, otherwise the image becomes blurred. Luckily, the average residence time in olefin polymerization reactors is in the order of several minutes (for solution processes) to a couple of hours (for most other processes), while the lifetime of a polymer chain is in the order of tenths of seconds to a few seconds. Therefore, we can apply the method of instantaneous distributions to the vast majority of industrial polyolefin reactors.

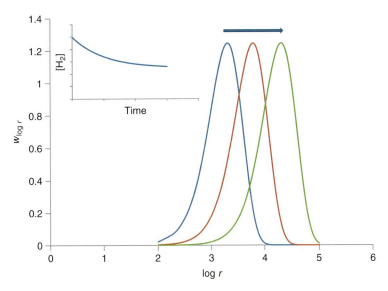

Figure 6.4 Instantaneous CLD as a function of [H$_2$] in the reactor (see insert). The three CLDs correspond to polymers with instantaneous r_n values of 1×10^3, 3×10^3, and 1×10^4, from left to right.

The cumulative CLD of polymer made during a given time interval Δt at non-steady-state operation is calculated with the equation

$$\overline{w_r} = \frac{\int_t^{t+\Delta t} w_r(t) R_p(t) \, dt}{\int_t^{t+\Delta t} R_p(t) \, dt} \tag{6.28}$$

where $R_p(t)$ is the instantaneous rate of polymerization (the mass of polymer made per unit of time). The instantaneous CLD is a function of time in Eq. (6.28) and is represented as $w_r(t)$ since τ varies during non-steady-state operation. In Figure 6.4, the dropping H$_2$ concentration causes τ to decrease and $w_r(t)$ to shift to the right.[5]

Example 6.1 illustrates the use of Eq. (6.28) with a simple case study.

Example 6.1: Effect of Non-Steady-State Reactor Operation on Chain Length Distribution and Averages

Suppose that the parameter τ varies during polymerization in a semibatch reactor according to the equation

$$\tau(t) = \tau_0 e^{-at}$$

where $a = 6.4 \times 10^{-4}$ s^{-1}, $\tau_0 = 1 \times 10^{-3}$, and t is the polymerization time. This change in τ may be caused by a decrease in hydrogen

---

5) For polymers made in CSTRs, the residence time distribution must be taken into consideration when calculating $\overline{w_r}$, as discussed in Chapter 8.

concentration in a semibatch reactor, when hydrogen is introduced into the reactor at the beginning and is consumed by chain transfer reactions during the polymerization. We use the equations derived above to calculate the variation of $\overline{w_r}$, $\overline{w_{\log r}}$, $\overline{r_w}$, and $\overline{PDI}$ with polymerization time, assuming that the rate of polymerization, $R_p(t)$, is constant.

First, we simplify Eq. (6.28) assuming that $R_p(t)$ is constant

$$\overline{w_r} = \frac{\int_0^t w_r(t)\,dt}{\int_0^t dt} = \frac{1}{t}\int_0^t w_r(t)\,dt$$

Then, we obtain the cumulative distribution $\overline{w_r}$ by substituting the expression for $\tau(t)$ into the definition of w_r, Eq. (6.17), and solving the resulting integral

$$\overline{w_r} = \frac{1}{t}\int_0^t r\tau^2 e^{-r\tau}\,dt = \frac{1}{t}\int_0^t r\left(\tau_0 e^{-at}\right)^2 e^{-r\tau_0 e^{-at}}\,dt$$

$$\overline{w_r} = \frac{e^{-r\tau_0 e^{-at}}\left(1 + r\tau_0 e^{-at}\right) - e^{-r\tau_0}(1 + r\tau_0)}{atr}$$

The cumulative average molecular weights and the PDI are calculated with the standard expressions

$$\overline{r_n} = \left(\int_0^\infty \frac{\overline{w_r}}{r}\,dr\right)^{-1} = \frac{at}{\tau_0(1 - e^{-at})}$$

$$\overline{r_w} = \int_0^\infty r\overline{w_r}\,dr = 2\frac{e^{at} - 1}{\tau_0 at}$$

$$\overline{PDI} = \frac{\overline{r_w}}{\overline{r_n}} = 2\frac{(1 - e^{-at})(e^{at} - 1)}{a^2 t^2}$$

Note that the limits

$$\lim_{a \to 0} \overline{r_n} = \frac{1}{\tau_0}$$

$$\lim_{a \to 0} \overline{r_w} = \frac{2}{\tau_0}$$

$$\lim_{a \to 0} \overline{PDI} = 2$$

prove that the non-steady-state expressions converge to the steady-state values for the CLD averages when the parameter τ does not vary during the polymerization.

Figure E.6.1.1 shows that $\overline{r_w}$ increases with polymerization time since the value of τ decreases according to $\tau(t) = 1 \times 10^{-3} e^{-6.4 \times 10^{-4} t}$ from 1×10^{-3}

to 1×10^{-4} after 1 h of polymerization. Because the polymer average chain lengths vary as a function of polymerization time, the CLD will broaden, as illustrated in Figure E.6.1.1, by the increasing value of $\overline{PDI}$.

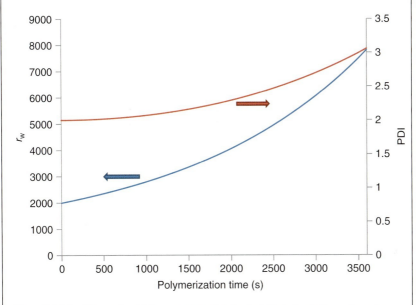

Figure E.6.1.1 Change in weight average molecular weight and polydispersity as a function of polymerization time when the parameter τ varies according to the equation $\tau(t) = 1 \times 10^{-3} e^{-6.4 \times 10^{-4} t}$.

It is also illustrative to plot the instantaneous and cumulative CLDs as a function of polymerization time. In this case, we plot the cumulative distribution in log scale, using the identity derived in Eq. (6.25) for instantaneous distributions

$$\overline{w}_{\log r} = 2.3026 \times r\overline{w}_r$$

Figure E.6.1.2 illustrates how the cumulative CLDs (solid lines) broaden with increasing polymerization time due to the variation of the parameter τ. The instantaneous distributions (dotted lines), on the other hand, remain narrow with PDI = 2, since they describe only the polymer produced instantaneously.

It is clear that this procedure can be repeated for more complex cases when several polymerization conditions, including the rate of polymerization, vary as a function of time. In general, it is not possible to find an analytical solution for these more general cases, and Eq. (6.28) needs to be integrated numerically to obtain the dynamic values for the CLD.

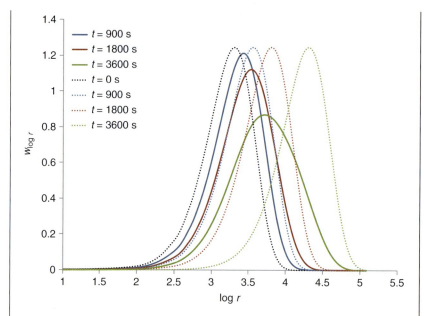

Figure E.6.1.2 Cumulative (solid lines) and instantaneous (dotted lines) chain length distributions produced at several polymerization times.

Thus far, we have restricted our model to the fairly simple polymerization mechanism described in Table 6.1. Fortunately, the inclusion of other chain transfer steps does not alter the final expression for Flory distribution, only the definition of the parameter τ. Therefore, if we include the additional chain transfer steps listed in Table 6.2, the parameter τ is redefined as

$$\tau = \frac{R_p}{R_t} = \frac{k_{tH}[H_2]}{k_p[M]} + \frac{k_{tM}}{k_p} + \frac{k_{t\beta}}{k_p[M]} + \frac{k_{tAl}[Al]}{k_p[M]} \qquad (6.29)$$

but all the equations we have derived for the CLDs remain the same. The beauty of this approach is that if more transfer reactions are added to the polymerization mechanism, we just need to include them in the definition of τ without modifying the expression for the CLD.

As formulated in Eqs. (6.13), (6.17), and (6.24), Flory distribution describes the CLD – r is the number of monomer molecules in the polymer chain – not the MWD. To obtain the MWD, we replace the number average chain length, r_n, by the number average molecular weight, M_n, and r by the molecular weight of the polymer chain, $MW = r \times mw$, where mw is the average molar mass of the repeating unit in the polymer chain

$$w_{MW} = \frac{MW}{M_n^2} \exp\left(-\frac{MW}{M_n}\right) = MW\hat{\tau}^2 \exp(-MW\hat{\tau}) \qquad (6.30)$$

Table 6.2 Additional chain transfer steps.

Description	Chemical equations	Rate constants
Transfer to monomer	$P_r + M \rightarrow P_0 + D_r$	k_{tM}
β-Hydride elimination	$P_r \rightarrow P_0 + D_r$	$k_{t\beta}$
Transfer to cocatalyst	$P_r + Al \rightarrow P_0 + D_r$	k_{tAl}

and the parameter $\hat{\tau}$ is defined as

$$\hat{\tau} = \frac{1}{M_n} = \frac{1}{r_n \text{mw}} = \frac{\tau}{\text{mw}} \tag{6.31}$$

Similarly, in logarithmic scale, Eq. (6.30) becomes

$$w_{\log MW} = 2.3026 \times MW^2 \hat{\tau}^2 \exp(-MW\hat{\tau}) \tag{6.32}$$

The CLD can be converted into the MWD by using the molar mass of the monomer for homopolymers, or the average molar mass of the repeating unit for copolymers. For instance, for an ethylene/1-hexene copolymer with the average molar fraction of ethylene equal to $\overline{F_E}$, mw is given by

$$\text{mw} = \overline{F_E} \text{mw}_E + \left(1 - \overline{F_E}\right) \text{mw}_H \tag{6.33}$$

where mw_E and mw_H are the molar masses of ethylene and 1-hexene, respectively.

6.2.1.2 Multiple-Site Catalysts

In Chapter 5, we modeled the polymerization rate of a multiple-site catalyst as a weighted sum of the polymerization rates of its several site types. We apply the same approach to describe the polymer microstructural distributions they generate. Therefore, the CLD of a polyolefin made with a catalyst having n site types is represented as

$$w_r = \sum_{j=1}^{n} m_j w_{rj} \tag{6.34}$$

where m_j is the mass fraction of polymer made on site type j

$$m_j = \frac{R_{pj}}{\sum_{i=1}^{n} R_{pi}} \tag{6.35}$$

It is evident that this approach will describe well the CLD of polyolefins made with a combination of several single-site catalysts, since we know *a priori* that the CLDs of polymers synthesized separately by each catalyst follow Flory distribution. Figure 6.5 shows that the MWD of a polyethylene sample made with Et(Ind)$_2$ZrCl$_2$ and Cp$_2$HfCl$_2$ supported on silica can be described with the superposition of two Flory distributions. The experimental MWD (line with the experimental noise) was determined by high-temperature GPC.

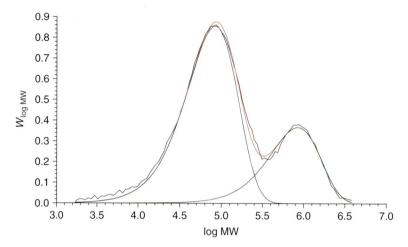

Figure 6.5 MWD of polyethylene made with Et(Ind)$_2$ZrCl$_2$ (low MW peak) and Cp$_2$HfCl$_2$ (high MW peak) supported on silica at 40 °C and ethylene partial pressure of 120 psi in a slurry reactor.

This modeling approach has been widely applied to heterogeneous Ziegler–Natta and Phillips catalysts, but it should be used with care. First, we must ensure that several requirements are met: (1) the polymerization reactor is operated at steady-state conditions, (2) the reactor conditions are spatially uniform, (3) there are no significant mass and heat transfer limitations, and (4) peak broadening in GPC is absent. Conditions 1–3 assure that all polymerization conditions are the same during different times and locations in the polymerization reactor and that they do not vary as a function of radial position in the polymer particles. These requirements are a must since, as we have studied in Example 6.1, the cumulative CLD of polymers made with a single-site catalyst (or site type, in the present case) under drifting conditions cannot be represented with only one Flory distribution. Requirement 4 guarantees that GPC instrumental broadening will not be identified as an extra active site.

If these conditions are met, we can estimate the *minimum* number of Flory distributions required to describe a given experimental MWD by minimizing the objective function

$$\chi^2 = \sum_{i=1}^{n_{GPC}} \left(w_{\log MW, i}^{GPC} - w_{\log MW, i} \right)^2$$

$$= \sum_{i=1}^{n_{GPC}} \left[w_{\log MW, i}^{GPC} - \sum_{j=1}^{n} m_j \left(2.3026 \times MW_i^2 \tau_j^2 e^{-MW_i \tau_j} \right) \right]^2 \quad (6.36)$$

where n_{GPC} is the number of points in the GPC distribution, $w_{\log MW}^{GPC}$. This parameter estimation process is called *MWD deconvolution*. At the end of the optimization, we find the "best" values for m_i, τ_i, and n by estimating $2 \times n - 1$ parameters ($\Sigma m_j = 1$). Often, four to five site types are required to represent the MWD of polyolefins made with heterogeneous Ziegler–Natta catalysts. More site types may be required for polymers made with Phillips catalysts because of their very broad MWDs.

Nonlinear optimization problems are frequently subject to multiple solutions, but because Flory distribution has a fixed width (PDI = 2.0), the MWD deconvolution procedure is generally quite robust. Had this not been the case, MWD deconvolution would be significantly less reliable. For instance, if we try to deconvolute MWDs with normal distributions, we will find that a few broader distributions may fit the data equally as well as several narrower distributions.

After the parameters m_j and τ_j have been estimated by MWD deconvolution, the overall molecular weight averages can be calculated with the following equations

$$M_n = \frac{1}{\sum_{j=1}^{n} m_j \hat{\tau}_j} = \frac{1}{\sum_{j=1}^{n} \frac{m_j}{M_{nj}}} \quad (6.37)$$

$$M_w = 2 \sum_{j=1}^{n} \frac{m_j}{\hat{\tau}_j} = \sum_{j=1}^{n} m_j M_{wj} \quad (6.38)$$

Example 6.2 applies this MWD deconvolution procedure to a polyethylene sample made with a heterogeneous Ziegler–Natta catalyst.

Example 6.2: MWD Deconvolution of a Polyethylene Sample Made with a Heterogeneous Ziegler–Natta Catalyst

Figure E.6.2.1 shows the MWD of a polyethylene sample made with a heterogeneous Ziegler–Natta catalyst, with $M_n = 19\,400$ and $M_w = 80\,500$. We need to determine the minimum number of Flory distributions required to describe this MWD by minimizing the objective function defined in Eq. (6.36).

We start with only two site types and gradually progress to more site types. The minimization of the objective function (using the Levenberg–Marquardt method) for the model with only two site types

$$\chi^2 = \sum_{i=1}^{n_{GPC}} \left[w_{\log MW,i}^{GPC} - m_1 \left(2.3026 \times MW_i^2 \hat{\tau}_1^2 e^{-MW_i \hat{\tau}_1} \right) \right.$$

$$\left. - (1 - m_1) \left(2.3026 \times MW_i^2 \hat{\tau}_2^2 e^{-MW_i \hat{\tau}_2} \right) \right]^2$$

leads to the estimation of m_1, $\hat{\tau}_1$, and $\hat{\tau}_2$ (or M_{n1} and M_{n2}), as shown in Figure E.6.2.2 and Table E.6.21. It is clear that two site types do not provide an adequate description of the MWD; additional site types need to be added.

The deconvolution with the three-site model

$$\chi^2 = \sum_{i=1}^{n_{GPC}} \left[w_{\log MW,i}^{GPC} - \sum_{j=1}^{3} m_j \left(2.3026 \times MW_i^2 \hat{\tau}_j^2 e^{-MW_i \hat{\tau}_j} \right) \right]^2$$

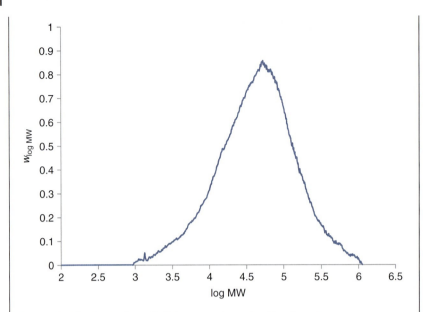

Figure E.6.2.1 MWD of a polyethylene sample made with a heterogeneous Ziegler–Natta catalyst measured by GPC ($M_n = 19\,400$ and $M_w = 80\,500$).

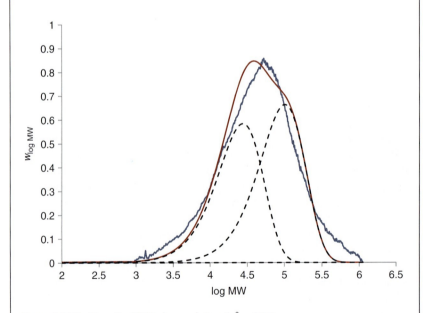

Figure E.6.2.2 Two-site MWD deconvolution ($\chi^2 = 3.35$).

Table E.6.2.1 MWD deconvolution parameters using two to six different site types.

No. of sites		1	2	3	4	5	6	All
2	m	0.467	0.533	–	–	–	–	1.000
	M_n	13 672	50 050	–	–	–	–	22 304
	M_w	27 343	100 099	–	–	–	–	66 087
	χ^2	–	–	–	–	–	–	3.35
3	m	0.197	0.602	0.201	–	–	–	1.000
	M_n	7 154	27 706	100 052	–	–	–	19 518
	M_w	14 308	55 412	200 104	–	–	–	76 408
	χ^2	–	–	–	–	–	–	0.29
4	m	0.146	0.462	0.304	0.088	–	–	1.000
	M_n	5 981	21 745	50 848	168 579	–	–	19 172
	M_w	11 961	43 491	101 696	337 159	–	–	82 306
	χ^2	–	–	–	–	–	–	0.086
5	m	0.020	0.156	0.481	0.263	0.080	–	1.000
	M_n	1 626	7 074	23 489	54 634	169 279	–	16 716
	M_w	3 252	14 148	46 977	109 269	338 557	–	80 782
	χ^2	–	–	–	–	–	–	0.040
6	m	0.020	0.156	0.290	0.191	0.263	0.080	1.000
	M_n	1 626	7 074	23 489	23 489	54 634	169 279	16 716
	M_w	3 252	14 148	46 977	46 977	109 269	338 557	80 782
	χ^2	–	–	–	–	–	–	0.040

leads to the estimation of m_1, m_2, M_{n1}, M_{n2}, and M_{n3}, and is presented in Figure E.6.2.3. The fit is considerably better than for the two-site model, and the value of χ^2 decreases from 3.35 to 0.29, but it still does not represent the MWD adequately, especially the high and low molecular weight shoulders of the distribution.

Repeating this procedure for four and five site types, we obtain the MWDs shown in Figures E.6.2.4 and E.6.2.5, respectively. While it is apparent that the four-site model is a significant improvement over the three-site model (χ^2 decreases from 0.29 to 0.086), the five-site model does not seem to be justified based on the deconvolution results, despite the slightly better fit of the low molecular weight shoulder. The additional fifth site type does not decrease the value of χ^2 significantly and accounts for only 2% of the total polymer mass.

Therefore, based on the MWD of only this sample, the four-site model seems to be the best option. This point is made clear by comparing the values for χ^2 shown in Table E.6.2.1. Notice that the values predicted for the overall M_n and M_w using Eqs. (6.37) and (6.38) are also very close to the values for the experimental MWD. Table E.6.2.1 also shows that if we try to fit a sixth site to the MWD, the optimization routine will simply make the values of the M_n for two sites identical, indicating that the model is overspecified.

By analyzing the MWDs of other samples made with the same catalyst at different polymerization conditions, we would be able to follow how each of these

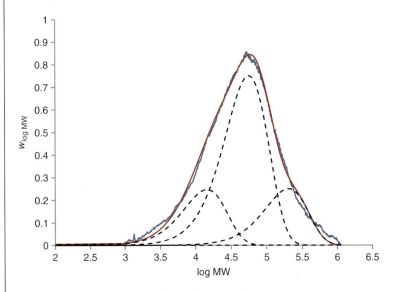

Figure E.6.2.3 Three-site MWD deconvolution ($\chi^2 = 0.29$).

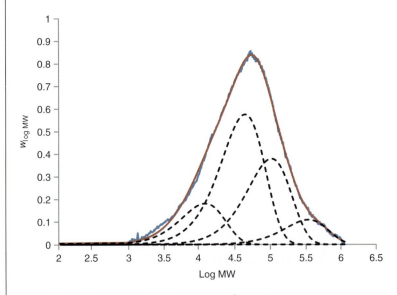

Figure E.6.2.4 Four-site MWD deconvolution ($\chi^2 = 0.086$).

site types responds to different polymerization conditions and make an even better decision regarding the number of site types required to model the MWD.

We can also use this example to point out a common shortcoming of several Ziegler–Natta models, particularly those that use the method of moments and only two site types. Parameter estimation with these models is commonly done by fitting only the molecular weight averages, M_n and M_w, for the whole polymer. If we minimize the objective function

$$\chi^2 = \left[M_n^{GPC} - M_n\right]^2 + \left[M_w^{GPC} - M_w\right]^2$$

$$= \left\{M_n^{GPC} - \left[\frac{m_1}{M_{n1}} + \frac{(1-m_1)}{M_{n2}}\right]^{-1}\right\}^2$$

$$+ \left\{M_n^{GPC} - [m_1 M_{n1} + (1-m_1) M_{n2}]\right\}^2$$

we will obtain the deconvolution results given in Figure E.6.2.6, where the predicted and experimental values for M_n and M_w have an almost perfect match ($\chi^2 = 2.3 \times 10^{-8}$), but the modeled MWD clearly does not represent the experimental MWD. Therefore, while a two-site model has the advantage of requiring only three independent parameters (m_1, M_{n1}, and M_{n2}), it is generally not adequate to describe the whole MWD of polyolefins made with Ziegler–Natta and Phillips catalysts.

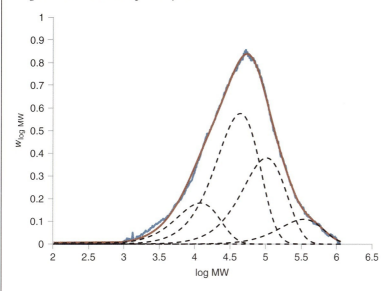

Figure E.6.2.5 Five-site MWD deconvolution ($\chi^2 = 0.040$).

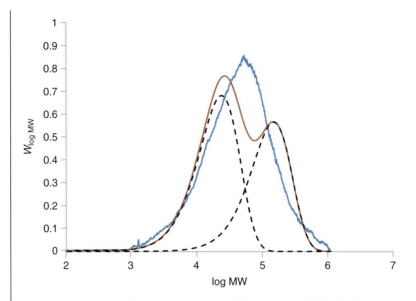

Figure E.6.2.6 Two-site fit using only $M_n = 19\,400$ and $M_w = 80\,500$. Model parameters: $m_1 = 0.546$, $M_{n1} = 12\,017$, $m_2 = 0.454$, and $M_{n2} = 74184$ ($\chi^2 = 2.3 \times 10^{-8}$). The molecular weight averages are very well represented, but the two site types are insufficient to capture all the MWD.

MWD deconvolution results should be interpreted with care. First, we must be sure that the polymer was produced under spatially and temporally uniform conditions. Second, we should ensure that peak broadening during MWD analysis by GPC is negligible. Third, and more importantly, MWD deconvolution can only retrieve the *minimum* number of Flory sites required to represent the measured MWD; more sites may be present, but not seen, because they produce polymers with very similar MWDs (peak superposition), but may, for instance, have different stereoregularities or comonomer reactivity ratios.

Despite its usefulness, considerable controversy lingers about the real meaning of the MWD deconvolution procedure. In Example 6.2, we found out that at least four site types would be required to describe the MWD. What would be the chemical natures of these four site types? Why should they be characterized by clearly distinct τ values that lead to their characteristic M_n and M_w averages?[6]

An alternative, and much less commonly used modeling approach, assumes that each chemically distinct site type is characterized by a distribution of values for τ.

6) It is, however, important to realize that the use of several Flory distributions to describe the MWD of polymer made with multiple-site catalysts is equivalent to using any other technique that assumes several site types with different propagation and transfer rates, such as population balances and the method of moments; they are just different ways of formulating the same problem.

Each *catalyst site* (not site type) still makes polymer that follows Flory distribution (as it must, if we accept the polymerization mechanism described in Tables 6.1 and 6.2), but the values of τ may vary from active site to active site (and as a function of polymerization time) within a given population of active sites of the same type. For instance, for a given active site type, τ may follow a normal distribution

$$N(\tau) = \frac{1}{\sigma\sqrt{2\pi}} \exp\left[-\frac{(\tau - \bar{\tau})^2}{2\sigma^2}\right] \tag{6.39}$$

with average value $\bar{\tau}$ and standard deviation σ.

We may speculate that within the same population of active site types, the values for τ may vary from site to site because of electronic and steric effects of neighboring sites, different degrees of coordination with counter ions, and the presence of electron donors, for instance.

In this case, the CLD of polymer made on each site type would be represented by the following equation

$$W_r = \int r\tau^2 e^{-r\tau} N(\tau) d\tau \tag{6.40}$$

where we used the capital W to indicate that this distribution will be broader than the original Flory distribution. It is easy to see that this approach will result in the identification of a smaller number of active site types since their individual MWDs will be broader than Flory distribution.

Unfortunately, even if this modeling approach may be a closer representation of reality than the direct use of Flory distributions, the form of the broadening function is unknown, adding another layer of uncertainty to the MWD deconvolution process.

From a merely mathematical point of view, the lognormal distribution is preferable to the normal distribution because it avoids the occurrence of physically impossible negative values for τ

$$N_{\ln}(\tau) = \frac{1}{\tau \sigma_{\ln}\sqrt{2\pi}} \exp\left[-\frac{(\ln \tau - \bar{\tau}_{\ln})^2}{2\sigma_{\ln}^2}\right] \tag{6.41}$$

where the parameters $\sigma_{\ln}$ and $\bar{\tau}_{\ln}$ are related to the normal values through the equations

$$\bar{\tau}_{\ln} = \ln \bar{\tau} - \frac{1}{2}\ln\left[1 + \left(\frac{\sigma}{\bar{\tau}}\right)^2\right] \tag{6.42}$$

$$\sigma_{\ln} = \sqrt{\ln\left[1 + \left(\frac{\sigma}{\bar{\tau}}\right)^2\right]} \tag{6.43}$$

We use these equations in Example 6.3 to show that the same MWD studied in Example 6.2 can be represented with fewer site types when we allow the values of τ to vary according to lognormal distributions.

Example 6.3: MWD Deconvolution of a Polyethylene Sample Made with a Heterogeneous Ziegler–Natta Catalyst Using Flory Distributions Broadened with Lognormal Distributions

We repeat the deconvolution exercise of Example 6.2 using Flory distributions broadened with lognormal distributions and search for the values of $\sigma_{\ln}$ and $\overline{\tau}_{\ln}$ (or the equivalent σ and $\overline{\tau}$) for each site type.

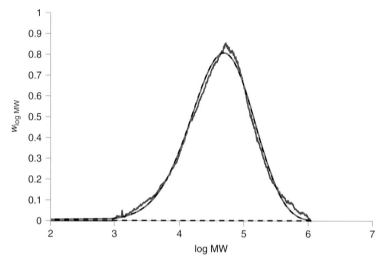

Figure E.6.3.1 One-site MWD deconvolution ($\chi^2 = 5.17 \times 10^{-1}$).

We start by using a single distribution

$$W_{\log MW} = 2.3026 \times \int MW^2 \hat{\tau}^2 e^{-MW\hat{\tau}} \frac{1}{\hat{\tau} \sigma_{\ln} \sqrt{2\pi}} \exp\left[-\frac{(\ln \hat{\tau} - \overline{\tau}_{\ln})^2}{2\sigma_{\ln}^2}\right] d\hat{\tau}$$

and minimizing the objective function

$$\chi^2 = \sum_{i=1}^{n_{GPC}} \left(W_{\log MW,i}^{GPC} - W_{\log MW,i}\right)^2$$

Figure E.6.3.1 and Table E.6.3.1 show the deconvolution results for a single-site catalyst that follows Flory distribution broadened with a lognormal distribution for the values of τ. The fit is much better than for two Flory distributions (Figure E.6.2.2) and comparable to that with three Flory distributions (Figure E.6.2.3). At this point, we may be tempted to look for different broadening functions that may result in even better fits, so that a single family of active sites would be able to describe the complete MWD. This may be mathematically feasible, but not desirable, since the bimodal CCDs commonly associated with polyolefins made with Ziegler–Natta catalysts imply the existence of at least two significantly different site types.

Table E.6.3.1 MWD deconvolution parameters using lognormal broadened site types.

No. of sites		1	2	3	4	5	6	All
1	m	1.000	–	–	–	–	–	1.000
	M_n	19 400	–	–	–	–	–	19 400
	M_w	74 900	–	–	–	–	–	74 900
	PDI	3.87	–	–	–	–	–	3.87
	σ	20 200	–	–	–	–	–	–
	χ^2	–	–	–	–	–	–	0.517
2	m	0.736	0.264	–	–	–	–	1.000
	M_n	14 600	27 900	–	–	–	–	16 700
	M_w	87 200	55 800	–	–	–	–	78 900
	PDI	5.95	2.00	–	–	–	–	4.70
	σ	24 200	798	–	–	–	–	–
	χ^2	–	–	–	–	–	–	0.0752
2 $(M_n/\sigma)_{max}$ = 0.5	m	0.373	0.627	–	–	–	–	1.000
	M_n	10 800	38 100	–	–	–	–	19 700
	M_w	26 800	94 300	–	–	–	–	60 100
	PDI	2.47	2.47	–	–	–	–	3.52
	σ	5 400	19 000	–	–	–	–	–
	χ^2	–	–	–	–	–	–	1.012
3 $(M_n/\sigma)_{max}$ = 0.5	m	0.160	0.695	0.145	–	–	–	1.000
	M_n	5 900	26 100	104 000	–	–	–	18 200
	M_w	14 700	61 600	257 300	–	–	–	82 400
	PDI	2.47	2.36	2.47	–	–	–	4.53
	σ	2 900	11 300	52 000	–	–	–	–
	χ^2	–	–	–	–	–	–	0.0696

Repeating the deconvolution for two site types by minimizing

$$\chi^2 = \sum_{i=1}^{n_{GPC}} \left\{ w_{\log MW,i}^{GPC} - \left[m_1\, W_{\log MW,1i} - (1-m_1)\, W_{\log MW,2i} \right] \right\}^2$$

we obtain the results shown in Figure E.6.3.2. Despite the fact that the fit is comparable to a fit with four Flory distributions, the MWDs for a polymer made by the individual site types seem counterintuitive, with one rather broad peak and one peak with PDI = 2. This result illustrates one of the shortcomings of this method: a good mathematical fit may lead to results without physical meaning. We may constrain the optimization by imposing, for instance, that the maximum value for M_n/σ for each site be equal to 0.5 and repeating the data fit procedure to obtain the results shown in Figure E.6.3.3. In this case, the fit looks more reasonable, even though the value of χ^2 is much higher than the one for the unconstrained optimization with two site types (but lower than for the two Flory distributions shown in Figure E.6.2.2). However, the choice of $(M_n/\sigma)_{max} = 0.5$ is arbitrary, and this is only one of many possible solutions for this deconvolution problem.

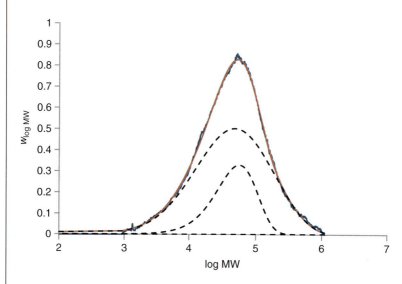

Figure E.6.3.2 Two-site MWD deconvolution ($\chi^2 = 7.51 \times 10^{-2}$).

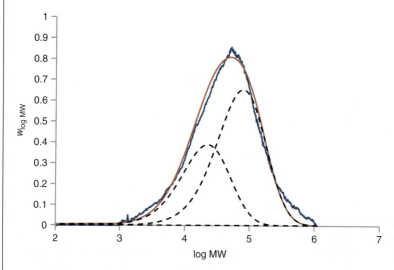

Figure E.6.3.3 Two-site constrained (($M_n/\sigma)_{max} = 0.5$) MWD deconvolution ($\chi^2 = 1.012$).

If we repeat this procedure for three-site-type families, by minimizing the objective function,

$$\chi^2 = \sum_{i=1}^{n_{GPC}} \left(W_{\log MW,i}^{GPC} - \sum_{j=1}^{3} m_j W_{\log MW,ji} \right)^2$$

still keeping the constraint $(M_n/\sigma)_{max} = 0.5$, we arrive at the results shown in Figure E.6.3.4, which are comparable to the final five-site deconvolution in Example 6.2. Therefore, we arrived at a very satisfactory MWD deconvolution result with only three site types, but at the cost of two arbitrary assumptions: (i) the values of τ follow a lognormal distribution and (ii) $(M_n/\sigma)_{max} = 0.5$ for each site type.

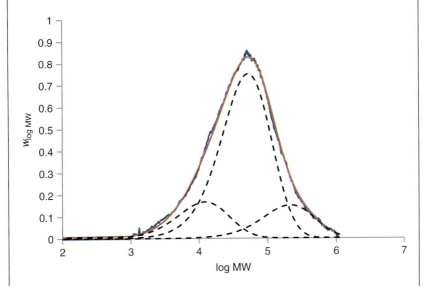

Figure E.6.3.4 Three-site constrained $((M_n/\sigma)_{max} = 0.5)$ MWD deconvolution $(\chi^2 = 6.97 \times 10^{-2})$.

What is then the preferred approach for MWD deconvolution, the direct use of Flory distribution illustrated in Example 6.2 or of the broadened Flory distribution described above? This question, unfortunately, has no straightforward answer, but we generally favor the former. It is only by the systematic analysis of a series of polymer samples made by the same catalyst under different polymerization conditions that we may be able to select the modeling approach that best describes that particular catalyst system and polymerization process. The deconvolution procedure used in Example 6.2, however, has the appeal of simplicity and does not require additional assumptions regarding the type of distribution for the parameter τ. In addition, the fact that Flory distribution has fixed width generally guarantees unique solutions. These properties make deconvolution with Flory distributions a very attractive option when modeling the MWD of polymers made with multiple-site catalysts.

6.2.2
Chemical Composition Distribution

6.2.2.1 Single-Site Catalysts

The models we considered above were developed for olefin homopolymerization with single- and multiple-site catalysts. However, most industrial polyolefins are copolymers of ethylene, propylene, and higher α-olefins. In this case, the kinetic rate constants generally depend on the type of monomer coordinating with the active site *and* on the type of monomer at the end of the chain coordinated to the transition metal site, as discussed in Chapter 5.

Table 6.3 lists the propagation rate steps for three models commonly used for olefin copolymerization. The Bernoullian model (also known as the *zeroth-order Markov model*) assumes that the last monomer added to the polymer chain does not influence the propagation rate of the coordinating monomer. Therefore, only two propagation rate constants, k_{pA} and k_{pB}, are needed, and the product of the reactivity ratios

$$r_A = \frac{k_{pAA}}{k_{pAB}} \tag{6.44}$$

$$r_B = \frac{k_{pBB}}{k_{pBA}} \tag{6.45}$$

is equal to one since we assume that $k_{pAA} = k_{pBA}$ and that $k_{pBB} = k_{pAB}$. The Bernoullian model is only suitable to describe the polymerization kinetics of some single-site catalysts that make ideal random copolymers.

The terminal model (*first-order Markov model*), involving the four propagation rate constants used in Eqs. (6.44) and (6.45), is the most common model for olefin copolymerization. It assumes that the monomers propagating and at the end of the polymer chain influence the polymerization rate. Finally, the penultimate model (*second-order Markov model*) postulates that the second to last monomer added to the polymer chain also influences the value of the propagation rate constant. Higher-order models, such as the pen-penultimate model (*third-order Markov model*), have also been proposed for other polymerization systems but are not commonly used for polyolefins.

Since the complexity added by using the penultimate model (eight propagation rate constants) is seldom justified for olefin polymerization, the terminal model is generally chosen. Table 6.4 lists chain transfer reactions associated with the terminal model; they are equivalent to the ones discussed for homopolymerization but now may depend on the type of monomer bonded to the chain end.

When we derived Flory distribution in Section 6.2.1, we saw that it described the CLD of polymers made under conditions where the chains could either propagate or

6.2 Instantaneous Distributions

Table 6.3 Copolymerization models.

Model type	Propagation steps[a]	Rate constants
Bernoullian	$P_r + A \rightarrow P_{r+1}$	k_{pA}
	$P_r + B \rightarrow P_{r+1}$	k_{pB}
Terminal	$P_r^A + A \rightarrow P_{r+1}^A$	k_{pAA}
	$P_r^A + B \rightarrow P_{r+1}^B$	k_{pAB}
	$P_r^B + A \rightarrow P_{r+1}^A$	k_{pBA}
	$P_r^B + B \rightarrow P_{r+1}^B$	k_{pBB}
Penultimate	$P_r^{AA} + A \rightarrow P_{r+1}^{AA}$	k_{pAAA}
	$P_r^{AA} + B \rightarrow P_{r+1}^{AB}$	k_{pAAB}
	$P_r^{AB} + A \rightarrow P_{r+1}^{BA}$	k_{pABA}
	$P_r^{AB} + B \rightarrow P_{r+1}^{BB}$	k_{pABB}
	$P_r^{BA} + A \rightarrow P_{r+1}^{AA}$	k_{pBAA}
	$P_r^{BA} + B \rightarrow P_{r+1}^{AB}$	k_{pBAB}
	$P_r^{BB} + A \rightarrow P_{r+1}^{BA}$	k_{pBBA}
	$P_r^{BB} + B \rightarrow P_{r+1}^{BB}$	k_{pBBB}

[a] A and B represent monomer type; the superscripts A and B indicate the type(s) of monomer(s) at the end of the polymer chain.

Table 6.4 Chain transfer steps for the terminal model.

Description	Chemical equations	Rate constants
Transfer to monomer	$P_r^A + A \rightarrow P_1^A + D_r$	k_{tAA}
	$P_r^A + B \rightarrow P_1^B + D_r$	k_{tAB}
	$P_r^B + A \rightarrow P_1^A + D_r$	k_{tBA}
	$P_r^B + B \rightarrow P_1^B + D_r$	k_{tBB}
β-Hydride elimination	$P_r^A \rightarrow P_0 + D_r$	$k_{t\beta A}$
	$P_r^B \rightarrow P_0 + D_r$	$k_{t\beta B}$
Transfer to hydrogen	$P_r^A + H_2 \rightarrow P_0 + D_r$	k_{tHA}
	$P_r^B + H_2 \rightarrow P_0 + D_r$	k_{tHB}
Transfer to cocatalyst	$P_r^A + Al \rightarrow P_0 + D_r$	k_{tAlA}
	$P_r^B + Al \rightarrow P_0 + D_r$	k_{tAlB}

terminate through chain transfer reactions. Since these conditions are still valid for copolymerization, Flory distribution also describes the CLD of olefin copolymers, and the same definition for the parameter τ applies; τ is still defined as the ratio between the total chain transfer rate and the monomer propagation rate.

The propagation rate for the terminal model is given by

$$R_p = k_{pAA} Y_0^A [A] + k_{pAB} Y_0^A [B] + k_{pBA} Y_0^B [A] + k_{pBB} Y_0^B [B] \tag{6.46}$$

where

$$Y_0^A = \sum_{r=1}^{\infty} P_r^A \tag{6.47}$$

$$Y_0^B = \sum_{r=1}^{\infty} P_r^B \tag{6.48}$$

Chain transfer rates, R_t, can be calculated in a similar manner to obtain the parameter τ for copolymerization

$$\tau = \frac{R_t}{R_p} \tag{6.49}$$

allowing the direct application of Eq. (6.13), (6.17), or (6.24) for the calculation of the CLD of olefin copolymers.

Alternatively, we may formulate the parameter τ in terms on pseudokinetic constants. As derived in Chapter 5, the overall copolymerization rate is given by

$$R_p = \left(k_{pAA}\phi_A f_A + k_{pAB}\phi_A f_B + k_{pBA}\phi_B f_A + k_{pBB}\phi_B f_B \right) Y_0 [M] = \hat{k}_p Y_0 [M] \tag{6.50}$$

We can now extend this approach to the chain transfer reactions shown in Table 6.4, by defining the following pseudokinetic constants

$$\hat{k}_{tM} = k_{tAA}\phi_A f_A + k_{tAB}\phi_A f_B + k_{tBA}\phi_B f_A + k_{tBB}\phi_B f_B \tag{6.51}$$

$$\hat{k}_{t\beta} = k_{t\beta A}\phi_A + k_{t\beta B}\phi_B \tag{6.52}$$

$$\hat{k}_{tH} = k_{tHA}\phi_A + k_{tHB}\phi_B \tag{6.53}$$

$$\hat{k}_{tAl} = k_{tAlA}\phi_A + k_{tAlB}\phi_B \tag{6.54}$$

which are used to define the parameter τ for copolymers

$$\tau = \frac{\hat{k}_{tH}[H_2]}{\hat{k}_p [M]} + \frac{\hat{k}_{tM}}{\hat{k}_p} + \frac{\hat{k}_{t\beta}}{\hat{k}_p [M]} + \frac{\hat{k}_{tAl}[Al]}{\hat{k}_p [M]} \tag{6.55}$$

Therefore, Flory distribution may be used to calculate the MWD of copolymers, provided we use pseudokinetic constants to define τ.

Besides the distribution of chain lengths, copolymers also have a distribution of comonomer molar fraction (called the *chemical composition distribution*). The joint distribution of chain length and chemical composition for binary copolymers[7] is

---

7) Unfortunately, no general analytical solution is available for the CCD of terpolymers and higher multicomponent copolymers. A solution for random multicomponent copolymers is available and is discussed in Further Reading. The CCDs for these copolymers can be predicted by the Monte Carlo simulation or by solving their complete population balances.

given by the Stockmayer bivariate distribution. This derivation is similar to that for Flory distribution, but it is much longer and is omitted here.

The Stockmayer distribution is given by the elegant equation[8]

$$w_{r y} = r\tau^2 e^{-r\tau} \sqrt{\frac{r}{2\pi\beta}} e^{-\frac{ry^2}{2\beta}} \qquad (6.56)$$

where y is the difference between the molar fraction of monomer A in a copolymer chain, F_A, and the *average* molar fraction of monomer A in the whole copolymer sample, $\overline{F_A}$

$$y = F_A - \overline{F_A} \qquad (6.57)$$

and the parameter β is defined as

$$\beta = \overline{F_A}\left(1 - \overline{F_A}\right)\sqrt{1 - 4\overline{F_A}\left(1 - \overline{F_A}\right)(1 - r_A r_B)} \qquad (6.58)$$

Finally, the variable r and the parameter τ have the same meanings as in Flory distribution.

The average molar fraction of comonomer A in the copolymer is calculated with the Mayo–Lewis equation, discussed in introductory textbooks of polymer science and engineering

$$\overline{F_A} = \frac{(r_A - 1)f_A^2 + f_A}{(r_A + r_B - 2)f_A^2 + 2(1 - r_B)f_A + r_B} \qquad (6.59)$$

The value of $\overline{F_A}$ can also be measured by a variety of analytical techniques, such as ^{13}C NMR and FTIR.

An alternative form of the Stockmayer distribution is

$$w_{r,F_A} = r\tau^2 e^{-r\tau}\sqrt{\frac{r}{2\pi\beta}} e^{-\frac{r(F_A - \overline{F_A})^2}{2\beta}} \qquad (6.60)$$

This representation is more convenient when we need to superimpose several CCDs with different $\overline{F_A}$ values.

Equation (6.60) may also be expressed as a function of molecular weight in logarithmic scale as

$$w_{\log MW, F_A} = 2.3026 \times MW^2 \hat{\tau}^2 e^{-MW\hat{\tau}} \sqrt{\frac{MW}{2\pi\hat{\beta}}} e^{-\frac{MW(F_A - \overline{F_A})^2}{2\hat{\beta}}} \qquad (6.61)$$

where

$$\hat{\beta} = \beta mw \qquad (6.62)$$

---

8) This form of Stockmayer distribution assumes that the molar masses of both comonomers are the same. This simplification has been proven to be accurate for most polyolefins of interest (see Further Reading at the end of this chapter).

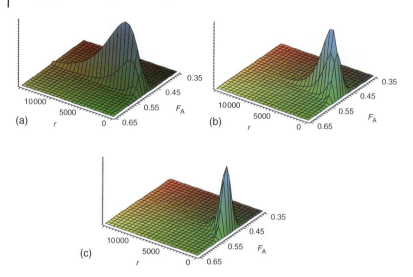

Figure 6.6 Effect of r_n on the bivariate CLD × CCD of a binary olefin copolymer with $\overline{F_A} = 0.5$ and $r_A r_B = 1$: (a) $r_n = 2000$; (b) $r_n = 1000$; and (c) $r_n = 500$.

It is interesting to notice that Stockmayer distribution reduces to Flory distribution if we integrate over all possible values of y

$$w_r = \int_{-\infty}^{\infty} r\tau^2 e^{-r\tau} \sqrt{\frac{r}{2\pi\beta}} e^{-\frac{ry^2}{2\beta}} dy = r\tau^2 e^{-r\tau} \tag{6.63}$$

demonstrating that the CLD of copolymers obeys Flory distribution, as we had inferred above on purely logical grounds.

Much can be learned about the microstructure of olefin copolymers from a careful inspection of the Stockmayer distribution. Figure 6.6 shows how the CCD of ideal random copolymers ($r_A r_B = 1$) is affected by chain length for resins with different r_n averages. As the chain length increases, the CCD becomes narrower, that is, shorter polyolefin chains will have a wider variation of comonomer fraction than longer chains; this phenomenon is observed even within a single population as the value of r increases. The reason for this behavior is the statistical nature of polymerization processes: as chain lengths tend to infinity, the composition of all chains in the polymer population will unavoidably approach $\overline{F_A}$.

A simple way to visualize this phenomenon is to imagine that we will make the decision between the polymerization of comonomers A or B by tossing a coin (a simple Monte Carlo simulation): faces correspond to monomer A and tails to monomer B. This process will produce polymer chains with an average molar fraction of A of 0.5, as illustrated in Figure 6.6. Now, imagine that we want to calculate the CCD of chains with only 10 monomer units. We do this by tossing the coin 10 times and recording the number of faces (A) and tails (B), repeating the process several times to generate the CCD for several chains of length 10. There is a small ($0.5^{10} \approx 9.765 \times 10^{-4}$) but not negligible probability that some chains will be composed only of monomer A or B; therefore, the CCD will span the whole range

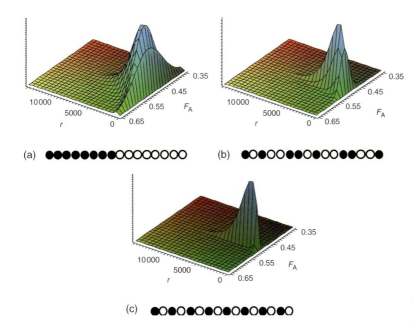

Figure 6.7 Effect of the $r_A r_B$ product on the bivariate CLD × CCD of a binary olefin copolymer with $\overline{F_A} = 0.5$ and $r_n = 1000$: (a) $r_A r_B = 100$, block copolymer; (b) $r_A r_B = 1$, random copolymer; and (c) $r_A r_B = 0.01$, alternating copolymer.

of compositions from $F_A = 0$ to $F_A = 1$. On the other hand, if we repeat this trial for chains with 1000 monomer units, the likelihood of finding only A or B units in the chains ($0.5^{1000} \approx 9.333 \times 10^{-302}$) is vanishingly small. This phenomenon is elegantly represented in Figure 6.6 by the Stockmayer distribution.

Figure 6.7 illustrates the effect of "blockines" on the CCD of chains with $r_n = 1000$. As expected, the CCD narrows when there is a tendency toward comonomer alternation ($r_A r_B \ll 1$) since, in a perfectly alternating copolymer, all chains have $F_A = 0.5$. Blocky copolymers ($r_A r_B \gg 1$), however, will have a broader CCD than random copolymers because once monomer A changes to monomer B in the chain (or vice versa), a long sequence of monomer B will be formed until another monomer A is polymerized, widening the CCD. Bear in mind, however, that the effect illustrated in Figure 6.7 has been exaggerated for illustration purposes. In general, the reactivity ratio product for polymerization with coordination catalysts does not differ from unity by two orders of magnitude, as shown in Figure 6.7. Single-site catalysts tend to form random, or near random, copolymers. Ziegler–Natta catalysts may have some site types that favor ethylene incorporation over α-olefins, but not to the extent shown in Figure 6.7a.

We can extract the CCD component of Stockmayer distribution by integrating it over all chain lengths

$$w_y = \int_0^\infty r\tau^2 e^{-r\tau} \sqrt{\frac{r}{2\pi\beta}} e^{-\frac{ry^2}{2\beta}} dr = \frac{3\tau^2 \beta^2}{(2\beta\tau + y^2)^{5/2}} \tag{6.64}$$

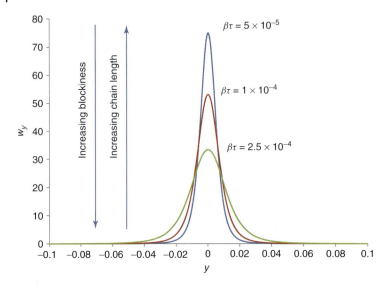

Figure 6.8 Effect of the product $\beta\tau$ on the CCD of a binary olefin copolymer.

or, alternatively

$$w_{F_A} = \frac{3\tau^2\beta^2}{\left[2\beta\tau + \left(F_A - \overline{F_A}\right)^2\right]^{5/2}} = \frac{3}{4\sqrt{2\beta\tau}\left[1 + \frac{\left(F - \overline{F}\right)^2}{2\beta\tau}\right]^{5/2}} \quad (6.65)$$

One interesting feature of Eqs. (6.64) and (6.65) is that the two parameters of the Stockmayer bivariate distribution appear as the product $\beta\tau$. As shown in Figure 6.8, this product regulates the width of the CCD. As the product $\beta\tau$ increases, the CCD becomes broader, as expected from our previous analysis of Stockmayer distribution: τ increases when r_n decreases and β increases when $r_A r_B$ increases (block copolymers) – both factors broaden the CCD.

Example 6.4 shows how Stockmayer distribution can be used to describe the CCD of a series of ethylene/1-hexene copolymers made with a metallocene catalyst.

Example 6.4: Effect of Average Comonomer Fraction on the CCD of Single-Site Ethylene/1-Hexene Copolymers

Figure E.6.4.1 shows the CRYSTAF profiles of a series of ethylene/1-hexene copolymers made with a metallocene catalyst, and Table E.6.4.1 gives some of their average properties. As the comonomer content increases, the CRYSTAF peaks move to lower crystallization temperatures and the profiles become broader. We use Stockmayer's distribution to explain this behavior.

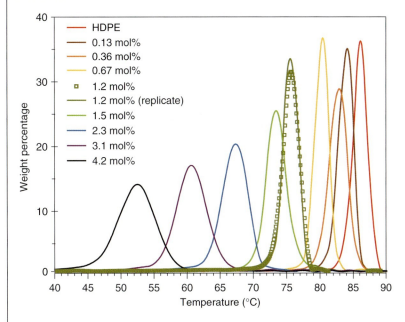

Figure E.6.4.1 CRYSTAF profiles of a series of ethylene/1-hexene copolymers made with a single-site catalyst.

Table E.6.4.1 Average properties of ethylene/1-hexene copolymers ($r_A r_B \approx 1$).

1-Hexene (%)	M_n	PDI	mw	τ	β	$\tau\beta$
0.13	38 800	2.6	28.08	7.23×10^{-4}	1.35×10^{-3}	9.76×10^{-7}
0.36	37 100	2.5	28.20	7.60×10^{-4}	3.58×10^{-3}	2.72×10^{-6}
0.67	37 200	2.5	28.38	7.63×10^{-4}	6.70×10^{-3}	5.11×10^{-6}
1.20	36 100	2.4	28.69	7.95×10^{-4}	1.21×10^{-2}	9.65×10^{-6}
1.50	36 300	2.4	28.85	7.95×10^{-4}	1.49×10^{-2}	1.18×10^{-5}
2.30	35 200	2.2	29.29	8.32×10^{-4}	2.26×10^{-2}	1.88×10^{-5}
3.10	34 300	2.2	29.76	8.68×10^{-4}	3.04×10^{-2}	2.64×10^{-5}
4.20	34 500	2.2	30.35	8.80×10^{-4}	4.02×10^{-2}	3.54×10^{-5}

As discussed in Chapter 2, calibration curves for CRYSTAF are generally linear; peak temperatures decrease linearly with increasing comonomer molar fractions. Therefore, the *x*-axis in Figure E.6.4.1 can be translated into molar fraction of ethylene in the copolymer without distorting the peak shapes.

First, we use the average properties listed in Table E.6.3.1 to calculate the Stockmayer parameters τ and β, assuming that the copolymers are random ($r_A r_B \approx 1$).

The product $\beta\tau$ increases with increasing comonomer content in the polymer, so we should expect the CRYSTAF peaks to become broader, as illustrated in Figure 6.8, in good agreement with the experimental data. As a consequence, when we plot the Stockmayer distributions predicted by Eq. (6.65) in Figure E.6.4.2, the same behavior observed in the CRYSTAF curves is replicated by the theoretical distributions.

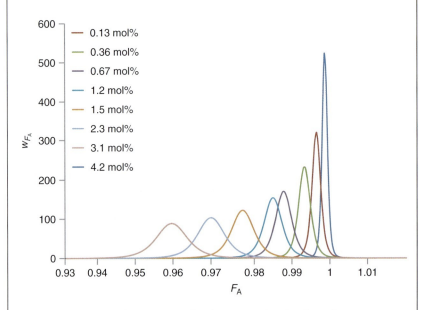

Figure E.6.4.2 Chemical composition distributions predicted with Stockmayer distribution.

We can draw a more general conclusion by plotting the value of β as a function of $\overline{F_A}$, as done in Figure E.6.4.3. We see that β increases, and therefore, the CCD broadens, when $\overline{F_A}$ decreases in a wide range of compositions. The blockier the copolymer, the higher will be this increase.

We should, however, be careful when using CCDs to describe CRYSTAF curves. As discussed in Chapter 2, the shape of CRYSTAF profiles are influenced by, but *are not*, CCDs. Several effects, such as cooling rates, cocrystallization, and, to a smaller extent, molecular weight, can influence the shape of CRYSTAF profiles. Therefore, the best we can expect is a qualitative agreement between CRYSTAF curves and theoretical CCDs such as the Stockmayer distribution.

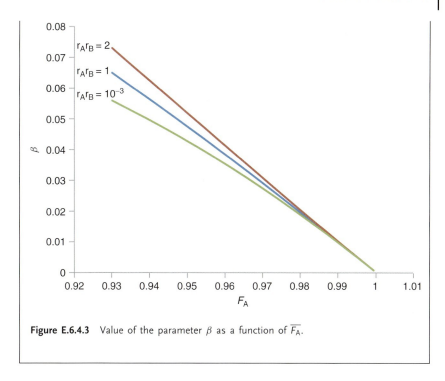

Figure E.6.4.3 Value of the parameter β as a function of $\overline{F_A}$.

We have seen in Chapter 2 that the use of multiple detectors with GPC, TREF, and CRYSTAF is becoming increasingly common in the polyolefin industry. When TREF is coupled with a light scattering detector, it is possible to measure how the weight average molecular weight varies as a function of the elution temperature. The Stockmayer distribution allows us to calculate this dependency from first principles, by solving the equation

$$r_{wF_A} = \frac{\int_0^\infty r w_{r,F_A} dr}{\int_0^\infty w_{r,F_A} dr} = \frac{5\beta}{2\beta\tau + \left(F_A - \overline{F_A}\right)^2} \tag{6.66}$$

Similarly, the number average chain length is obtained with the expression

$$r_{nF_A} = \left(\frac{\int_0^\infty \frac{w_{r,F_A}}{r} dr}{\int_0^\infty w_{r,F_A} dr}\right)^{-1} = \frac{3\beta}{2\beta\tau + \left(F_A - \overline{F_A}\right)^2} \tag{6.67}$$

that leads to the very interesting theoretical result for the PDI of chains with the same comonomer fraction F_A

$$PDI_{F_A} = \frac{r_{wF_A}}{r_{nF_A}} = \frac{5}{3} \tag{6.68}$$

Figure 6.9 shows the predicted variation of r_n, r_w, and PDI for a model copolymer. It is interesting to see that the average chain lengths increase as the comonomer

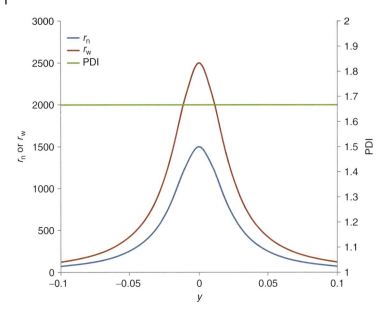

Figure 6.9 Variation of chain length averages with comonomer fraction. Model parameters: $\beta = 0.25$ and $\tau = 1 \times 10^{-3}$.

fraction approaches $\overline{F_A}$. This is another consequence of the phenomenon we discussed above: longer chains have a narrower CCD and, therefore, will approach the average comonomer fraction for the whole copolymer.

This interesting theoretical result is confirmed experimentally in Figure 6.10, where the M_n, M_w, and PDI of an ethylene/1-octene copolymer made with a single-site catalyst are plotted as a function of TREF elution temperature.[9] The values for M_n and M_w pass through a maximum as they approach the peak elution temperature (average copolymer composition), as predicted in Figure 6.9. The PDI values are relatively constant throughout the TREF profile but do not reach the theoretical limit of 1.67 predicted in Eq. (6.68), likely due to nonidealities during CFC fractionation.

6.2.2.2 Multiple-Site Catalysts

The approach to modeling the microstructure of binary copolymers made with multiple-site catalysts is analogous to the one applied to homopolymers; the bivariate CLD × CCD for the whole resin is considered a weighted sum of the Stockmayer distributions for each site type j

$$w_{ry} = \sum_{j=1}^{n} m_j w_{ryj} \tag{6.69}$$

---

9) This plot was obtained by projecting the cross-fractionation (CFC) profile of the resin onto the composition (elution temperature) plane, as explained in Chapter 2.

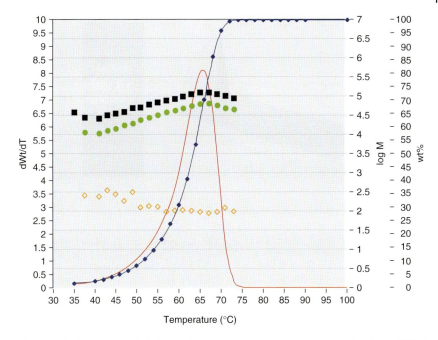

Figure 6.10 Plot of M_n (circles), M_w (squares), and PDI (diamonds) as a function of TREF elution temperature of a single-site ethylene/1-octene copolymer measured by CFC.

or when only the CCD component is being considered

$$w_y = \sum_{j=1}^{n} m_j w_{yj} \qquad (6.70)$$

Finally, the average comonomer molar fraction in the chains is given by the expression

$$\overline{F_A} = \sum_{j=1}^{n} m_j \overline{F_{Aj}} \qquad (6.71)$$

Figure 6.11 shows the Stockmayer bivariate distribution of a polymer made with a model catalyst having three site types. We observe the characteristic broad and bimodal CLD × CCD distribution of Ziegler–Natta resins, as measured by GPC–TREF cross-fractionation, showing that the model provides a good qualitative representation for these resins.

Because the CCD component of the Stockmayer distribution, Eq. (6.65), has variable width defined by the product $\beta \tau$, the deconvolution of the CCD of copolymers made with multiple-site catalysts, even if we neglect cocrystallization effects and other nonidealities during TREF or CRYSTAF analysis, may lead to multiple solutions. Two techniques may be used to eliminate this problem. In the first approach, the MWD is deconvoluted before the CCD to determine the number of active site types and polymer mass fractions made on them (n, m_j);

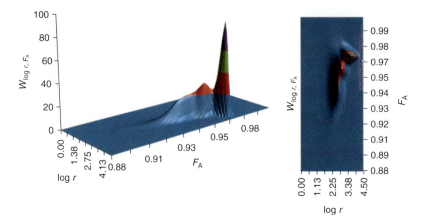

Figure 6.11 Stockmayer bivariate CLD × CCD for a model polymer made with a catalyst having three site types. Model parameters: $w_1 = 0.4$, $r_{n,1} = 4000$, $\overline{F}_{A1} = 0.97$, and $\beta_1 = 0.04719$; $w_2 = 0.3$, $r_{n,2} = 800$, $\overline{F}_{A2} = 0.96$, and $\beta_2 = 0.04879$; and $w_3 = 0.3$, $r_{n,3} = 400$, $\overline{F}_{A3} = 0.945$, and $\beta_3 = 0.0184$. Overall polymer properties: $r_n = 820$, PDI = 4.8, and $\overline{F}_A = 0.96$.

these parameters are kept constant during CCD deconvolution, which is only used to search for the best values of the parameters β_j and $\overline{F}_{Aj}$. In the second approach, the joint CLD × CCD is deconvoluted simultaneously to find out n, m_j, τ_j, β_j, and $\overline{F}_{Aj}$. However, it must be stressed again that TREF or CRYSTAF profiles are not true CCDs, albeit they are related to them. Consequently, the estimates for β obtained with CCD deconvolution are only apparent values and should not be used to calculate the product $r_A r_B$ using Eq. (6.58), as they will result in incorrect estimates.

Figure 6.12 shows the deconvolution results for an ethylene/1-butene copolymer made with a heterogeneous Ziegler–Natta catalyst. Four site types were required to represent the copolymer MWD. The values for m_j identified during the MWD deconvolution were kept constant during the CCD deconvolution, which was used to estimate the values of β_j and $\overline{F}_{Aj}$. Table 6.5 shows the values estimated for the parameters needed to describe the MWD and CCD for this resin.

For this particular sample, the MWD is better represented than the CCD; this is generally the case since TREF profiles, when converted to CCDs via a calibration curve, are just approximations of the true CCD due to crystallization kinetic effects and cocrystallization phenomena taking place in TREF (or CRYSTAF). Nonetheless, the CCD approximation is very good, and we notice that as the M_n for each site type increases, their CCDs get narrower, as theoretically expected from the Stockmayer distribution.

The simultaneous deconvolution of these two profiles would result in the estimation of very similar model parameters.

An interesting problem occurs when trying to deconvolute the CCD of a copolymer having a sizeable TREF or CRYSTAF soluble fraction. In this case, part (or all) of the polymer chains made in the active sites having high α-olefin reactivity

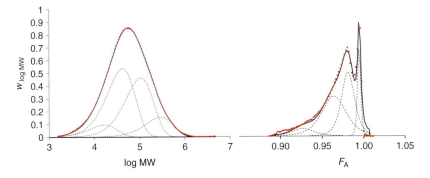

Figure 6.12 MWD and CCD deconvolution for an ethylene/1-butene resin made with a heterogeneous Ziegler–Natta catalyst. Model parameters are reported in Table 6.5.

Table 6.5 MWD and CCD deconvolution parameters for an ethylene/1-butene copolymer made with a heterogeneous Ziegler–Natta catalyst.

Site	m	M_n	$\overline{F_A}$	β
1	0.076	8 380	0.926	0.1114
2	0.429	21 200	0.964	0.3356
3	0.371	53 400	0.981	0.2513
4	0.124	146 000	0.994	0.0381
All	1.00	26 700	0.971	–

ratios is separated at room temperature and cannot be considered in the CCD deconvolution process. For these resins, the joint MWD × CCD deconvolution procedure must be modified to account for the soluble fraction, as illustrated in Example 6.5.

> **Example 6.5: Simultaneous MWD × CCD Deconvolution Considering the Soluble Fraction of an Ethylene/1-Octene Copolymer Made with a Ziegler–Natta Catalyst**
>
> Figure E.6.5.1 shows the TREF profile (inset) and cumulative CCD for an ethylene/1-octene resin made with a multiple-site catalyst. A significant fraction of this copolymer is soluble in TCB below 25 °C, the lowest temperature used in the TREF analysis. Since the "soluble peak" is not a chromatographic peak, but instead a purge peak, it cannot be described with Eq. (6.65). Consequently, the CCD deconvolution procedure illustrated in Figure 6.12 cannot be used, without modifications for this resin. Since several LLDPE resins have TREF soluble peaks, we need to find an alternative way to model these copolymers.

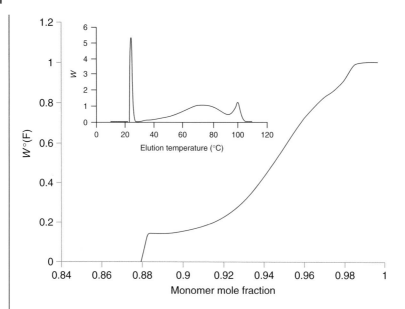

Figure E.6.5.1 TREF profile (inset) and cumulative CCD for an ethylene/1-octene sample made with a Ziegler–Natta catalyst.

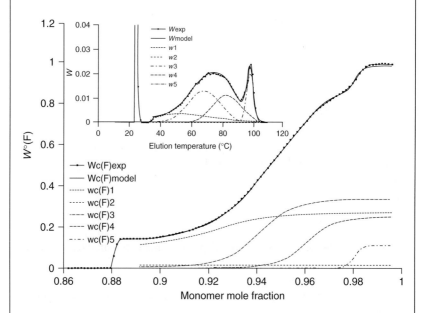

Figure E.6.5.2 Deconvolution of TREF cumulative profile into five site types. Model parameters are shown in Table E.6.5.1.

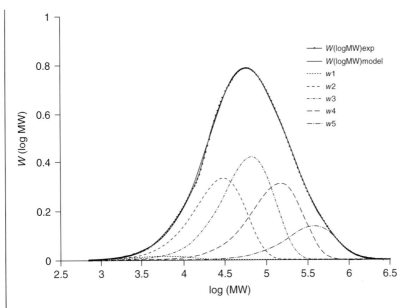

Figure E.6.5.3 MWD deconvolution into five site types. Model parameters are shown in Table E.6.5.1.

It is possible to overcome this problem by using the cumulative form of Eq. (6.65) during deconvolution

$$w_{F_A}^C = \int_{-\infty}^{F_A} w_{F_A}\, dF_A$$

which has the analytical solution

$$w_{F_A}^C = \frac{2\left[2\hat{\beta}\tau+(F_A-\overline{F_A})^2\right]^{5/2}+3(F_A-\overline{F_A})(2\hat{\beta}\tau)^2+5(2\hat{\beta}\tau)(F_A-\overline{F_A})^3+2(F_A-\overline{F_A})^5}{4\left[2\hat{\beta}\tau+\left(F_A-\overline{F}\right)^2\right]^{5/2}}$$

However, to include the soluble fraction in the deconvolution, m_j in Eq. (6.70) needs to be redefined as

$$m_j = m_j^s + m_j^{ns}$$

where m_j is the total mass fraction of polymer made on site type j and m_j^s and m_j^{ns} are the mass fractions of polymer that are soluble and not soluble at the lowest TREF temperature (commonly, the room temperature), respectively. Only the insoluble polymer fraction can be described with Stockmayer distribution. Therefore, the cumulative CCD for the whole resin is given by

$$w_{FA}^C = \sum_{j=1}^{n} m_j^{ns} w_{FA,i}^C \quad \text{if } F_A \geq F_{A,\text{crit}}$$

$$w_{FA}^C = \sum_{i=j}^{n} m_j^s \quad \text{if } F_A < F_{A,\text{crit}}$$

and

$$m_j^s = \int_{-\infty}^{F_{A,\text{crit}}} w_{FA,j}\, dF_A$$

where $F_{A,\text{crit}}$ is the critical ethylene mole fraction below which the polymer is soluble at room temperature. The critical ethylene mole fraction depends on solvent type, comonomer type, and analysis conditions; it is easily determined during the TREF calibration procedure by extrapolating the curve to room temperature; for this example, $F_{A,\text{crit}} = 0.89$.

The next step in the deconvolution procedure is to use a nonlinear least squares optimization routine to minimize the squares of the difference between the measured and predicted distributions. Different from what was done for the resin illustrated in Figure 6.12, we now use the cumulative CCD measured by TREF, not its differential profile.

The objective function for the simultaneous MWD and cumulative CCD deconvolution is given by the expression

$$\chi^2 = \sum_{i=1}^{n_{\text{GPC}}} \left(w_{\log \text{MW},i}^{\text{GPC}} - w_{\log \text{MW},i}\right)^2 + \sum_{i=1}^{n_{\text{TREF}}} \left(w_{FA,i}^{C,\text{TREF}} - w_{FA,i}^{C}\right)^2$$

Figures E.6.5.2 and E.6.5.3 show the best fit for the MWD and CCD of this resin, when the simultaneous deconvolution procedure outlined above was followed. Five site types were necessary to describe these distributions adequately.

In the case of LLDPE resins made with heterogeneous Ziegler–Natta catalysts, the sites that produce polymer with lower average comonomer fractions also have higher number average molecular weights. We added this constraint to the optimization routine to make sure that the results are consistent with this experimentally observed behavior for heterogeneous LLDPE resins.

Table E.6.5.1 shows that site type 1 makes polymer that is completely soluble at room temperature. Because it is not detected as a chromatographic peak, it cannot be described by Eq. (6.65) (or its equivalent cumulative version); consequently, no information is obtained for $\overline{F_A}$ and β for site type 1. Polymer made on site 5, with the highest M_n, is not present in the soluble fraction, while the fraction of polymer made in the other sites increases in the order $4 < 3 < 2$, as also illustrated in Table E.6.5.1. These results correspond to our expectations of how a heterogeneous Ziegler–Natta catalyst makes ethylene/α-olefin copolymers.

Table E.6.5.1 Model parameters for MWD × CCD simultaneous deconvolution.

n	1	2	3	4	5
m	0.0166	0.3005	0.3519	0.2185	0.1124
m^{ns}	0	0.2061	0.3459	0.2181	0.1124
m^s	0.0166	0.0945	0.0060	0.0005	0
M_n	3 370	15 882	37 513	80 361	198 730
τ	2.97×10^{-4}	6.30×10^{-5}	2.67×10^{-5}	1.24×10^{-5}	5.03×10^{-6}
$\overline{F_A}$	–	0.9254	0.9417	0.9608	0.9801
β	–	0.5209	0.3544	0.3541	0.0678

An alternative modeling strategy involves the use of broadening functions for the CCD, in analogy to the approach we used for MWD in Eq. (6.40)

$$W_y = \int w_y N(\beta\tau) d(\beta\tau) \quad (6.72)$$

where the capital W is again used to indicate that the site distributions are broader than the one predicted by Eq. (6.65), according to the function $N(\beta\tau)$. This approach, despite requiring fewer number of site types, suffers from the same limitations discussed for broadened MWD distributions above.

The use of online FTIR detectors with GPC is becoming increasingly popular for polyolefin characterization. This relatively simple combination permits the detection of the average chemical composition (generally reported as molar fraction of α-olefin or short-chain branching (SCB) frequency) as a function of molecular weight, as already discussed in Chapter 2. Figure 6.13 shows the GPC-FTIR plot for an ethylene/1-butene LLDPE, with the typical behavior of a heterogeneous

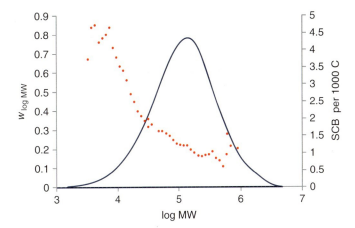

Figure 6.13 MWD and average fraction of 1-butene in an LLDPE resin made with a heterogeneous Ziegler–Natta catalyst.

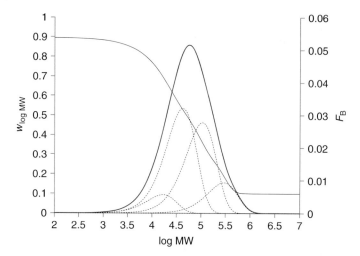

Figure 6.14 MWD and average fraction of 1-butene for the copolymer described in Figure 6.12 and Table 6.5.

Ziegler–Natta resin: as the molecular weight increases, the fraction of 1-butene in the sample decreases.

The molar fraction of ethylene as a function of molecular weight for a copolymer made with a multiple-site catalyst is calculated with the equation

$$F_{Br} = \frac{\sum_{j=1}^{n} m_j \overline{F_{Bj}} w_{rj}}{\sum_{j=1}^{n} m_j w_{rj}} \qquad (6.73)$$

provided that we know the MWDs, average copolymer composition, and masses of polymer made on each site type. For instance, for the polymer shown in Figure 6.12 and Table 6.5, the predicted FTIR–MWD curve is given in Figure 6.14.

FTIR copolymer composition results are generally reported in number of SCBs per 1000 carbon atoms, instead of comonomer molar fractions. Equation (6.73) still applies in this case, by simply replacing $\overline{F_{Bj}}$ with the equivalent $\overline{SCB_j}$.[10]

This simple representation of GPC-FTIR profiles permits a systematic interpretation of results observed in several academic and industrial polyolefin analytical laboratories. Furthermore, if the MWD deconvolution procedure is combined with fitting of the FTIR profile, the average comonomer molar fraction per site type can be estimated, as shown in Example 6.6.

10) Branching frequencies reported as SCB/1000 C can be converted into comonomer molar fractions with the equation $F_B = \frac{2 SCB}{1000 + (2 - n_C) SCB}$, where n_C is the number of carbons in the α-olefin.

For the low molecular weight region, the effect of methyl end groups on the experimental FTIR data may also have to be considered.

Example 6.6: Simultaneous MWD-FTIR Deconvolution of an Ethylene/1-Butene Copolymer made with a Ziegler–Natta Catalyst

We fit the MWD-FTIR profiles shown in Figure 6.13 by combining the MWD deconvolution procedure seen above with the approach proposed in Eq. (6.73). First, the MWD deconvolution is performed to find the minimum number of site types required to describe the polymer and the values for the parameters m_j and τ_j. In a second step, we minimize the square of the differences between predicted and FTIR-measured copolymer composition, as given by the objective function

$$\chi^2 = \sum_{i}^{n_{FTIR}} \left[SCB_{exp,i} - SCB_{model,i}\right]^2$$

$$= \sum_{i}^{n_{FTIR}} \left(SCB_{exp,i} - \frac{\sum_{j=1}^{n} m_j \overline{SCB_j} w_{i,j}}{\sum_{j=1}^{n} m_j w_{i,j}} \right)^2$$

where n_{FTIR} is the number of data points collected by the FTIR detector, $\overline{SCB_j}$ is the average SBC/1000 C for site type j, m_j is the mass fraction of the polymer made on site type j, and $w_{i,j}$ is the value of Flory distribution at site type j for the molecular weight corresponding to FTIR sampling point i.

Figure E.6.6.1 shows that this procedure leads to an excellent fit for both MWD and the copolymer composition curve. The estimated model parameters are shown in Table E.6.6.1.

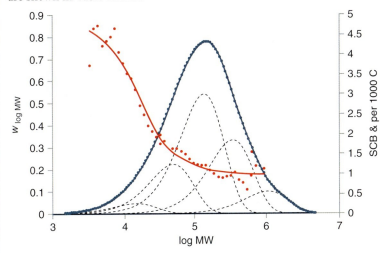

Figure E.6.6.1 MWD-FTIR plot of an ethylene/1-butene copolymer made with a heterogeneous Ziegler–Natta catalyst.

Table E.6.6.1 MWD × FTIR deconvolution parameters for the sample shown in Figure E.6.6.1.

j	1	2	3	4	5
m	0.0378	0.1800	0.4333	0.2674	0.0815
M_n	7 400	23 900	64 900	164 000	507 000
SCBs per 1000 C	7.05	1.98	1.10	0.99	0.99
$\overline{F_B}$	0.0143	0.0040	0.0022	0.0020	0.0020

6.2.3
Comonomer Sequence Length Distribution

Comonomer sequence length distributions (CSLDs) are measured by ^{13}C NMR analysis as dyads, triads, or higher sequences. They are complementary to CCDs measured by TREF, CRYSTAF, CEF, or MWD-FTIR detection. For polyolefins made with single-site catalysts, they provide direct information on the catalyst reactivity ratio; for multiple-site catalysts, the analysis is complicated because the detected sequences are averages for copolymer chains made on different site types that may have distinct reactivity ratios.

We will develop complete sequence length distribution expressions for triads considering Bernoullian and terminal model statistics. Extrapolation to higher-order models, or longer sequences, is straightforward.

There are six possible triads in a binary copolymer: AAA, AAB, BAB, ABA, BBA, and BBB. In the Bernoullian model, the probability of an ABA triad being formed is given by

$$\mathrm{ABA} = \overline{F_A} P_B P_A \tag{6.74}$$

where $\overline{F_A}$ is the molar fraction of the monomer A in the copolymer (given by the Mayo–Lewis equation, for instance) and P_A and P_B are the probabilities of adding monomers A and B, respectively, to the polymer chain; in the case of Bernoullian statistics, these probabilities do not depend on chain-end type, just on catalyst type and monomer molar fraction in the reactor (f_A or f_B). $\overline{F_A}$ is used as the probability to find the first monomer in the sequence, since it is equal to average molar fraction of monomer A in the polymer chain.

Similarly, for the tetrad ABAA, we write

$$\mathrm{ABAA} = \overline{F_A} P_B P_A^2 \tag{6.75}$$

Table 6.6 Triad expressions for Bernoullian and terminal models.

Triad	Bernoullian model	Terminal model
AAA	$\overline{F_A} P_A^2$	$\overline{F_A} P_{AA}^2$
AAB	$\overline{F_A} P_A P_B + \overline{F_B} P_A^2$	$\overline{F_A} P_{AA} P_{AB} + \overline{F_B} P_{BA} P_{AA}$
BAB	$\overline{F_B} P_A P_B$	$\overline{F_B} P_{BA} P_{AB}$
ABA	$\overline{F_A} P_B P_A$	$\overline{F_A} P_{AB} P_{BA}$
BBA	$\overline{F_B} P_B P_A + \overline{F_A} P_B^2$	$\overline{F_B} P_{BB} P_{BA} + \overline{F_A} P_{AB} P_{BB}$
BBB	$\overline{F_B} P_B^2$	$\overline{F_B} P_{BB}^2$

P_i, propagation probability for monomer A (Bernoullian model); P_{ij}, conditional probability for monomer j adding to a chain terminated in monomer i (terminal model).

The probability of adding monomer A is expressed as

$$P_A = \frac{R_{pA}}{R_{pA} + R_{pB}} = \frac{k_{pA}[A]}{k_{pA}[A] + k_{pB}[B]} = \frac{f_A}{f_A + \rho f_B} \tag{6.76}$$

and

$$\rho = \frac{k_{pB}}{k_{pA}} \tag{6.77}$$

$$P_B = 1 - P_A \tag{6.78}$$

The tetrad AAB can be constructed starting from monomer A or B; therefore, the triad expression is

$$AAB = \overline{F_A} P_A P_B + \overline{F_B} P_A^2 \tag{6.79}$$

This procedure can be applied to all six triads, to obtain the formulas summarized in the first column of Table 6.6.

The procedure used to derive equations for the terminal model is analogous, but we also need to consider the type of monomer last added to the chain end. For instance, for the triad AAB

$$AAB = \overline{F_A} P_{AA} P_{AB} + \overline{F_B} P_{BA} P_{AA} \tag{6.80}$$

where P_{AA} and P_{AB} are the conditional probabilities of monomer A and B adding to a chain terminated in monomer A, respectively. These probabilities are given by the equations

$$P_{AB} = \frac{R_{pAB}}{R_{pAA} + R_{pAB}} = \frac{k_{pAB}[B]}{k_{pAA}[A] + k_{pAB}[B]} = \frac{f_B}{r_A f_A + f_B} \tag{6.81}$$

$$P_{AA} = 1 - P_{AB} \tag{6.82}$$

Similarly,

$$P_{BA} = \frac{R_{pBA}}{R_{pBB} + R_{pBA}} = \frac{f_A}{f_A + r_B f_B} \qquad (6.83)$$

$$P_{BB} = 1 - P_{BA} \qquad (6.84)$$

For polymers made with multiple-site catalysts, the sequence length distribution measured by ^{13}C NMR spectroscopy results from the weighted superposition of the sequences made on each site type. For instance, for the AAB triad

$$\overline{AAB} = \sum_{j=1}^{n} m_j (AAB)_j = \sum_{j=1}^{n} m_j \left(\overline{F_{A,j}} P_{AA,j} P_{AB,j} + \overline{F_{B,j}} P_{BA,j} P_{AA,j} \right) \qquad (6.85)$$

where the probabilities refer to each active site type j.

Parameter identification, in this case, is much more involved. Example 6.7 examines this topic in more details.

Example 6.7: Triad Distributions for Copolymers Made with a Model Multiple-Site Catalyst

Consider the catalyst for which model parameters are given in Table E.6.7.1. We use the equations derived above to predict the triad distribution for a copolymer made with this catalyst under varying comonomer molar fractions in the reactor. The catalyst has three site types: site type 1 makes polymer with the lowest molecular weight average and the highest α-olefin (B) molar fraction and follows Bernoullian statistics; site type 3 makes polymer with the highest molecular weight averages and lowest comonomer incorporation, with a more blocky structure ($r_A r_B > 1.0$); and site type 2 has intermediate behavior between sites 1 and 2. This is the pattern observed for most heterogeneous Ziegler–Natta catalysts, even though the values shown in Table E.6.7.1 are arbitrary. The M_n values are not required for this example but shown here for completeness.

Table E.6.7.1 Parameters for a three-site model catalyst.

Site type	1	2	3
m	0.35	0.4	0.25
M_n	7400	23 900	64 900
r_A	5	10	20
r_B	0.2	0.2	0.15
$r_A r_B$	1.0	2.0	3.0

Figure E.6.7.1 shows the triad distribution when the molar fraction of ethylene in the reactor is $f_A = 0.8$. To obtain this distribution, $\overline{F_A}$ is estimated using the Mayo–Lewis expression, Eq. (6.59), and the probabilities P_{AA}, P_{AB}, P_{BA}, and P_{BB} for each site type are computed using the expressions given Eqs. (6.81–6.84) for the terminal model. Finally, the triad distribution for each site type is calculated with the formulas shown in Table 6.6, and the overall triad distribution is estimated with expressions analogous to Eq. (6.85). Since the copolymer has a high ethylene molar fraction (0.970), the triad AAA predominates in its structure, with a small fraction of ABA and AAB triads, followed by negligible amounts of the other triads. When the fraction of ethylene in the reactor is reduced to $f_A = 0.6$ (Figure E.6.7.2), the overall ethylene fraction in the copolymer drops to $\overline{F_A} = 0.924$, and the overall contribution of triads containing comonomer B increases, as expected. For both cases, comonomer B is more readily incorporated by site type 1, as expected from its reactivity ratio values.

Table E.6.7.2 Triad distributions for the polymers shown in Figures E.6.7.1 and E.6.7.2.

Triad	$f_A = 0.6$	$f_A = 0.8$
AAA	7.95×10^{-1}	9.14×10^{-1}
AAB	1.23×10^{-1}	5.48×10^{-1}
BAB	5.99×10^{-3}	1.03×10^{-3}
ABA	5.97×10^{-2}	2.71×10^{-2}
BBA	1.55×10^{-2}	2.64×10^{-3}
BBB	1.01×10^{-3}	6.46×10^{-5}

Table E.6.7.2 lists the value for the triad fractions for the two model polymers shown in Figures E.6.7.1 and E.6.7.2.

In this example, we assumed that the model parameters were known and used them to predict the triad distributions for two different comonomer molar fractions in the reactor. Alternatively, if the triad distributions were measured by ^{13}C NMR, we could have used them to estimate the reactivity ratios and number of active site types in the catalyst. Since the signal for each triad is an average of the signals for each site type, multiple solutions are often encountered when this approach is applied, especially when the number of active site types is not known (or estimated) *a priori*. For instance, the alternative parameters given in Table E.6.7.3 for a two- and a four-site catalyst will predict essentially the same triad distribution given in Table E.6.7.2.

This problem can be avoided, at least partially, if we use MWD deconvolution to identify the minimum number of active site types and then apply the comonomer sequence length equations to estimate the reactivity ratios for each site type. This method also assures that there will be consistency among MWD and CSLD modeling for the same catalyst.

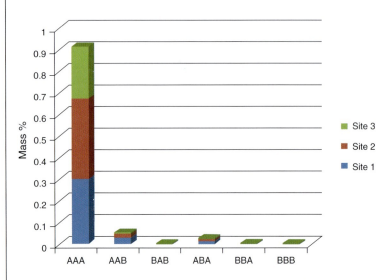

Figure E.6.7.1 Comonomer sequence length distribution when $f_A = 0.8$ ($\overline{F_A} = 0.970$).

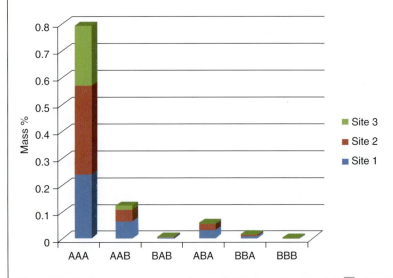

Figure E.6.7.2 Comonomer sequence length distribution when $f_A = 0.6$ ($\overline{F_A} = 0.924$).

Table E.6.7.3 Parameters for alternative two- and four-site catalysts that predict essentially the same triad distribution shown in Table E.6.7.2.

Site type	Two sites		Four sites			
	1	2	1	2	3	4
m	0.174	0.826	0.208	0.743	0.036	0.013
r_A	3.776	10.795	4.065	10.77	19.99	19.99
r_B	0.170	0.206	0.175	0.206	0.185	0.247

CSLDs for terpolymer (or multicomponent copolymers) can be derived in a similar way using the propagation probabilities of all comonomers involved in the copolymerization.

6.2.4
Long-Chain Branching Distribution

In Chapter 3, we saw that LCBs are formed during olefin polymerization with coordination catalysts by the insertion of a vinyl-terminated polymer chain (macromonomer) onto a growing polymer chain. Thus, LCBs result from the copolymerization of ethylene and α-olefins with macromonomers, either formed *in situ* or added to the reactor.

Following the approach used to derive Flory and Stockmayer distributions, we can find analytical solutions for the bivariate distribution of molecular weight and LCB and for the trivariate MWD × CCD × LCB distribution of polymers made with a single-site catalyst assuming *in situ* macromonomer formation in a CSTR. The derivation of these equations is long and is omitted herein; it can be found in the references discussed at the end of this chapter.

The instantaneous frequency and weight distributions of chain length and LCB for polymer populations with k LCBs per chain made in a CSTR operated at steady state are given by the expressions

$$f_{rk} = \frac{1}{(2k)!} r^{2k} \tau_B^{2k+1} \exp(-\tau_B r) \qquad (6.86)$$

$$w_{rk} = \frac{1}{(2k+1)!} r^{2k+1} \tau_B^{2k+2} \exp(-\tau_B r) \qquad (6.87)$$

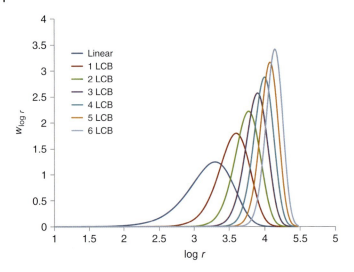

Figure 6.15 MWD for polymer populations with increasing number of LCBs per chain, for a polymer with $\tau_B = 10^{-3}$.

or in logarithmic scale

$$f_{\log r,k} = 2.3026 \times \frac{1}{(2k)!} r^{2k+1} \tau_B^{2k+1} \exp(-\tau_B r) \tag{6.88}$$

$$w_{\log r,k} = 2.3026 \times \frac{1}{(2k+1)!} r^{2k+2} \tau_B^{2k+2} \exp(-\tau_B r) \tag{6.89}$$

The parameter τ_B closely resembles the parameter τ in Flory and Stockmayer distributions

$$\tau_B = \frac{R_t + R_{LCB}}{R_p} = \tau + \frac{R_{LCB}}{R_p} \tag{6.90}$$

Equations (6.86–6.89) describe the MWDs of polymer populations with different numbers of LCBs per chain that compose the MWD of the whole polymer. When $k = 0$, $\tau_B = \tau$ and these expressions are reduced to Flory distribution.

Figure 6.15 shows the MWDs of chain populations for a polymer with $\tau_B = 10^{-3}$ and increasing number of LCBs per chain. The MWDs shift to higher averages and become narrower as the number of LCB per chain increases. The narrowing of the MWD with increasing LCB per chain is a consequence of the size averaging process that takes place when macromonomers are incorporated onto the polymer chains as LCBs; the more highly branched the population, the more likely they will have similar chain lengths. The MWD for the whole polymer, however, becomes broader with increasing LCB frequency. This distinction will be made clearer when we present the equations for PDI later in this section.

In Figure 6.15, the MWDs are normalized for each population. Another set of equations is necessary to calculate the fractions of these populations that compose the whole polymer. Because of the way LCBs are formed with coordination catalysts, the molar fraction of linear chains is always greater than that of chains with one LCB, which, in turn, is greater than the molar fraction of chains with two LCBs, and so on. These fractions, y_k, are given by

$$y_k = \phi_k \frac{(\mu f^=)^k}{(1 + \mu f^=)^{2k+1}} = \phi_k \frac{\alpha^k}{(1 + \alpha)^{2k+1}} \tag{6.91}$$

The molar fraction of macromonomers in the reactor, $f^=$, measured with respect to the total number of dead polymer molecules is given by the ratio

$$f^= = \frac{K_t^=}{K_t^= + K_t} \tag{6.92}$$

where $K_t^=$ and K_t are the frequencies of macromonomer and saturated dead chain formation, respectively. Finally, the parameters α, μ, and ϕ_k (Catalan numbers) are defined by the expressions

$$\alpha = \mu f^= \tag{6.93}$$

$$\mu = \frac{1}{1 + \dfrac{s}{k_{LCB} Y_0}} \tag{6.94}$$

$$\phi_k = \frac{(2k)!}{k!(k+1)!} \tag{6.95}$$

where s is the reciprocal of the average residence time in the CSTR, k_{LCB} is the rate constant for LCB formation, and Y_0 is the concentration of catalyst active sites.

The equivalent mass fractions, m_k, for the polymer populations are given by

$$m_k = \phi_k \frac{\alpha^k (1 - \alpha)(2k + 1)}{(1 + \alpha)^{2k+2}} \tag{6.96}$$

The parameter α is the key to understanding what regulates the branching structures of these polyolefins. First, notice that $0 \leq \alpha \leq 1$. All chains are linear when $\alpha = 0$ since this requires that either the fraction of macromonomers be zero, $f^= = 0$, or that $\mu = 0$. The last condition is obeyed if $s/k_{LCB} Y_0 \to \infty$, which is true when either $k_{LCB} = 0$ (no LCB formation) or when $s \to \infty$ (reactor residence time equal to zero; no macromonomer is formed). On the other hand, LCB formation is maximum when $\alpha = 1$, a condition only obeyed when all dead chains have terminal unsaturations, $f^= = 1$, and the residence time in the reactor is infinite, $s \to 0$, or the rate constant for LCB formation is infinite, $k_{LCB} \to \infty$.

An alternative definition for the parameter μ can be used to shed more light on the mechanism of LCB formation. The molar balance for the concentration of living macromonomer, $P^=$, in the reactor is

$$\frac{dP^=}{dt} = K_t^= Y_0 - k_{LCB}[P^=]Y_0 - s[P^=] \tag{6.97}$$

At steady state, we can write

$$\frac{s}{k_{LCB} Y_0} = \frac{K_t^=}{k_{LCB}[P^=]} - 1 \tag{6.98}$$

Substituting Eq. (6.98) into Eq. (6.94), we reach the equivalent, but very instructive, definition for the parameter μ

$$\mu = \frac{k_{LCB}[P^=]}{K_t^=} \tag{6.99}$$

as the ratio between the frequencies for macromonomer insertion (LCB formation) and macromonomer formation. Notice that $0 \leq \mu \leq 1$ since macromonomers can only be inserted at a rate lower than, or equal to, that they are formed. As $\mu \to 1$, the rate of macromonomer formation approaches the rate of macromonomer insertion (a limit not possible to be reached in practice), and LCB formation is maximized in the reactor.

The elegant set of expressions defined by Eqs. (6.86–6.99) serve as a path for understanding the microstructure of polyolefins containing LCBs that lies beyond what can be found using currently available characterization techniques. These equations exemplify very well how mathematical models can be used to overcome experimental limitations and open the door to a deeper comprehension of polymer microstructures. Besides, they are beautiful.[11]

Table 6.7 lists the molar and weight fractions for a polymer with $\tau_B = 10^{-3}$ and $\alpha = 0.1$. It is interesting to see that the molar fraction of linear chains is higher than their mass fractions since they are present in high concentration, but weigh less than the more branched species.

The weight fractions shown in Table 6.7 are used in Figure 6.16 to calculate the overall MWD for the polymer. The polymer is composed mostly of linear chains (74.4 wt%), a relatively large fraction of chains with one LCB (18.4 wt%), and decreasing fractions of more branched species, which explains why these materials are called *essentially linear polyolefins*. The average number of LCBs per chain, also displayed in Figure 6.16, is computed with the expression

$$\overline{LCB}_r = \frac{\sum_{k=0}^{\infty} k m_k w_{\log r, k}}{\sum_{k=0}^{\infty} m_k w_{\log r, k}} \tag{6.100}$$

11) Quoting Paul Dirac once more, "The research worker, in his effort to express the fundamental laws of Nature in mathematical form, should strive mainly for mathematical beauty."

Table 6.7 Molar and mass fractions for polymer populations with $\tau_B = 10^{-3}$ and $\alpha = 0.1$.

LCBs per chain	y	m
0	9.09×10^{-1}	7.44×10^{-1}
1	7.51×10^{-1}	1.84×10^{-1}
2	1.24×10^{-2}	5.08×10^{-2}
3	2.57×10^{-3}	1.47×10^{-2}
4	5.94×10^{-4}	4.37×10^{-3}
5	1.47×10^{-4}	1.33×10^{-3}
6	3.82×10^{-5}	4.07×10^{-4}

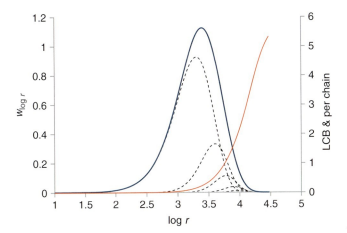

Figure 6.16 MWD for overall and individual polymer populations, for a model polymer with $\tau_B = 10^{-3}$ and $\alpha = 0.1$. The MWD of individual populations are indicated with dashed lines and their mass fractions are listed in Table 6.7. The red curve shows the average number of LCBs per chain.

where the horizontal bar is used to indicate that the LCB value applies to the whole polymer. This notation is used to specify whole polymer properties for the rest of this section.

The fractions calculated with Eqs. (6.91) and (6.96) can be combined with the distributions given by Eqs. (6.86) and (6.87), respectively, to obtain the CLDs for the whole polymer

$$\overline{f_r} = \sum_{k=0}^{\infty} y_k f_{rk} = \frac{\exp(-r\tau_B)}{r\sqrt{\alpha}} I_1\left(2\frac{r\tau_B\sqrt{\alpha}}{1+\alpha}\right) \tag{6.101}$$

$$\overline{w_r} = \sum_{k=0}^{\infty} m_k w_{rk} = \frac{(1-\alpha)\tau_B \exp(-r\tau_B)}{(1+\alpha)\sqrt{\alpha}} I_1\left(2\frac{r\tau_B\sqrt{\alpha}}{1+\alpha}\right) \tag{6.102}$$

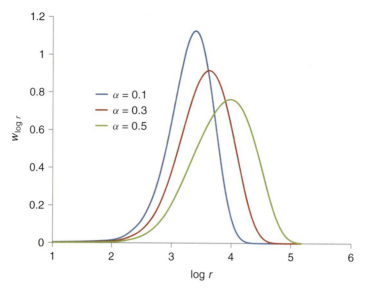

Figure 6.17 Overall MWD for $\tau_B = 10^{-3}$ and $\alpha = 0.1$, 0.3, and 0.5.

where I_1 is the modified Bessel function of the first kind and order 1.[12] The logarithmic form of these equations can be found by applying the transformation shown in Eq. (6.25). Figure 6.17 shows how the MWD broadens and moves to higher averages as the value of α increases because of the formation of more branched populations.

Analytical solutions are also available for chain length averages as a function of the number of LCBs per chain. Interestingly, the averages for the branched chains are related to those for the linear chains through concise expressions

$$r_{nk} = (1 + 2k)r_{n,0} = (1 + 2k)\frac{1}{\tau_B} \tag{6.103}$$

$$r_{wk} = (1 + k)r_{w,0} = (1 + k)\frac{2}{\tau_B} \tag{6.104}$$

$$r_{zk} = \left(1 + \frac{2}{3}k\right)r_{z,0} = \left(1 + \frac{2}{3}k\right)\frac{3}{\tau_B} \tag{6.105}$$

$$\text{PDI}_k = \left(\frac{1+k}{1+2k}\right)\text{PDI}_0 = 2\left(\frac{1+k}{1+2k}\right) \tag{6.106}$$

---

12) Bessel functions of the first kind and order 1 are defined as $I_1(x) = \sum_{k=0}^{\infty}\frac{\left(\frac{x}{2}\right)^{1-2k}}{k!\Gamma(k+2)}$; they are readily available in most scientific software packages and spreadsheets.

Table 6.8 Chain length averages for the populations shown in Figure 6.15 with $\tau_B = 10^{-3}$.

LCBs per chain	r_n	r_w	r_z	PDI
0	1 000	2 000	3 000	2.00
1	3 000	4 000	5 000	1.33
2	5 000	6 000	7 000	1.20
3	7 000	8 000	9 000	1.14
4	9 000	10 000	11 000	1.11
5	11 000	12 000	13 000	1.09
6	13 000	14 000	15 000	1.08

where r_{nk}, r_{wk}, and r_{zk} are the number, weight, and z-average chain lengths for populations with k LCBs, respectively, and PDI_k is their PDI indices. It is instructive to point out that as the number of LCBs per chain increases, chain length averages increase, but PDI indices decrease and tend to a limiting value of 1.0 for chains with an infinite number of LCBs, as already illustrated in Figure 6.15. For instance, according to Eq. (6.106), all populations with more than five LCBs per chain have PDIs that are less than 1.1. As discussed above, the PDI of the branched chains decrease because LCB formation is a size averaging process that consists of randomly inserting macromonomers, independent of their sizes, onto the growing chains.

The chain length averages for the populations shown in Figure 6.15 are listed in Table 6.8, where the variation pattern from population to population clearly emerges. It is difficult not to feel a certain esthetic pleasure in the regular pattern displayed by the chain length averages of these polymer populations.

On the other hand, the PDI for the overall polymer increases with increasing branching frequencies. Expressions for the chain length averages for the whole polymer are given below

$$\bar{r}_n = \frac{1}{\tau_B} \times \frac{1+\alpha}{1-\alpha} \tag{6.107}$$

$$\bar{r}_w = \frac{2}{\tau_B} \times \frac{1+\alpha}{(1-\alpha)^2} \tag{6.108}$$

$$\overline{PDI} = \frac{2}{1-\alpha} \tag{6.109}$$

Equation (6.109) shows that the PDI for the whole polymer is always greater than 2 and approaches infinity (gel point) as α tends to 1.

Equations for average LCB frequencies are equally easy to use

$$\lambda = 500 \frac{\alpha}{\bar{r}_n} \tag{6.110}$$

Table 6.9 Chain length and branching averages for the polymers shown in Figure 6.17 with $\tau_B = 10^{-3}$.

α	$\tilde{r}_n$	PDI	n_{LCB}	λ
0.1	1222	2.22	0.11	0.0455
0.3	1857	2.86	0.43	0.115
0.5	3000	4.00	1.00	0.117

$$n_{LCB} = \frac{\alpha}{1-\alpha} \tag{6.111}$$

where λ is the average number of LCBs per 1000 carbon atoms (for polyethylene), n_{LCB} is the average number of LCBs per chain, and $\tilde{r}_n$ is the number average chain length that would be obtained *in the absence of LCB-forming reactions*.[13]

Table 6.9 lists the chain length and branching averages for the polymers illustrated in Figure 6.17, quantifying how the chain length averages increase, and the CLD broadens, as the branching frequency increases.

Using Eq. (6.111) to solve for α, and replacing the resulting expression into Eqs. (6.107–6.109), a useful set of expressions relating chain length averages and PDI to the average number of LCBs per chain is obtained

$$\bar{r}_n = \frac{1}{\tau_B}(1 + 2n_{LCB}) \tag{6.112}$$

$$\bar{r}_w = \frac{2}{\tau_B}(1 + 2n_{LCB})(1 + n_{LCB}) \tag{6.113}$$

$$\overline{PDI} = 2(1 + n_{LCB}) \tag{6.114}$$

Since the number of LCBs reaches a maximum when $\alpha = 1$, the limiting branching averages become

$$\lambda_{max} = \frac{500}{\tilde{r}_n} \tag{6.115}$$

$$n_{LCBmax} \to \infty \tag{6.116}$$

---

13) Notice that $\tilde{r}_n$ is not equal to $r_{n,0}$, the number average of the linear chains in the population; instead, $\tilde{r}_n = \frac{1}{\tau} = \frac{1}{\tau_B - \frac{R_{LCB}}{R_p}}$, that is, the number average chain length that would be obtained if all LCB-formation reactions were suppressed. Some additional mathematical manipulations will show that $\tau = \frac{\tau_B}{1+\alpha}$. The demonstration of this identity is left as an exercise to the interested reader.

These two branching indices lead to very different numerical values: while λ tends to a finite limit, n_{LCB} becomes infinite, an observation that can be explained by the steady-state hypothesis made in the derivation of these equations. As explained above, a value of unity for α is interpreted as an infinite residence time in the CSTR. Under steady-state conditions, this implies that all macromonomers produced must be reincorporated into the growing chains, leading to chains with infinite number of LCBs. This condition is not attainable in practice, but it is still useful as the limiting behavior of LCB formation with these catalysts.

Examples 6.8 and 6.9 illustrate some additional applications for these LCB modeling equations.

Example 6.8: Investigation of LCB Microstructure from Average Polymer Properties

Table E.6.8.1 shows the molecular weight and branching averages of eight polyethylene resins having different LCB frequencies made with a constrained geometry catalyst. PDI increases with LCB frequency, as predicted by Eq. (6.109) or (6.114). We combine this information with the model equations shown above to obtain more details on the microstructure of these resins.

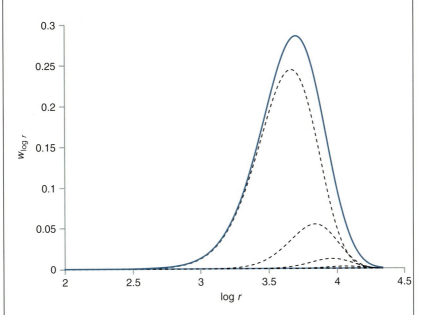

Figure E.6.8.1 CLD predicted for sample 2 in Table E.6.8.1 ($\lambda = 0.026$, PDI $= 2.13$).

6 Polyolefin Microstructural Modeling

Table E.6.8.1 Molecular weight and LCB averages for a series of polyethylene resins.

Sample	M_w	PDI	λ (LCBs per 1000 C)
1	94 000	2.00	0
2	77 000	2.13	0.026
3	82 000	2.20	0.037
4	86 000	2.23	0.042
5	96 000	2.45	0.080
6	79 000	2.50	0.090
7	68 000	2.80	0.190
8	70 000	3.10	0.330

Equation (6.109) can be used to estimate the values of the parameter α for each resin

$$\alpha = 1 - \frac{2}{\text{PDI}}$$

and Eq. (6.108) to get the values for τ_B

$$\tau_B = \frac{2}{\overline{r}_w} \times \frac{1+\alpha}{(1-\alpha)^2} = \frac{2}{\overline{M}_w/28(\text{g/mol})} \times \frac{1+\alpha}{(1-\alpha)^2}$$

Applying these equations to the data in Table E.6.8.1, we recover the model parameters shown in Table E.6.8.2, which can be used to calculate the CLDs of the individual populations and overall polymer using Eqs. (6.87), (6.96), and (6.102).

Table E.6.8.2 Model parameters for samples shown in Table E.6.8.1.

Sample	α	τ_B
1	94 000	2.00
2	77 000	2.13
3	82 000	2.20
4	86 000	2.23
5	96 000	2.45
6	79 000	2.50
7	68 000	2.80
8	70 000	3.10

The CLD distributions for samples 2 and 8, depicted in Figures E.6.8.1 and E.6.8.2, respectively, illustrate how the overall CLD is composed of polymer populations with different branching frequencies. This type of in-depth CLD × LCB analysis can only be achieved through the application of microstructural models such as the ones discussed in this section.

It is also illustrative to compare the LCB frequencies measured by ^{13}C NMR with the ones predicted by the model. According to Eq. (6.110)

$$\lambda = 500 \frac{\alpha}{\bar{r}_n}$$

Since (see footnote 13)

$$\tau = \frac{1}{\bar{r}_n} = \frac{\tau_B}{1+\alpha}$$

Then

$$\lambda_{predicted} = 500 \frac{\alpha \tau_B}{1+\alpha}$$

Figure E.6.8.3 shows that the agreement between measured and predicted LCB frequencies is very good for the sample set analyzed in this example. More interestingly, this figure shows that the model is capable to predict LCB averages from molecular weight information alone, confirming that its main assumptions are valid.

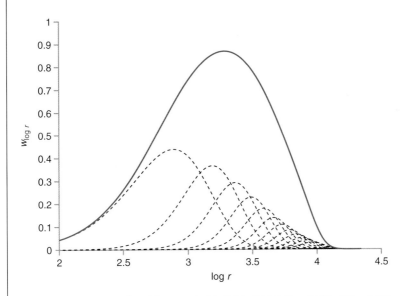

Figure E.6.8.2 CLD predicted for sample 8 in Table E.6.8.1 ($\lambda = 0.330$, PDI = 3.10).

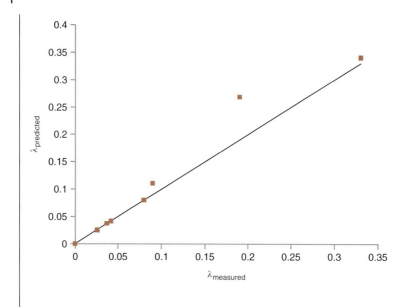

Figure E.6.8.3 LCBs per 1000 C measured × predicted for the samples shown in Table E.6.8.1.

Example 6.9: Prediction of Experimental MWD Profiles Measured by GPC Using Zimm–Stockmayer Correction Factors for the Radius of Gyration of Branched Molecules

The Zimm–Stockmayer correction factors for the radius of gyration of trifunctional, randomly branched polymer chains, are given by the equation

$$g = \frac{[\eta]_k}{[\eta]_0} = \left[\left(1 + \frac{k}{7}\right)^{1/2} + \frac{4n}{9\pi}\right]^{-1/2}$$

where $[\eta]_0$ and $[\eta]_k$ are the intrinsic viscosities of linear and branched chains, respectively.

We apply these correction factors to sample 8, listed in Table E.6.8.1, to predict its apparent CLD, as it would be measured by GPC. Table E.6.9.1 lists the correction factors for polymers chains with up to 11 LCBs per chain.

Since chains with the same hydrodynamic volume, $[\eta]MW$, elute at the same time from a GPC column, we can write

$$[\eta]_0 MW_0 = [\eta]_k MW_k$$

Table E.6.9.1 Zimm–Stockmayer correction factors for branched chains.

No. of LCBs	g	No. of LCBs	g
0	1.000	6	0.672
1	0.909	7	0.645
2	0.840	8	0.621
3	0.786	9	0.599
4	0.741	10	0.580
5	0.704	11	0.563

where MW_0 is the molecular weight measured by GPC using linear calibration standards and MW_k is the real molecular weight of polymer chains with k LCBs per chain. Therefore, the apparent MW_k of the branched chains is

$$MW_k = \frac{[\eta]_0}{[\eta]_k} MW_0 = \frac{MW_0}{g}$$

Figure E.6.9.1 compares the actual and apparent CLDs for sample 8 shown in the previous example. As expected, the apparent CLD is narrower and shifted toward lower averages, than the actual one, due to the size contraction experienced by the more highly branched polymer chains.

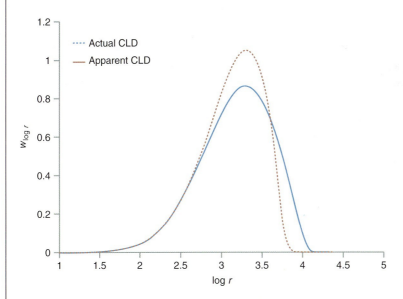

Figure E.6.9.1 Corrected and apparent CLDs for sample 8 in Table E.6.8.1.

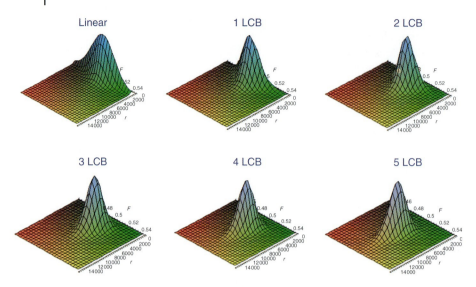

Figure 6.18 CLD × CCD × LCB trivariate distribution for a model polymer with $\tau_B = 2 \times 10^{-3}$, $\overline{F_A} = 0.5$, and $r_A r_B = 1.0$.

The most complete microstructural distribution for polyolefins containing LCBs is given by the trivariate CLD × CCD × LCB distribution

$$w_{r,F_A,k} = \frac{1}{(2k+1)!} r^{2k+1} \tau_B^{2k+2} e^{-r\tau_B} \sqrt{\frac{r}{2\pi\beta}} e^{-\frac{r(F_A - \overline{F_A})^2}{2\beta}} \tag{6.117}$$

which extends the Stockmayer distribution to nonlinear chains. Figure 6.18 shows CLD × CCD plots for chain populations with different number of LCBs. As the chains become more branched, their CCDs become narrower because chains with more LCBs also have higher chain length averages and, consequently, converge toward the average comonomer composition $\overline{F_A}$, as already seen when we discussed Stockmayer distribution for linear copolymers.

Figure 6.18 illustrates the inherent order and structural beauty of these branched polyolefins.

6.2.5
Polypropylene: Regio- and Stereoregularity

All the methods described above for modeling the microstructure of ethylene/α-olefin copolymers apply to propylene homopolymers and copolymers as well and do not need to be discussed any further.

Polypropylene, however, has an additional microstructural dimension characterized by its distributions of regio- and stereoregularity, as discussed in the first three chapters of this book. These distributions can be modeled in a way similar to that used to describe the CCD of polyolefins, if we assume that each monomer orientation is equivalent to a comonomer type. For instance, considering for simplicity that all insertions are regioregular (1-2, or primary insertions), we have only two possible orientations for the coordination of propylene molecules to the active sites: meso (m) or racemic (r). If all insertions are meso, a perfectly isotactic polypropylene will be produced; if they are all racemic, syndiotactic polypropylene will result; and, if they vary randomly throughout the polymer chain, atactic polypropylene will be synthesized. The m and r insertions can be treated as the copolymerization of comonomers A and B to predict the stereoregularity distribution of polypropylene using Bernoullian or Markov models of varying orders, where

$$k_{pAA} = k_{pBB} = k_{pm} \tag{6.118}$$

$$k_{pAB} = k_{pBA} = k_{pr} \tag{6.119}$$

where k_{pm} and k_{pr} are the propagation rate constants for meso and racemic insertions, respectively.

If regioselectivity also needs to be considered, the model becomes analogous to a multicomponent copolymerization problem with four comonomers: m_{1-2}, m_{2-1}, r_{1-2}, and r_{2-1}. No instantaneous distributions are available for these multicomponent problems, but they can be treated using Monte Carlo simulation, which is briefly described in the next section, or by population balances.

In the case of multiple-site catalysts, each site type will be characterized by its own regio- and stereoregularity parameters. We have seen in Chapter 3 how the regio- and stereoselectivity of Ziegler–Natta catalysts can be changed by the addition of internal and external electron donors. This effect is well understood qualitatively, but still no fundamental mathematical models are available to describe it quantitatively.

6.3
Monte Carlo Simulation

It is difficult to do justice to Monte Carlo methods in this short section, but this subject has such importance for polymer microstructural modeling that it could not be completely omitted from this book. Rather, we describe a few simple Monte Carlo examples to give the reader a flavor of this powerful modeling technique. We also hope to show that Monte Carlo models are relatively easy to develop and apply and that they provide a wealth of information that is very difficult to obtain by other methods.

One of the advantages of Monte Carlo simulation is that there is no need to solve differential or algebraic equations to predict the microstructures of polymers made in batch, semibatch, or continuous reactor, operated at steady state or dynamically.

At least in its simplest implementation forms,[14] as they are treated in this section, no knowledge of advanced mathematics is required to develop Monte Carlo models. What is required is a polymerization mechanism composed of several elementary steps (that we may also call *events*), probabilities associated to each of these steps, and a random number (RND) generator. Armed with these concepts, and a basic knowledge of programing techniques, we can develop Monte Carlo programs to describe a variety of polymerization systems, varying from very simple to rather complex.

We start by describing steady-state Monte Carlo models and then give a brief overview of dynamic Monte Carlo simulations.

6.3.1
Steady-State Monte Carlo Models

For illustration purposes, let us consider the simple polymerization mechanism described in Table 6.1: the monomer can either propagate or terminate by transfer to hydrogen. We have seen that the analytical solution for the CLD of these polymers is Flory distribution; it will serve as a check for the Monte Carlo model we are about to develop.

First, we need to define the probabilities for each event, propagation or termination, in the proposed mechanism. The probability of propagation is given by

$$P_p = \frac{R_p}{R_p + R_t} = \frac{k_p[M]}{k_p[M] + k_{tH}[H_2]} \tag{6.120}$$

Since there are only two events in the mechanism, $P_p + P_t = 1$, the chain transfer probability is given by

$$P_t = 1 - P_p \tag{6.121}$$

The propagation probability may also be related to the parameter τ, or the number average chain length, in Flory distribution by the equation

$$P_p = \frac{R_p}{R_p + R_t} = \frac{1}{1 + \dfrac{R_t}{R_p}} = \frac{1}{1 + \tau} = \frac{r_n}{r_n + 1} \tag{6.122}$$

Second, we need to define how many chains will be generated in the simulation; the higher the number of chains, the smoother the CLD will be, at the cost of longer computation times. The simulation starts with the length of the first polymer chain set to zero, $r = 0$; a RND is generated in the interval [0,1] to select the next event

---

14) Which may be inefficient from the point of view of computation time, but still generate detailed polymer microstructural information.

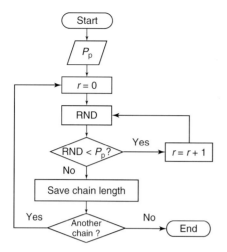

Figure 6.19 Flowchart for CLD Monte Carlo simulation.

that will take place in the simulation.[15] Propagation is selected if RND $< P_p$; the chain length is increased by one monomer unit, and another RND is generated to select the next event. This process is repeated until RND $\geq P_p$, when the chain is terminated. The length of the chain is stored in a CLD vector, and a new chain is started. The complete process is repeated until the desired number of chains is obtained. Figure 6.19 shows the flowchart for this simple model.

Figure 6.20 compares the CLDs predicted by Monte Carlo simulations using different number of polymer chains with the exact solution given by Flory distribution; the Monte Carlo predictions improve as more chains are simulated to filter out statistical variations.

The power of Monte Carlo modeling is apparent even in this trivial application. Since chains are generated one by one during the simulation, *all* microstructural information can be computed. In the example studied above, this amounted to only the CLD, but for more complex examples, Monte Carlo simulation becomes an attractive modeling approach. For instance, no analytical solution is available for the CCD of terpolymers that follow first- or higher-order Markov models. For these copolymers, the probabilities of propagation and termination will depend on both chain-end type and comonomer type. For instance, for chains terminated in monomer A

$$P_{pA} = \frac{R_{pA}}{R_{pA} + R_{tA}} = \frac{k_{pAA}[A] + k_{pAB}[B] + k_{pAC}[C]}{k_{pAA}[A] + k_{pAB}[B] + k_{pAC}[C] + K_{TA}} \quad (6.123)$$

$$P_{tA} = 1 - P_{pA} \quad (6.124)$$

where P_{pA} and P_{tA} are the probabilities of propagation and termination, respectively, for chains terminated in monomer A and K_{TA} is the chain transfer frequency for chains terminated in monomer A. Considering transfer to hydrogen, β-hydride

---

15) Most scientific software and spreadsheets come equipped with a RND generator.

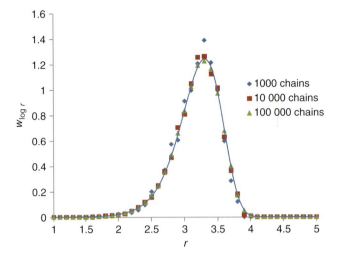

Figure 6.20 CLDs predicted by Monte Carlo simulation for a polymer with $r_n = 1000$ using different polymer population sizes. The continuous line is the exact solution given by Flory distribution.

elimination, transfer to cocatalyst, and transfer to monomer, the expression for K_{TA} is

$$K_{TA} = k_{tH,A}[H_2] + k_{t\beta,A} + k_{tAl,A}[Al] + \left(k_{tM,AA}[A] + k_{tM,AB}[B] + k_{tM,AC}[C]\right)$$
(6.125)

Analogous to Eq. (6.122) for homopolymerization, the probability of propagation for chains terminated in monomer A may be related to the average chain length of the copolymer by the expression

$$P_{pA} = \dfrac{\dfrac{R_{pA}}{\sum_j R_{pj}}}{\dfrac{R_{pA}}{\sum_j R_{pj}} + \dfrac{R_{tA}}{\sum_j R_{tj}}\dfrac{\sum_j R_{tj}}{\sum_j R_{pj}}} = \dfrac{\overline{F_A}}{\overline{F_A} + \dfrac{R_{tA}}{\sum_j R_{tj}}\dfrac{1}{r_n}} = \dfrac{\overline{F_A} r_n}{\overline{F_A} r_n + \rho_{tA}}$$
(6.126)

where $\sum_j R_{pj}$ and $\sum_j R_{tj}$ are the overall rates of propagation and chain transfer, respectively, and the parameter ρ_{tA}

$$\rho_{tA} = \dfrac{R_{tA}}{\sum_j R_{tj}}$$
(6.127)

is the fraction of chains that undergo chain transfer reactions when their last monomer unit is A. ^{13}C NMR chain-end analysis can be used to measure this parameter.

If propagation is selected (RND $< P_{pA}$), the type of monomer that will propagate is selected according to their propagation probabilities. For chains terminated in monomer A

$$P_{AA} = \frac{k_{pAA}[A]}{k_{pAA}[A] + k_{pAB}[B] + k_{pAC}[C]}$$
$$= \frac{k_{pAA}f_A}{k_{pAA}f_A + k_{pAB}f_B + k_{pAC}f_C} = \frac{f_A}{f_A + \dfrac{f_B}{r_{AB}} + \dfrac{f_C}{r_{AC}}} \qquad (6.128)$$

where

$$r_{AB} = \frac{k_{pAA}}{k_{pAB}} \qquad (6.129)$$

$$r_{AC} = \frac{k_{pAA}}{k_{pAC}} \qquad (6.130)$$

Similarly,

$$P_{AB} = \frac{\dfrac{f_B}{r_{AB}}}{f_A + \dfrac{f_B}{r_{AB}} + \dfrac{f_C}{r_{AC}}} \qquad (6.131)$$

$$P_{AC} = \frac{\dfrac{f_C}{r_{AC}}}{f_A + \dfrac{f_B}{r_{AB}} + \dfrac{f_C}{r_{AC}}} \qquad (6.132)$$

In general,

$$P_{ij} = \frac{\dfrac{f_j}{r_{ij}}}{\sum_j \dfrac{f_j}{r_{ij}}} \quad i = A,B,C \qquad (6.133)$$

$$r_{ij} = \frac{k_{pii}}{k_{pij}} \quad i = A,B,C \qquad (6.134)$$

Notice that because of the way we defined these probabilities, they must add up to one: $\sum_j P_{ij} = 1$.

Similar expressions can be obtained for the propagation probabilities of chains terminating in monomers B and C.

Finally, $\overline{F_A}$ (needed in Eq. (6.126) to calculate P_{pA} from average polymer properties) can be found from the reactor comonomer molar fractions and reactivity ratios using the Alfrey–Goldfinger equation for terpolymers[16]

$$\overline{F_A} : \overline{F_B} : \overline{F_C} = f_A \lambda_A : f_B \lambda_B : f_C \lambda_C \qquad (6.135)$$

16) The Alfrey–Goldfinger equation is an extension of the Mayo–Lewis equation for terpolymers.

where

$$\lambda_A = \left(\frac{f_A}{r_{CA}r_{BA}} + \frac{f_B}{r_{BA}r_{CB}} + \frac{f_C}{r_{CA}r_{BC}}\right) \times \left(f_A + \frac{f_B}{r_{AB}} + \frac{f_C}{r_{AC}}\right) \quad (6.136)$$

$$\lambda_B = \left(\frac{f_A}{r_{AB}r_{CA}} + \frac{f_B}{r_{AB}r_{CB}} + \frac{f_C}{r_{CB}r_{AC}}\right) \times \left(f_B + \frac{f_A}{r_{BA}} + \frac{f_C}{r_{BC}}\right) \quad (6.137)$$

$$\lambda_C = \left(\frac{f_A}{r_{AC}r_{BA}} + \frac{f_B}{r_{BC}r_{AB}} + \frac{f_C}{r_{AC}r_{BC}}\right) \times \left(f_C + \frac{f_A}{r_{CA}} + \frac{f_B}{r_{CB}}\right) \quad (6.138)$$

We can solve for the average comonomer molar fractions in the copolymer by rearranging Eq. (6.135)

$$\overline{F_A} = \left(1 + \frac{\lambda_B f_B + \lambda_C f_C}{\lambda_A f_A}\right)^{-1} \quad (6.139)$$

$$\overline{F_B} = \left(1 + \frac{\lambda_A f_A + \lambda_C f_C}{\lambda_B f_B}\right)^{-1} \quad (6.140)$$

$$\overline{F_C} = 1 - \overline{F_A} - \overline{F_B} \quad (6.141)$$

The flowchart for a Monte Carlo program that describes olefin terpolymerization with the terminal model is presented in Figure 6.21.

We first test our algorithm with a binary copolymer, for which we have the CCD analytical solution given by the Stockmayer distribution. Figure 6.22 shows that the agreement between the CCDs predicted with the Monte Carlo and the analytical solution is excellent when 5×10^5 chains are simulated, indicating that the algorithm is accurate.[17]

Extending the model to terpolymerization, Figure 6.23 shows the CCD for comonomers B and C of a model terpolymer (the model parameters are listed in the caption). The CCD for monomer A is obtained from the difference of these two and was omitted for clarity. We have also generated 5×10^5 chains to obtain these CCDs. No general analytical solution is possible for the CCD of this terpolymer.

Since the Monte Carlo program can be designed to record all information on the polymer chain architecture, other microstructural details may also be recovered, such as the joint comonomer CCD, shown in Figure 6.24, or the CSLDs.

For multiple-site catalysts, similar simulations could be repeated for each site type to generate microstructural distributions for the whole polymer.

17) This is another useful application of Monte Carlo simulation: it can be used to confirm analytical solutions for microstructural distributions and vice versa. Since Monte Carlo programs are essentially "math free" and depend only on logic, they provide very good confirmation for results that may involve complex derivations.

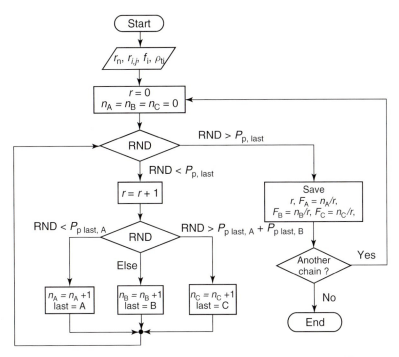

Figure 6.21 Flowchart for Monte Carlo simulation of terpolymerization with the terminal model. The variable *last* keeps track of the last monomer added to the growing polymer chain.

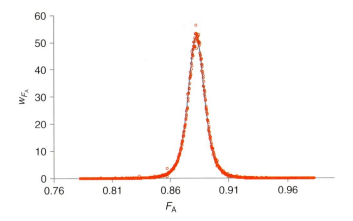

Figure 6.22 Monte Carlo simulation (dots) for the CCD of a polymer with $r_A = 5.0$, $r_B = 0.2$, $f_A = 0.6$, and $r_n = 1 \times 10^3$ ($\overline{F_A} = 0.882$). The continuous line is the CCD predicted by Stockmayer distribution.

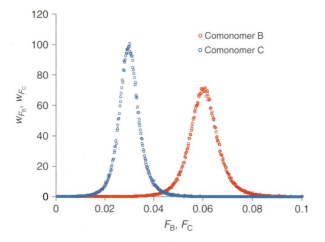

Figure 6.23 Monte Carlo simulation for the CCD of a polymer with $r_{AB} = 5.0$, $r_{AC} = 10$, $r_{BA} = 0.2$, $r_{BC} = 2.0$, $r_{CA} = 0.1$, $r_{CB} = 0.5$, $f_A = 0.6$, $f_B = f_C = 0.2$, $\rho_{tA} = \rho_{tB} = 0.33$, $\rho_{tC} = 0.34$, and $r_n = 1 \times 10^3$ ($\overline{F_A} = 0.9091$, $\overline{F_B} = 0.0606$, and $\overline{F_C} = 0.0303$).

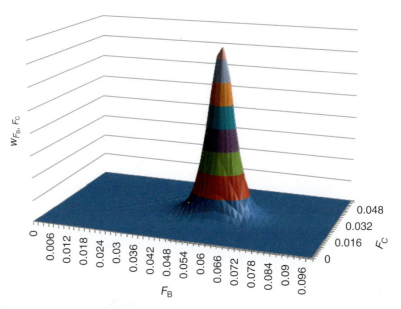

Figure 6.24 Monte Carlo simulation for the joint $F_B \times F_C$ CCD of the terpolymer shown in Figure 6.23 ($\overline{F_A} = 0.9091$, $\overline{F_B} = 0.0606$, and $\overline{F_C} = 0.0303$).

Example 6.10 shows another interesting application of the Monte Carlo methods for the simulation of the branching structure of polyethylenes made with Ni-diimine catalysts that undergo chain walking during the polymerization.

Example 6.10: Monte Carlo Simulation of Polyethylene SCB Distribution Formed by Chain Walking

In Chapter 3, we discussed how polyethylene with SCBs could be produced by polymerizing only ethylene with Ni-diimine catalysts because of the chain-walking mechanism. In these polymerizations, the active site "walks" away from the chain end through an isomerization step that starts with a β-hydride elimination step followed by a reinsertion; if a monomer insertion takes place when the catalyst is coordinated to a nonterminal carbon, an SCB is formed, with the length equal to the number of carbon atoms between the catalyst location and the nearest chain end in the polymer.

Monte Carlo techniques are well suited to describe these systems since the SCB distribution results from a one-dimensional random (or near random) walk process along the polymer backbone. A simplified flowchart for the model is shown in Figure E.6.10.1.

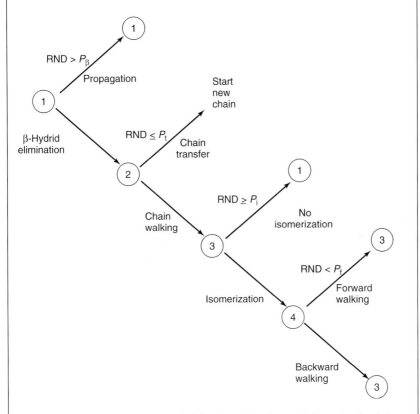

Figure E.6.10.1 Monte Carlo simplified flowchart describing SCB formation by chain walking.

First, the model decides whether propagation or β-hydride elimination takes place. If propagation happens, the chain grows by one unit, and the program returns to step (1); if β-hydride is selected, the next event may be chain transfer or chain walking. In the case of chain transfer, the chain length and SCB details are stored and a new chain is started; in the case of chain walking, isomerization may take place. When the isomerization event is selected, the program needs to decide whether the active site will move forward or backward along the backbone.

For simplicity, we have omitted from the flowchart the probability for methyl SCB favoring, P_m, which accounts for the preference that late transition metal catalysts have for the formation of methyl branches. In essence, this probability ensures that the active site is more likely to stay on the β-carbon than in any other nonterminal carbon atom in the backbone.

This set of five probabilities (P_m, P_β, P_t, P_i, and P_f) uniquely define the polymerization in the Monte Carlo model. Figure E.6.10.2 shows that the model fits well the experimental SCB distributions of two polyethylenes made with a Ni-diimine catalyst. The model probabilities for the two resins are shown in Table E.6.10.1.

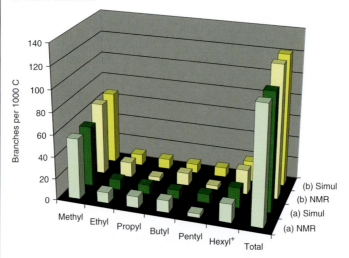

Figure E.6.10.2 Experimental and Monte Carlo SCB distributions. Model parameters are shown in Table E.6.10.1.

Figure E.6.10.2 demonstrates that the Monte Carlo model describes adequately the ^{13}C NMR results for these two polymers. More interestingly, the model can tell us much more about the polymer microstructure than can be measured by ^{13}C NMR. The model can also be applied to predict the relative position of SCBs in the chains, following the convention illustrated for methyl branches in Figure E.6.10.3, using the same parameters estimated to fit the experimental data shown in Figure E.6.10.2.

Table E.6.10.1 Monte Carlo model probabilities.

Probabilities	Sample (a)	Sample (b)
P_t	8.85×10^{-4}	7.78×10^{-4}
P_β	0.226	0.257
P_i	0.940	0.940
P_f	0.700	0.700
P_m	0.797	0.814

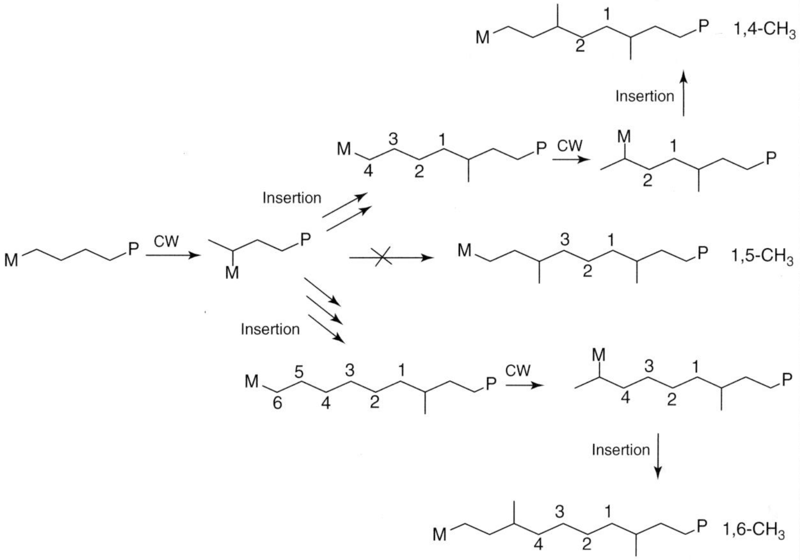

Figure E.6.10.3 Relative SCB positions for methyl branches.

Figure E.6.10.4 shows the distribution of relative SCB positions up to decyl+ (C_{10} and longer branches) for sample (a). The decyl+ limit is arbitrary; we could have kept track of SCBs of any length in the simulation, but their frequency decreases fast as they become longer. Since SCBs result from the random or near random walk (the β-position is favored) of the catalyst site along the backbone, the likelihood of forming SCBs decreases with increasing branch size. We can come to this conclusion intuitively since there is no reason why a catalyst molecule should move preferentially in one direction along the backbone, particularly when it is a few carbon chain atoms removed from the terminal carbon and is no longer influenced by chain-end effects.

It is generally only possible to obtain this type of detailed microstructural information by Monte Carlo simulation; it is, unquestionably, the most powerful modeling tool for linking polymerization mechanisms to polymer microstructures.

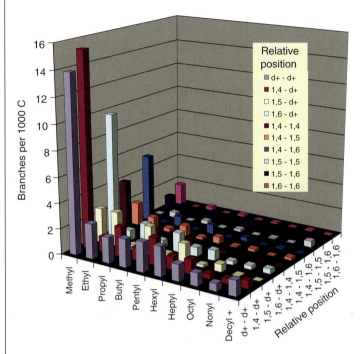

Figure E.6.10.4 Monte Carlo SCB distribution for relative SCB positions in sample (a). Model parameters are shown in Table E.6.10.1.

6.3.2
Dynamic Monte Carlo Models

If the polymerization conditions vary in time, the Monte Carlo probabilities will also change, and the method discussed in Section 6.3.1 is valid only instantaneously. We could still adopt a procedure similar to the one used for instantaneous distributions in Example 6.1, with probabilities varying as a function of polymerization time, but this approach may be too time consuming since each Monte Carlo simulation may take several minutes to hours to be completed.

For dynamic simulations, it is more convenient to use a Monte Carlo method that allows for the variation of the model probabilities as the conditions change during the polymerization. The procedure explained in this section requires the selection of a suitable simulation volume containing a certain number of reactant molecules (such as monomer, hydrogen, catalyst, and cocatalyst), the transformation of

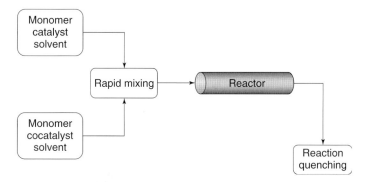

Figure 6.25 Stopped-flow reactor.

experimental reaction rate constants to Monte Carlo rate constants, and the selection of the several reaction steps using randomly generated numbers. We summarize the basic ideas behind this method below; further information can be found in the references discussed in Further Reading.

We develop a transient model for homopolymerization in a stopped-flow reactor in the remaining of this section. A stopped-flow reactor is a tubular reactor with very short residence time (in the order of hundredths to a couple of seconds) and is used for mechanistic studies on olefin polymerization. The reactor is operated under non-steady-state conditions, generally under varying monomer and polymer species concentrations. Since residence time in the reactor is comparable to, or shorter than, the average lifetime of the polymer chains, the polymer CLD does not get fully established and cannot be described with Flory distribution. Typically, PDI is smaller than two if a single-site catalyst is used. Figure 6.25 shows a schematic for a stopped-flow reactor. A solution (or slurry) containing monomer and catalyst is rapidly mixed with another solution, composed of monomer and cocatalyst, before flowing through a tubular reactor where most of the polymerization takes place. The polymerization occuring in the reactor effluent is rapidly quenched by mixing it with an acidified alcohol solution. As the reactive mixture flows through the reactor, monomer concentration decreases and the concentrations of monomer-free sites and growing and dead chains of different lengths vary, constituting an ideal example for dynamic Monte Carlo simulation.

We again use the simple mechanism illustrated in Table 6.1 to develop the dynamic Monte Carlo method in this section, assuming that the site activation reaction between catalyst precursor and cocatalyst is instantaneous.

For a generic reaction, Monte Carlo reaction rates and rate constants, R^{MC} and k^{MC}, respectively, are defined as

$$R^{MC} = k^{MC} N_C \tag{6.142}$$

where N_C is the number of unique combinations between reactant molecules. The Monte Carlo reaction rates and reaction constants have units of reciprocal time; they are, in fact, frequencies proportional to molecular collisions in the control volume.

The number of combinations between reactant molecules depends on the type of reaction; for bimolecular reactions between different molecules, N_C is equal to the product of the number of molecules of each type participating in the reaction. For instance, for propagation reactions

$$N_C = n_M n_P \tag{6.143}$$

where n_M is the number of monomer molecules and n_P is the number of living chains in the reactor at a given time.

The number of type i molecules in the simulation volume V is calculated from the concentration of the reactant i, C_i, in the reactor

$$n_i = C_i V N_A \tag{6.144}$$

where N_A is the Avogadro number.

The size of the simulation volume determines how smooth (that is, how free of stochastic noise) the results for a given simulation will be, but otherwise it does not affect the simulation. The simulation volume should be large enough to generate statistically valid results.

The Monte Carlo rate constant for a bimolecular reaction between different reactants is given by

$$k^{MC} = \frac{k}{V N_A} \tag{6.145}$$

This transformation is used to eliminate the units of volume from k^{MC} since we are now dealing with number of molecules in the simulation volume, not reactor concentrations.

For a polymerization model that involves site initiation (first monomer insertion, assumed to have the same kinetic constant as propagation, k_p), monomer propagation, and chain transfer, the three Monte Carlo rates are defined by the following equations

$$R_i^{MC} = \frac{k_p}{V N_A} n_M n_{C*} \tag{6.146}$$

$$R_p^{MC} = \frac{k_p}{V N_A} n_M n_P \tag{6.147}$$

$$R_t^{MC} = \frac{k_{tH}}{V N_A} n_H n_P \tag{6.148}$$

where n_{C*} and n_H are the number of monomer-free catalyst sites and hydrogen molecules in the reactor, respectively.

The overall rate of reaction is, therefore

$$R^{MC} = R_i^{MC} + R_p^{MC} + R_t^{MC} \tag{6.149}$$

The algorithm for the dynamic Monte Carlo program requires the generation of two random numbers, RND_1 and RND_2, for each reaction step. The first number is used to determine the time elapsed between two consecutive reactions and is given by the expression

$$\Delta t = \frac{1}{R^{MC}} \ln \frac{1}{RND_1} \qquad (6.150)$$

The derivation of Eq. (6.150) was shown by Gillespie in the reference given in Further Reading.

The total reaction time at iteration j is calculated with the equation

$$t_j = t_{j-1} + \Delta t_{j-1} \qquad (6.151)$$

The next simulation step selects the reaction that will take place: this is done by multiplying RND_2 by the overall rate of reaction and comparing this result with the rates for each reaction in the mechanism

Initiation: $\qquad RND_2 R^{MC} \leq R_i^{MC}$

Propagation: $\qquad R_i^{MC} < RND_2 R^{MC} \leq R_i^{MC} + R_p^{MC}$

Transfer: $\qquad R_i^{MC} + R_p^{MC} < RND_2 R^{MC} \leq R_i^{MC} + R_p^{MC} + R_t^{MC}$

Starting with the initial concentration of active sites, monomer, and chain transfer agent, we calculate the initial number of each molecule type in the simulation volume using Eq. (6.144). The following actions should be carried out when one of the elementary steps described in Eqs. (6.146–6.148) is selected:

- **Initiation:** Decrease the number of molecules of active sites ($n_{C^*} = n_{C^*} - 1$) and increase the number of living polymer chains ($n_P = n_p + 1$) by one. Set the length of this chain to one ($r = 1$).
- **Propagation:** Randomly select one living chain in the control volume using another random number, RND_3, increase its length ($r = r + 1$), and decrease the number of monomer molecules ($n_M = n_M - 1$).
- **Chain transfer:** Randomly select one living chain and store its chain length in the chain length vector for dead chains, decrease the number of living chains ($n_p = n_p - 1$), decrease the number of hydrogen molecules ($n_H = n_H - 1$), and increase the number of monomer-free active sites ($n_{C^*} = n_{C^*} + 1$).

The computation steps described above are repeated until the simulation time reaches the required residence time in the reactor. Notice that it is not necessary to assume that the concentrations of monomer and hydrogen are constant, as required when deriving Flory distribution or using the steady-state Monte Carlo models of the previous section.

Figure 6.26 describes schematically this dynamic Monte Carlo procedure.

Figure 6.27 tracks how the CLD changes as a function of time inside a stopped-flow reactor (the same results would be obtained for batch polymerization in an autoclave reactor). It is fascinating to see how the CLD for the living polymer "moves forward" in the time dimension, leaving behind a trail of dead polymer chains in its wake.

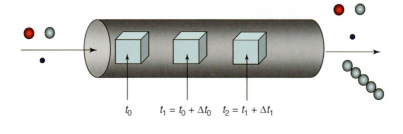

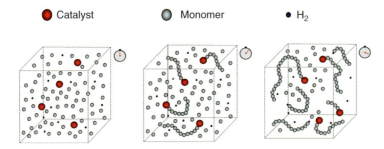

Figure 6.26 Schematic for the simulation of a stopped-flow reactor with a dynamic Monte Carlo model.

Incidentally, if the monomer concentration can be assumed constant, these CLDs have analytical solutions given by

$$f_r^L = \frac{P_p^{r-1}}{1-e^{-\psi t}}\left\{P_t\left[1-\frac{\Gamma(r,\psi t)}{\Gamma(r)}\right]+P_p\frac{(\psi t)^r}{r!}e^{-\psi t}\right\} \quad (6.152)$$

$$f_r^D = \frac{P_p^{r-1}}{\psi t + e^{-\psi t} - 1}$$
$$\times \left\{ P_t\left[\psi t - r + \frac{\Gamma(r-1,\psi t)(1-r)(\psi t - r) + e^{-\psi t}(\psi t)^{r-1} r}{\Gamma(r)}\right] + P_p\left[1-\frac{\Gamma(r,\psi t)}{\Gamma(r)}\right] \right\} \quad (6.153)$$

where

$$\psi = k_p[M] + k_{tH}[H_2] \cong k_p[M] \quad (6.154)$$

$$P_p = \frac{k_p[M]}{\psi} \quad (6.155)$$

$$P_t = \frac{k_{tH}[H]}{\psi} \quad (6.156)$$

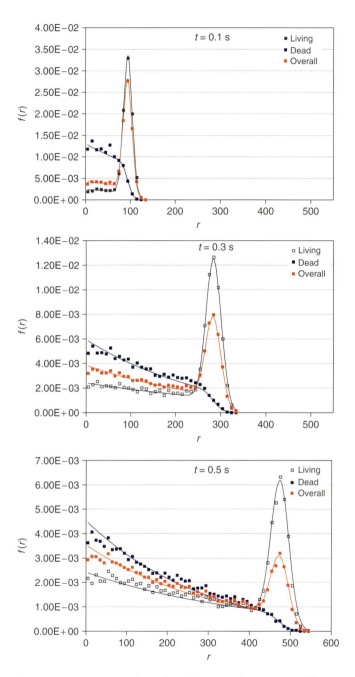

Figure 6.27 CLD time evolution for polymers made in a stopped-flow reactor. Model parameters: $k_p = 3800$ l (mol s)$^{-1}$, $k_H[H_2] = 2.3$ s^{-1}, $[M] = 0.25$ mol l^{-1}, and $n_P = 1 \times 10^4$. The continuous lines are analytical results, and the square dots are Monte Carlo predictions.

and the gamma and complementary gamma functions are defined as

$$\Gamma(r) = \int_0^\infty e^{-x} x^{r-1} dx \qquad (6.157)$$

$$\Gamma(r, \psi t) = \int_{\psi t}^\infty e^{-x} x^{r-1} dx \qquad (6.158)$$

The derivations of these solutions are beyond the scope of this book, but their results are also shown for comparison in Figure 6.27.

Further Reading

A short overview of polymer reaction engineering that includes the history of several instantaneous distributions described in this chapter was written by Ray, W.H., Soares, J.B.P., and Hutchinson, R.A. (2004) *Macromol. Symp.*, **206**, 1.

Applications of these instantaneous distributions to polymer characterization techniques were reviewed by Soares, J.B.P. (2007) *Macromol. Symp.*, **257**, 1.

Polyolefin MWD deconvolution was originally proposed by Vickroy, V.V., Schneider, H., and Abbott, R.F. (1993) *J. Appl. Polym. Sci.*, **50**, 551 and has been widely used by several groups to describe polymers made with heterogeneous Ziegler–Natta catalysts.

Soares, J.B.P. and Hamielec, A.E. (1995) *Polymer*, **36**, 2257.

Kissin, Y.V., Mink, R.I., Nowlin, T.E., and Brandolini, A.J. (1999) *Top. Catal.*, **7**, 69.

Kissin, Y.V., Rishina, L.A., and Vizen, E.I. (2002) *J. Polym. Sci., Part A: Polym. Chem.*, **40**, 1899.

Kissin, Y.V., Mirabella, F.M., and Meverden, C.C. (2005) *J. Polym. Sci., Part A: Polym. Chem.*, **43**, 4351.

Broad Flory distributions for MWD deconvolution were introduced by Soares, J.B.P. (1998) *Polym. React. Eng.*, **6**, 225.

In the same reference, an extension to CCD modeling with broad Stockmayer distributions is suggested.

The use of Stockmayer distribution to describe the CCD of polyolefins made with multiple-site was first proposed catalyst by Soares, J.B.P. and Hamielec, A.E. (1995) *Macromol. Theory Simul.*, **4**, 305, and later applied to Ziegler–Natta catalysts, da Silva Filho, A.A., de Galand, G.B., and single-site catalysts, Soares, J.B.P. (2000) *Macromol. Chem. Phys.*, **201**, 1226.

Sarzotti, D.M., Soares, J.B.P., and Penlidis, A. (2002) *J. Polym. Sci., Part B: Polym. Phys.*, **40**, 2595.

CCD deconvolutions considering the TREF soluble fraction were introduced by Al-ghyamah, A.A. and Soares, J.B.P. (2009) *Macromol. Rapid Commun.*, **30**, 384.

GPC-FTIR profiles of polymers made with multiple-site catalysts were first modeled by Soares, J.B.P., Abbott, R.F., Willis, J.N., and Liu, X. (1996) *Macromol. Chem. Phys.*, **197**, 3383 and were later used for combinations of metallocene resins.

Faldi, A. and Soares, J.B.P. (2001) *Polymer*, **42**, 3057.

Mathematical models for the CSLD of copolymers made with multiple-site catalysts have been extensively studied by Cheng, H.N. (1989) *Computer Applications in Applied Polymer Science II*, ACS, p. 174; Cheng, H.N. (1990) *Polym. Bull.*, **23**, 589; Cheng, H.N. (1991) *Polym. Bull.*, **26**, 325; Cheng, H.N. (1993) *New Advances in Polyolefins*, Plenum Press, New York, p. 15.

A new approach combining MWD and CSLD deconvolution has been proposed by Al-Saleh, M., Soares, J.B.P., and Duever, T.A. (2010) *Macromol. React. Eng.*, **4**, 578; (2011) *Macromol. React. Eng.*, **5**, 587.

An analytical solution for the longest ethylene sequence per chain for ethylene/α-olefin copolymers, including some Monte Carlo studies, has been published by Costeaux, S., Anantawaraskul, S., Wood-Adams, P.M., and Soares, J.B.P. (2002) *Macromol. Theory Simul.*, **11**, 326.

The analytical solution for the CCD of random multicomponent copolymers is also available in the literature.

Anatawaraskul, S., Soares, J.B.P., and Wood-Adamns, P. (2003) *Macromol. Theory Simul.*, **12**, 229.

Instantaneous LCB distributions for polyolefins made with single-site catalysts were derived by Soares, J.B.P. and Hamielec, A.E. (1996) *Macromol. Theory Simul.*, **5**, 547; Soares, J.B.P. and Hamielec, A.E. (1997) *Macromol. Theory Simul.*, **6**, 591, including full derivations not covered in this chapter. An extensive review of LCB models, including Monte Carlo applications, was given by Soares, J.B.P. (2004) *Macromol. Mater. Eng.*, **289**, 70.

Read and Soares have also shown how the LCB distribution for polymers made with two catalysts was described with an analytical solution that could be used to predict rheological properties.

Read, D.J. and Soares, J.B.P. (2003) *Macromolecules*, **36**, 10037.

Instantaneous distributions have been used to model the microstructures of polymers made with a combination of two or more single-site catalysts. These systems provide good microstructural control and are useful as models for the behavior or multiple-site Ziegler–Natta and Phillips catalysts. These applications were not covered in this chapter (except in passing, in Figure 6.5) but can be found in the literature for MWD (Kim, J.D., Soares, J.B.P., and Rempel, G.L. (1998) *Macromol. Rapid Commun.*, **19**, 197; Kim, J.D. and Soares, J.B.P. (2000) *J. Polym. Sci., Part A: Polym. Chem.*, **38**, 1408; Kim, J.D. and Soares, J.B.P. (2000) *J. Polym. Sci., Part A: Polym. Chem.*, **38**, 1417), CCD (Kim, J.D. and Soares, J.B.P. (2000) *J. Polym. Sci., Part A: Polym. Chem.*, **38**, 1427), LCB (Beigzadeh, D., Soares, J.B.P., and Duever, T.A. (1999) *Macromol. Rapid Commun.*, **20**, 541) control.

Steady-state Monte Carlo models to describe the formation of LCBs by macromonomer incorporation (Beigzadeh, D., Soares, J.B.P., and Penlidis, A. (1999) *Polym. React. Eng.*, **7**, 195), and of SCBs by chain walking (Simon, L.C., Soares, J.B.P., and de Souza, R.F. (2000) *AIChE J.*, **45**, 1234; Simon, L.C., Soares, J.B.P., and de Souza, R.F. (2001) *Chem. Eng. Sci.*, **56**, 4181) have also been published in the literature. An analytical solution, not discussed in this chapter, has also been developed to model SCB formation by chain walking.

Soares, J.B.P., Simon, L.C., and de Souza, R.F. (2001) *Polym. React. Eng.*, **9**, 199.

The fine molecular structure of LCB polymers made with two single-site catalysts simultaneously, including the classification of the LCB populations in families subdivided according to their branching topology, was described by Simon, L.C. and Soares, J.B.P. (2002) *Macromol. Theory Simul.*, **11**, 184; Simon, L.C. and Soares, J.B.P. (2005) *Ind. Eng. Chem. Res.*, **44**, 2461.

Haag *et al.* showed how a Monte Carlo model could be used to describe the microstructure of a thermoplastic polyolefin elastomers made with two metallocene catalysts.

Haag, M.C., Simon, L.C., and Soares, J.B.P. (2003) *Macromol. Theory Simul.*, **12**, 142.

The dynamic Monte Carlo method we introduced in this chapter was proposed by Gillespie, D.T. (1977) *J. Phys. Chem.*, **81**, 2340 and applied to stopped-flow reactors for single- and multiple-site catalysts by Soares, J.B.P. and Hamielec, A.E. (2007) *Macromol. React. Eng.*, **1**, 53; Soares, J.B.P. and Hamielec, A.E. (2008) *Macromol. React. Eng.*, **2**, 115.

A modification of the same model, which accounts for variation of model probabilities from site to site within the same site type (similar to using broad Flory distributions) was proposed by Soares, J.B.P. and Nguyem, T. (2007) *Macromol. Symp.*, **260**, 189.

7
Particle Growth and Single Particle Modeling

> We can't solve problems by using the same kind of thinking we used when we created them.
> *Albert Einstein (1879–1955)*

7.1
Introduction

As we have already discussed in the previous chapters, the attractiveness of polyolefins is based in large part on both the simplicity of the raw materials and the fact that they can be assembled to give very useful final products with a tremendous range of properties. One of the reasons for this diversity in product structures is that the catalysts used to make polyolefins are carefully designed to reach specific polymer yields, polymerization rate profiles, and molecular architecture. As reaction engineers, the major part of our job is to ensure that these carefully designed catalysts are exposed to the reaction conditions required to achieve their desired performance. To a certain extent, this is done through informed reactor design and operation and judicious catalyst selection and formulation. However, once the reactor is built and the catalyst is chosen, the only means that we have to control the productivity of our system and the polymer properties is by acting on parameters associated with the bulk properties of the reactor such as polymerization temperatures, monomer pressures, and reactant flow rates. As shown in Figure 7.1, a supported catalyst particle will act as a filter between the bulk properties of the reactor and the actual conditions under which the polymer is made at the active sites. Thus, if we are to develop a quantitative description of the relationship between reactor conditions and polymer properties, it is necessary to understand how the catalyst particle behaves and changes during polymerization.

The act of supporting active sites on the surfaces of magnesium dichloride ($MgCl_2$) or silica (SiO_2) will obviously have an impact on how the catalyst functions, generally affecting catalyst activity and polymer microstructure, as discussed in Chapter 3. Chemical interactions between the support and the transition metal sites are still poorly understood and their discussion is beyond the scope of this book but has been extensively discussed in an excellent book edited by Chadwick

Polyolefin Reaction Engineering, First Edition. João B. P. Soares and Timothy F. L. McKenna.
© 2012 Wiley-VCH Verlag GmbH & Co. KGaA. Published 2012 by Wiley-VCH Verlag GmbH & Co. KGaA.

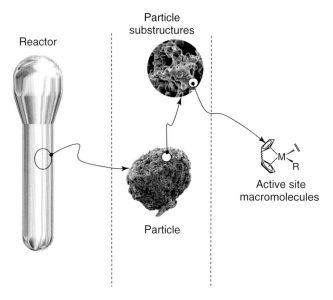

Figure 7.1 The particle as a filter between the bulk conditions in the reactor and the active sites. The particle structure will have an impact on the concentration of reactants and temperature at the active sites, and thus on the properties of the polymer.

and Severn [1]. In addition to the effects supporting may have on the behavior of the active sites, one must also deal with transport resistances that are encountered in all heterogeneously catalyzed reactions. It is this aspect that we are addressing in this chapter.

When the active sites are situated inside a catalyst particle, monomer must be transported through the continuous phase of the reactor into and through the boundary layer surrounding the particles, then into and through the particle pore space, and finally through the polymer phase surrounding the active sites before the polymerization takes place. Conversely, any heat generated at the active sites should be transported in the opposite direction. Each of these steps will be subject to resistances, which might be significant. If mass transfer resistances cannot be neglected, the concentrations of the transported species will not necessarily be the same inside and outside of the particles. Any resistances to energy transport from the particle could lead to particle overheating, thermal deactivation of the catalyst, loss of reactor control, and eventually production of molten polymer. If mass and/or heat transfer resistances cannot be neglected, concentration and/or temperature gradients will form inside of the particles, as illustrated in Figure 7.2. If these gradients are significant, then local reaction rates and polymer properties will vary inside the growing particles. One of the objectives of this chapter is to discuss how we can develop mathematical models to predict polymerization rates, concentration, and temperature gradients in order to use the models presented in Chapters 5 and 6 to predict polymer microstructural distributions.

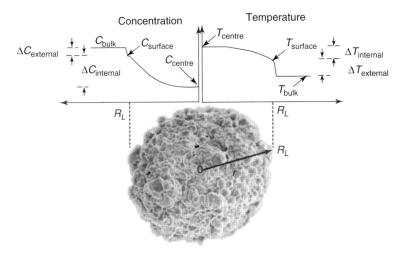

Figure 7.2 Schematic representation of the potential temperature and concentration gradients that need to be considered when dealing with heterogeneous systems. In particular, we have to consider eventual concentration and temperature gradients in the boundary layers around the particle (external), as well as internal gradients.

Important steps in model building include

1) identification of the model purpose;
2) identification of the model mechanisms;
3) definition of a physical framework for the model that defines the geometry of the problem;
4) identification of the mathematical expressions that describe the model mechanisms in the proper coordinate system;
5) exploration of how the model can be simplified without compromising any key modeling objective.

The purpose of the models discussed in this chapter is to describe polymer particle growth and to use our understanding of how the particle grows to develop a model that helps us calculate external and internal temperature and concentration gradients as a function of time and polymerization conditions. The mechanisms involved are heat transfer and mass transfer phenomena coupled to polymerization reactions. Before writing the actual mathematical expressions, we begin our discussion by looking at the initial geometry or morphology of the catalyst particles and at how they evolve in the reactor during the polymerization. Particle morphology evolution involves two steps, namely, fragmentation and growth. We then use these concepts to propose mathematical expressions for a generalized single particle model (SPM), the Multigrain Model (MGM), and its simpler derivative the polymer flow model (PFM). We also discuss the necessary simplifying assumptions required to set up a mathematical model linking bulk reactor conditions, polymer particle structural characteristics, and polymerization rate. Finally, we also look at the impact of some of these simplifying assumptions have on model validity.

7.2
Particle Fragmentation and Growth

Supported catalyst particles are highly porous and typically have diameters in the order of 10–100 μm, depending on the type of polymerization process for which they are intended. While interest in other support types is growing, $MgCl_2$ and SiO_2 are essentially the only commercially used supports at present. The catalyst particle shown in Figure 7.3a is a typical example of an Ziegler–Natta catalyst, $TiCl_4$

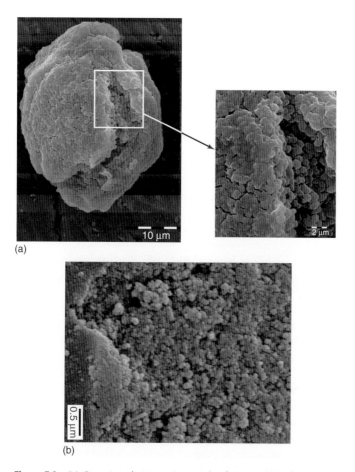

Figure 7.3 (a) Scanning electron micrograph of a typical Ziegler–Natta catalyst particle. Close up on the right: the roughly spherical particle is a porous assembly of smaller, spherical grains. (b) A scanning electron micrograph of a silica support; note that the spherical substructures are much smaller in the case of this particular silica support.

supported on an $MgCl_2$ support. This image shows us that the supported catalyst particle (macrograin or macroparticle) is composed of an assembly of smaller structures, often referred to as *micrograins* (also called *microparticles*). Figure 7.3b shows that similar observations can be made for silica supports. Even though SiO_2 has different material properties from $MgCl_2$ and is manufactured in a different way, the same macrograin/micrograin structure is often used to describe the structure of support particles made from it. In the rest of this chapter, we initially focus on particle morphology with these two levels of organization, macrograins and micrograins, which is at the core of the MGM that is discussed below. It has been suggested that the micrograins seen in Figure 7.3 are themselves organizations of even smaller structures. However, it would appear that using an extra level of spatial organization does nothing more than renders the mathematical description of these systems even more complex without providing any extra information about particle growth and polymer properties.

Experimental observations such as these tell us that the macroscopic structure of most catalyst particles is (roughly) spherical, and that we probably need to account for at least two levels of spatial organization in our model-building exercise. However, this is only the beginning, since the structure of the particles changes once the supported catalysts are injected into the reactor and polymer starts to form on the surface of the micrograins, as explained in the following sections.

7.2.1
The Fragmentation Step

In an ideal polymerization process, one catalyst particle is transformed into a much larger polymer particle with a diameter on the order of several hundred microns up to 1 or 2 mm. This transformation is complex to say the least, and begins with a step that is referred to as *particle fragmentation*.

When the particles are injected into the reactor, monomer, hydrogen, and any other species present in the continuous phase of the reactor begin to diffuse through the boundary layer surrounding the particle and into the particle pores. As soon as the reactive species reach the active sites, they start to react, forming polymer layers inside the pores of the catalyst particle, and the structure, or morphology, of the particle begins to evolve.

For a short period of time (on the order of 10^{-1} to 10^2 s, depending on the catalytic system in question and the nature of the support), the particle remains as a continuous inorganic phase with polymer chains growing on active sites located on the surface of its pores. As shown in Figure 7.4, as polymer accumulates at the active sites, the inorganic phase suffers a local buildup of stress at different points, and very quickly fragments into a series of unconnected mineral substructures held together by a polymer phase. This process continues throughout the entire support as monomer keeps reaching the active sites and polymer builds up. The end result of the fragmentation step is that the original two-phase structure (continuous solid support and pore space) is converted into a three-phase structure composed of micrograins of the original support, on which are situated the active

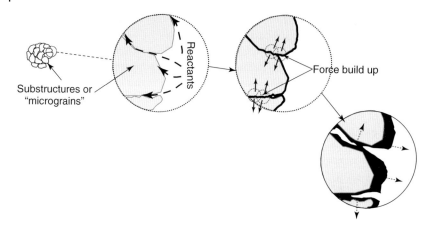

Figure 7.4 Particle morphology evolution and growth: monomers diffuse into the pores where they form polymer (black); this builds up stress at the weak points holding two micrograins together; fragmentation occurs at the weak point when the hydraulic forces pushing the neighboring grains apart exceed the intrinsic strength of the support material (gray).

sites, a continuous polymer phase that covers the micrograins, and pores through which the reactants are transported. New polymer is continually generated at the active sites covered with older material, and the pressure caused by the formation of this new material causes the particle to continue to expand. This process of expansion continues until the particles are removed from the reactor or the catalyst is completely deactivated. Because we can produce 50–100 kg of polymer per gram of catalyst (in 1–2 h), the catalyst fragments are not recovered after the end of the polymerization. Ideally, the fragmentation step proceeds in a controlled manner and a single supported catalyst particle is transformed into one polymer particle with the same overall form. This is referred to as the *replication phenomenon*: the particle size distribution of the polymer particles at the end of batch or semibatch polymerization closely approximates the particle size distribution of the catalyst at the beginning of polymerization. Good replication is assumed to occur when there is an adequate balance between the mechanical strength of the particle and catalytic activity.

For the polymerization to proceed, monomer needs to be transported through the boundary layer around the particle, then into the particle pores. One of the more important aspects of particle fragmentation and growth is that pores are generated, facilitating monomer access to the active sites inside the particles. If the catalyst support is resistant to the build up of stress and does not fragment, either because the support is too strong or the stress is too low, the pores will eventually become plugged with polymer. Since effective diffusion coefficients are several orders of magnitude greater in the pore space than in the polymer phase (up 4–6 orders of magnitude), the polymerization will rapidly shut down in this case. On the contrary, if the particle is too weak to properly dissipate the stress caused by polymer formation in the pores, then fragmentation will occur before enough

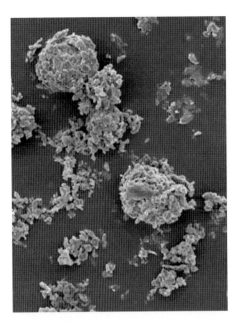

Figure 7.5 High-density polyethylene particles produced in a gas-phase reactor with a Ziegler–Natta catalyst. Poor morphological control has resulted in the generation of a significant portion of fines, as well as in loosely formed polymer particles with low bulk densities. The large particle in the left upper corner of the photo has a diameter of approximately 300 μm.

polymer has accumulated and the original particle will burst into fine particles that are often detrimental to good reactor operation and product quality. It should be noted that although fines can be generated through poor fragmentation, they have other causes as well, such as particle breakup, that can take place in later stages of the particle growth process.

Even though the time it takes for fragmentation to occur is in the order of $10^{-1} - 10^2$ s and the fraction of polymer produced during this step is insignificant with respect to the total mass of material made during the polymerization, this initial step has a significant impact on how the particles grow and behave in the reactor. Good morphology of the final polymer is one of the more important properties required for a successful polymerization process. Process engineers often use the term *"good morphology"* in different ways, but in the context of this chapter it refers to compact, well-defined polymer particles that do not sinter or break up into fines during the polymerization and that form a powder with high-apparent bulk density.[1] An example of poor morphology, displaying a significant fraction of fines, is shown in Figure 7.5. Experience (and logic) dictates

---

1) Note that polymer density and bulk density are not the same. Bulk density refers to the density of the polymer particles, including pore, and interstitial volumes. Polymer density is measured for polymer pellets or films.

that good morphological control requires controlled initial fragmentation. Note, however, that properly controlled fragmentation is a necessary but not sufficient step to obtaining desirable final particle morphologies. In summary, a good support should be porous enough to be active, be fragile enough to fragment once polymer is formed yet still rigid enough to handle easily, and give a high-polymer bulk density.

Experience shows that it is not always easy to do a controlled initial fragmentation step under full reactor conditions. As we see later in this chapter, particle overheating and the risk of mass transfer limitations that might lead to inhomogeneous particle growth are the most severe during the initial stages of industrial polymerization, especially for active catalysts. High reaction rates during the fragmentation step can also lead to rapid accumulation of forces in the fragmenting particles that can also cause uneven fragmentation and loss of control over the morphology during this critical stage of the reaction. One common way of avoiding loss of control over the reaction and the morphology at this point is the use of a prepolymerization step. Doing this allows us to produce polymer at a reasonable rate, generating just enough stress that the particles fragment but not so much and not too quickly that they disintegrate. In addition, it has the benefit that particles are grown to a size large enough that heat and mass transfer limitations are eliminated, or at least reduced to manageable levels. Finally, it appears that prepolymerization also helps to increase the activity of the catalyst in the main reactor with respect to nonprepolymerized ones.

Prepolymerization refers to the act of injecting the catalyst powder into a reactor that operates under relatively mild conditions (a few bars of monomer at most and occasionally a lower temperature) and produces in the order of 10–100 g of polymer per gram of catalyst. The prepolymerized powder is then injected into the main reactor train. Prepolymerization is typically carried out in slurry conditions even if the main reactor is a gas-phase one and is done in a reactor that is significantly smaller than the main reactor(s). Presumably this is to avoid producing sticky powder in the gas phase and to control the temperature better. As we mention below, temperature control is usually much less of a problem when the continuous phase in the reactor is a liquid. Of course, performing a prepolymerization step requires an additional unit operation in the reactor train, which adds capital and operating costs to the overall process, as well as potential sources of errors. However, given the importance of controlling the particle growth within predefined limits during the initial phase of the reaction, it is very common to find a reactor dedicated to this in a large number of processes, in particular processes using active Ziegler–Natta catalysts and gas-phase reactors.

While we are reasonably certain of the qualitative nature of particle fragmentation process, there remain a certain number of imprecisions in our understanding of the fundamental aspects of this important phenomenon. For instance, can we reasonably consider it to be instantaneous? How is the break up of the support influenced by polymerization conditions and/or polymer properties? How does the fragmentation process influence the long-term catalytic behavior of the active sites, or intraparticle heat and mass transfer rates? In large part this uncertainty is due to the difficulties associated with the experimental studies at the fragmentation scale:

7.2 Particle Fragmentation and Growth

- Initial polymerization rates are extremely fast and difficult to control, and the particles are relatively small (on the order of 10–100 microns in diameter). In addition, under industrial polymerization conditions, the characteristic dimensions of the macrograins can change rapidly.
- The highly exothermic nature of the polymerization makes it difficult to get precise estimates of the temperature of the particles, and large temperature excursions can influence the state of the polymer and the intrinsic activity of the active sites.
- In the case of $MgCl_2$-supported catalysts, trace quantities of humidity at levels present in the atmosphere can lead to the hydrolysis of the support and thus its destruction. This makes it quite challenging to use techniques such as electron microscopy or porosimetry to investigate particle morphological changes during particle fragmentation without special precautions.
- The dissipation of forces that provokes particle rupture will depend to a large extent on the polymerization conditions – fragmentation under mild conditions proceeds very differently from fragmentation under typical industrial reactor conditions – and on the properties of the polymer produced. For instance, it appears that polymer produced at very short times is much less crystalline than polymer produced after a few seconds or minutes of polymerization.

This said, it is still possible to investigate certain aspects of the fragmentation process using techniques including scanning electron microscopy (SEM) or transmission electron microscopy (TEM) to investigate the changes in particle porosity and pore size distribution, and energy dispersive X-ray spectrometry (EDX) to perform intraparticle elemental analyses. Particle fragmentation studies have appeared in the literature since the late 1960s, beginning with the investigation of the silica-supported catalysts, then $MgCl_2$-supported Ziegler catalysts in the early 1970s, and similar investigations are still being done at present since we still do not have all the answers.

Some of the earliest studies on particle fragmentation were done using silica-supported chromium catalysts. Polymer powders with different polymer/SiO_2 ratios were burned at high temperatures to leave behind only the silica fragments, and then nitrogen absorption/desorption or mercury intrusion porosimetry were used to investigate the change in pore size and size distribution of the recovered catalyst fragments. These early studies led to a qualitative description of the fragmentation of silica-supported catalysts that is shown in Figure 7.6a. The support is said to fragment first along the largest macropores; in other words, on the pores most easily accessible to the monomer during polymerization. According to this model, mesopores and micropores do not play a role in the fragmentation process during the very first instants of polymerization. Subsequently, as the porosity of the particle increases and the smaller pores become exposed to monomer, polymer forms at the active sites on the interior surfaces of the mesopores, then on the micropores, until reaching the size of the constitutive elements of the macrograins.

Other researchers proposed the alternative fragmentation model shown in Figure 7.6b. Controlled experiments on the fragmentation of silica supports

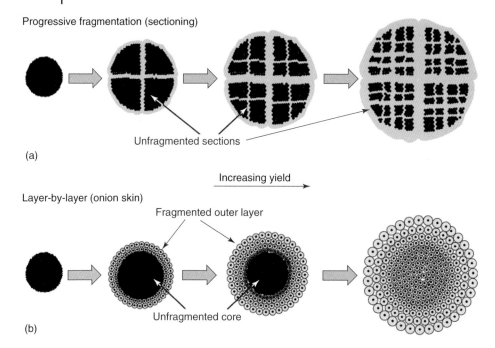

Figure 7.6 (a) Sectioning or progressive support fragmentation during olefin polymerization. Fragmentation occurs first on the largest macropores, then on successively smaller pores. (b) Layer-by-layer, or "onion skin" representation of fragmentation that assumes the particle is composed of roughly concentric layers of primary particles and that fragmentation begins at the outermost layer and progresses toward the center.

suggest that when the catalyst particles are spatially homogeneous, fragmentation follows a layer-by-layer mechanism. The active sites located closer to the center of the particle are not as accessible to the monomer as those at the outer edges because of the small pore size of the catalyst carrier and/or the blocking effect of polymer produced at the outer layer of the particle. Only once enough polymer is produced to fragment the outer layers do the inner parts more accessible to monomer molecules. This mechanism can be observed in silica-supported catalyst systems, or eventually in $MgCl_2$-supported systems. So, which of these two opposing views is correct?

To answer this question (in so far as is possible), let us consider the computed X-ray tomography (CXRT) images shown in Figure 7.7. X-ray tomography is a powerful noninvasive tool that allows one to visualize the inside of a particle without mechanically cutting it or treating it chemically. Briefly explained, the images are made by placing the particles to be examined in a small capillary that is fixed on a rotating plate in front of a synchrotron X-ray beam. The sample is scanned across a single transverse slice. Detectors on the opposite side of the sample from the X-ray source measure the intensity of the scattered beam and software is used to reconstruct the data into a map of the density variation inside the slice. The major limitation of this technique is its resolution of several hundred nanometers at best. This means that

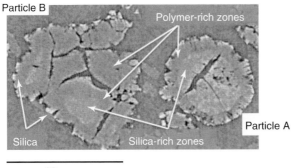

Figure 7.7 Computed X-ray tomograph of two fragmented silica-supported metallocene (Et[Ind]$_2$ZrCl$_2$) particles.

we can "see" the macroporosity, but not the microporosity. The image in Figure 7.7 is of particles produced by the polymerization of ethylene on silica-supported metallocene particles at 80 °C in a heptane slurry for approximately 10 s. We can see a number of interesting features, the most relevant to the present discussion being the presence of silica-rich zones near the center of the particles and polymer-rich zones near the exterior (the densest material – silica – is bright white, and the least dense – pore space – is the darkest gray). This is particularly evident in the case of Particle A. On the basis of the above discussion, one might be tempted to say that the layer-by-layer model is a reasonable representation of the particle morphology evolution during fragmentation. However, the situation is different for Particle B. The major difference between these two particles lies in fact that Particle B seems to have large pores that connect the interior of the particle to the surrounding bulk reactor phase. However, if one looks closely, it can be seen that the layer-by-layer representation works here, but only on the relatively homogeneous substructures within the particle. The boundaries of these substructures are either the particle exterior surfaces and/or the large pores crisscrossing the macroparticle. In some cases, the substructures are so small that it appears that the fragmentation is complete on the corresponding local scale. For larger ones, the fragmentation front seems to be moving toward the center of the substructures. The conclusion of this analysis is that the fragmentation model shown in Figure 7.6b seems to be valid for structures that are spatially homogeneous, such as particle A and some portions of particle B.

All things considered, the "true" fragmentation mechanism involves aspects of the two-limiting cases shown in Figure 7.6; which one dominates will likely depend strongly on the support particle pore structure, due in part to the fact that not all support particles are created spatially equal (even within a given batch). In other words, the more homogeneous the particle, the more uniform the fragmentation process and the morphology of the polymerizing particles will be. Figure 7.8 shows a CXRT image of a sample of untreated silica. Clearly, we should expect a certain range of different fragmentation behavior from catalyst particles made with this silica, given the differences in internal structure that can be seen in the picture.

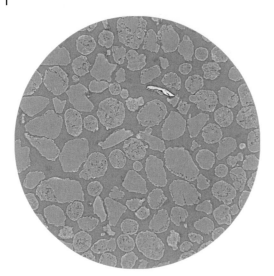

Figure 7.8 A CXRT image of a sample of silica support. Note the wide range of internal structures from particle to particle.

Another issue to consider in fragmentation is the timescale needed to completely fragment the support. The rate at which the particles fragment can have an influence on the initial rate of polymerization (and thus heat and mass transfer as we see below). Obviously, this will depend to a large extent on the type of support, with $MgCl_2$ fragmenting much faster than silica since it is more friable. Even under mild conditions, it would appear that $MgCl_2$ supports fragments in at most a few seconds, whereas it can take silica particles several tens of seconds, if not more depending on the reaction conditions.

As shown above, fragmentation will occur first where the polymer forms most quickly, either inside large pores (at least a few hundred nanometers in diameter) and near the catalyst surface. However, the role that smaller pores plays in the fragmentation process is not clear. In the example shown in Figure 7.7, once the substructures are relatively isotropic, fragmentation appears to be uniform within that substructure and to proceed in a roughly layer-by-layer manner. This leads us to suspect that while micropores and nanopores are essential to create the high surface areas needed to achieve commercially viable catalyst loadings, they do not seem to play a direct role in particle fragmentation. It is possible that if the smallest pores are extremely numerous (resulting in a catalysts with very high specific surface area) then the support structure will be more fragile than a similar support with a lower specific surface area. An interesting thought on how little the micropores of the catalyst directly influence the fragmentation process comes from a third type of support. Nonporous supports, such as those made via an emulsification process where the active sites and relevant activators are dispersed inside droplets that are then solidified, can still polymerize efficiently, despite having no initial porosity at all. It can be postulated that the fragmentation of these solid catalysts proceeds via a layer-by-layer mechanism similar to the one shown in Figure 7.6b. It is possible that monomer molecules penetrate a short depth into the solid support and polymerize when they reach an active site. The formed polymer breaks the support, creating enough porosity to render sites at a slightly deeper level available for polymerization, and the process continues until the support has been completely ruptured, as it is the case with the more traditional SiO_2-supported and $MgCl_2$-supported catalysts. Thus, if initially nonporous catalysts can polymerize olefins, this leads one to think that micropores and nanopores do not play an essential role in the fragmentation process. However, this type of fragmentation is not yet well understood, and this discussion should be considered speculative for the time being.

We have argued that fragmentation will create or maintain the pore space necessary for the polymerization to progress reasonably quickly and suggested that properly controlled fragmentation is a *sine qua non* condition to obtain polymer particles with good morphology. The possibility that the fragmentation step also influences the polymerization rate by exposing new active sites to monomer molecules as the inorganic support material breaks up has also been raised (as opposed to enlarging the pore space and making sites already on the surface of the pores more accessible to monomer). There is no consensus about this point in the literature, but if one reflects on how the active catalysts are made, the exposure of new active sites (by exposure, we mean active sites embedded in the support structure and not accessible to monomer unless the support is physically ruptured) through fragmentation does not seem highly probable for silica supports. In principle, silica is an inert support, and the active sites are anchored in a variety of different ways onto the surface of already-formed silica particles. For instance, in the case of a chromium oxide catalyst, the active particle is prepared by impregnating a chromium compound onto a silica support and then calcining it in the presence of oxygen to activate the catalyst. In other words, the structure of the support is fixed before the active sites are added. In this case, it would be reasonable to assume that no active sites are embedded in the structure to be exposed when the bridges of support material rupture during the fragmentation process. The same argument would appear to be valid in the case of metallocenes, anchored onto the inert silica support with methaluminoxane (MAO). In the case of a fourth generation Ziegler–Natta catalyst, $TiCl_4$ molecules and the $MgCl_2$ form a surface-activated crystal structure. Depending on the technology used to create the catalyst, rupturing the structure might allow some additional $TiCl_4$ sites to be exposed and eventually form new catalytic sites. However, it does not appear obvious that this would lead to the creation of a significant number of new sites, not to mention that the chemistry of Ziegler–Natta catalysts is not understood in detail, so new sites can be activated or deactivated for any number of reasons, making it difficult to test the idea. Finally, one might argue that fragmentation exposes all of the sites in the case of a nonporous support.

Before moving to the next section, we can conclude with some final thoughts on particle fragmentation:

- It appears reasonable to assume that conventional porous supports are organized into at least two levels of morphology: the overall particle (macrograin) and the smaller primary substructures (micrograins).

- Fragmentation occurs locally when enough polymer is formed, and the forces generated by its accumulation are greater than the cohesive forces binding the primary substructures together.
- The timescales for fragmentation will vary from support to support. This is due not only to different mechanical strengths of the supports but also to the structure of the particles themselves and the accessibility of the active sites.
- If a supported catalyst particle contains macropores or large cracks, fragmentation will take place first on these highly accessible parts of the catalyst.
- If a catalyst particle (or a portion of one) has a relatively homogeneous pore structure[2] then a moving fragmentation front will cause this structure to break up concentrically, following a layer-by-layer mechanism.

7.2.2
Particle Growth

Once fragmentation occurs, the polymer particle will grow by expansion: the polymer formed at the active sites displaces the previously formed polymer, causing the particle to grow. While this might seem like a simple process at first glance, the real situation is more complex than that. On the one hand, it is logical to assume that the particles expand as polymer forms at the active sites, but, on the other hand, it is very difficult to predict a priori how this expansion step leads to the final particle morphology and particle size distribution, how it influences pore structure and polymer powder bulk density, how the evolution of particle morphology influences transport rates, observed polymerization kinetics, and polymer microstructure (and vice versa).

We have already mentioned the replication phenomenon, where one polymer particle is generated from one catalyst particle, and the final particle retains the same overall shape and structure of the original catalyst. With a good catalyst and a well-run process, it should be possible to avoid fines generation or particle agglomeration, and get close to a one to one mapping of catalyst particles into polymer particles. However, the second part of the replication concept is not well supported by the experimental evidence available. If the particles were perfectly replicated, we would have an evolution of the overall particle morphology as the one proposed in Figure 7.6b. However, what one often finds are particles that have grown into shapes as those shown in Figure 7.9, which depicts a range of polyethylene and polypropylene particles, all made using $MgCl_2$-supported Ziegler–Natta catalysts (except Figure 7.9d which was made with a silica-supported chromium catalyst). These images represent a random selection of particles collected from a number of different polymer powders, some of which were cut open (or broken in liquid

---

2) By relatively homogeneous pore structure, we mean pores with average diameters that are less than a few hundred nanometers, although this is an admittedly arbitrary limit.

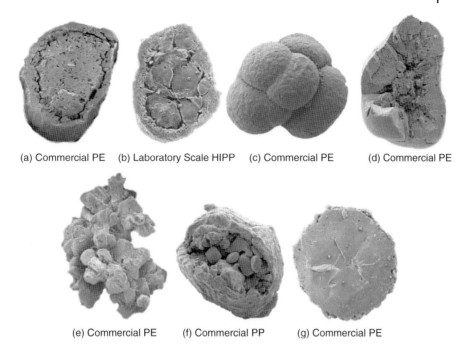

(a) Commercial PE (b) Laboratory Scale HIPP (c) Commercial PE (d) Commercial PE

(e) Commercial PE (f) Commercial PP (g) Commercial PE

Figure 7.9 Different commercial- and laboratory-scale polyolefin particles. All, with the exception of (d) are Ziegler–Natta products. Sample (e) was made with a silica-supported chromium catalyst. HIPP, high-impact polypropylene; PP, polypropylene; PE, polyethylene.

nitrogen). While most of them show some level of secondary morphological organization, the final shapes do not resemble the more idealized image presented in Figure 7.6. Clearly, one needs to take the concept of replication, at least in so far as the internal structure of the particles is concerned, with a grain of salt.

What are the reasons for this discrepancy? If we think about it, there are certain similarities between particle fragmentation and growth: polymerization creates forces inside the particles that initially cause them to fragment, and then expand. This implies that the internal structure of the particles, as well as their integrity, will be a function of the rate of generation of internal mechanical forces and the rate of dissipation of energy by the solid phase. Therefore, one can reasonably say that the means by which energy is dissipated at a given location within the particle will depend on local mechanical properties and polymerization rates. In fact, some efforts have been made to link polymer properties such as Young's modulus, the initial morphology of a prepolymer or catalyst particle, and the polymerization rate to polymer morphology evolution. Several articles describing these efforts can be found in the Further Reading section at the end of this chapter. These models can help explain certain types of anomalous replication behavior. For instance, some morphology models support the idea evoked above that the particles fragment first along the large pores, followed by increasingly smaller ones. They also demonstrate that if the polymer is less rigid, the supports will fragment more slowly and that the substructures might fuse together. Nevertheless, these models remain, at least for the time being, far too speculative and computationally intensive to be of use in predicting polymerization rates, polymer microstructure, and heat-generation rates to be of any practical use in the context of this book.

Furthermore, depending on the mechanical properties of the polymer being made and on polymerization conditions (including local temperature, pressure, type and intensity of mixing, and polymerization rate), we can run into situations where agglomeration and/or generation of fine particles occur. Agglomeration can have different causes, among them the generation of static electricity in gas-phase reactors, overheating and softening of neighboring particles, or even capillary forces in instances where liquids are injected into gas-phase reactors. Occasionally, these sources of agglomeration can be regulated to a certain extent by using antistatic agents and controlling the temperature and rate of injection of liquids into the reactor.

Fines generation is more problematic as its source is difficult to identify. As we mentioned above, it is possible that fines can be generated by poor fragmentation such as when the polymerization rate is too high and/or the support too fragile, causing the catalyst particle to break up before enough polymer accumulates to maintain its integrity. Fines can also be created when parts of a large particle break off during the polymerization for different reasons. In this last instance, the morphology of the particles might evolve in an unfavorable direction, and fragments could be knocked off due to collisions with another particles, pumps, or some other structure in the reactor (this is likely at the origin of the fines in Figure 7.5). In the case of ethylene polymerization in gas-phase reactors, fines are more likely to be observed for high-density polyethylene (HDPE) than for LLDPE, supposedly because HDPE is more brittle than LLDPE. It turns out that we can reduce fines formation during particle growth by slightly increasing the reactor temperature; it is thought that increasing the temperature will soften slightly the HDPE particles, making them less fragile and more ductile.

This example of reducing fines generation by adjusting the reactor temperature also illustrates the concept that the mechanical properties of the polymer itself can have an impact on how the particle morphology evolves during the polymerization. It does not take a leap of faith to accept that if the interior of a particle is hot then its substructures might fuse more easily than if the particle remains cool, thus leading to less or more porous particles, respectively, depending on the polymerization temperature. In other words, the structure of the growing particles will depend on more than just the morphology of the original support and the fragmentation step; it will evolve during the polymerization in a rather complex manner that is still not accurately quantified by any model.

7.3
Single Particle Models

Regardless of the type of olefin polymerization catalyst used, quantifying how heat and mass transfer occur is essential if we are to have confidence in our olefin polymerization reactor models. Efficient heat removal from the particle is evidently important to prevent thermal runaway, polymer melting, and particle agglomeration, while intraparticle temperature profiles will certainly impact local polymerization kinetics. On the other hand, mass transfer is important because the monomer concentration at the active sites will ultimately determine polymerization rates and the evolution of molecular weight and chemical composition distributions. In the following section, we look at the basis of building SPMs that describe the rates of heat and mass transfer inside growing polymer particles.

A SPM is simply a set of combined mass and energy balances around a polymerizing particle, coupled by a reaction term. Once we have solved the balances, we will obtain the temperate and concentration of reactants at every point inside the particle, and we will then use this information to predict the polymerization rate and polymer properties.

The first realistic effort to develop a model describing polymer particle growth, accounting for the two levels of structural organization discussed above (micrograins and macrograins), and that assumed that particle growth was only accomplished by expansion is often referred to as the *multigrain model*. For all intents and purposes, the MGM is based on the interpretation shown in Figure 7.6b. This model is the basis of most single particle modeling efforts at the current time, although a simplified version of the MGM, the PFM is also very popular for a number of reasons that are discussed later in this chapter.

While the MGM can be adapted to solve a number of problems, it has some shortcomings that make it difficult to model the phenomena already discussed in this chapter. The major limitation of the MGM is that it is built on the concept that the internal structure of the polymer particle replicates that of the catalyst particle, which we have seen is not always the case with supported olefin polymerization catalysts. However, we are obliged to use an MGM-based approach to calculate the growth rates of the polymerizing particles and polymer molecular properties because there are no tractable alternatives at the moment. As we see in the next section, such an approach is nonetheless a very practical tool for model building, and with judicious choices of model parameters and boundary conditions, the MGM representation of particle structure can be used to predict certain quantities and to control polymerization reactors and product quality.

7.3.1
Particle Mass and Energy Balances: the Multigrain Model (MGM)

Let us begin by looking at the mass balance equations. As noted above, we are more or less constrained to using the MGM representation of particle morphology, so the obvious choice is to write this mass balance assuming that the macroparticle is a sphere. While this is not strictly true in all cases, it is a very reasonable first approximation and any error introduced by making this assumption is likely to be small, especially with respect to uncertainties in other important model parameters. Furthermore, we also assume that the particle is isotropic so that only gradients in the radial direction will be of concern. We also need to assume a shape for

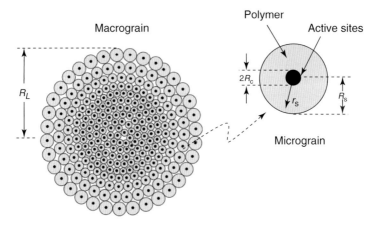

Figure 7.10 Geometry and length scales used in the standard MGM approach to single particle modeling: R_C is the radius of the catalyst fragments, R_S the radius of the microparticles, r_s the radial coordinate in a micrograin, and R_L and r_L are the radius and radial coordinates of the macrograin.

the micrograins in order to define the model equations and specify appropriate boundary conditions and will by necessity decide that they are also spherical. The geometry and principal length scales of the MGM approach are shown in Figure 7.10.

It can be seen from Figure 7.10 that there are 2 levels of resistance to mass transfer that are considered in the MGM: transport into and through the pores of the macrograin; and absorption of monomer into, followed by diffusion through the polymer layer that surrounds the micrograins. Heat transfer occurs in the opposite direction following a similar two-level mechanism. The characteristic length scale for transport in the particles can therefore be considered to be the radii of the macrograins and of the micrograins.

The general balance equation for transport and chemical reaction around a particle is

$$\frac{\partial [M_i]}{\partial t} = -(\nabla \times N_i) - R_i \tag{7.1}$$

where $[M_i]$, N_i, and R_i are the concentration and flux of species i inside the pores of the macroparticle, and the local volumetric reaction rate of monomer i, respectively. For inert components such as nitrogen, the same balance is used, but with $R_i = 0$.

The initial and boundary conditions for this equation can be written as

$$t = 0, [M_i] = [M_i^0] \tag{7.2}$$

$$r = 0, \quad \frac{\partial [M_i]}{\partial r_L} = 0 \tag{7.3}$$

$$r_L = R_L, \begin{cases} -N_i = k_s \left([M_i]_{\text{bulk}} - [M_i]_{r_L=R_L}\right) & (7.4) \\ \text{or} \\ [M_i]_{\text{bulk}} = [M_i]_{r_L=R_L} & (7.5) \end{cases}$$

In these equations, k_s is the boundary layer mass transfer coefficient and the subscript bulk refers to the concentration of monomer in the continuous phase outside the growing particle. Assuming that the mass transfer resistance across the particle boundary layer is negligible, then the boundary condition expressed in Eq. (7.5) can be used; otherwise, we need to use Eq. (7.4). The initial concentration in the secondary particle, defined in Eq. (7.2), may be set to zero for a monomer-free catalyst particle, but this generally leads to stiff differential equations that may be very hard to solve. It is common to assume a pseudo steady-state concentration at $t = 0$ to obtain the initial condition for Eq. (7.2). Unless one is interested in the intraparticle monomer profiles for the very first seconds of polymerization, this approximation generally leads to a system of partial differential equations that is easier to solve.

In most traditional SPMs, the flux term in Eq. (7.1) is taken to be purely diffusional, expressed using Fick's law with an effective diffusivity D_{eff}, thus rendering the equation

$$\frac{\partial [M_i]}{\partial t} = -(\nabla \times N_i) - R_i = \frac{1}{r_L^2} \frac{\partial}{\partial r_L} \left(r_L^2 \frac{\partial}{\partial r_L} (D_{\text{eff}} [M_i]) \right) - R_i \qquad (7.6)$$

Finally, another key assumption that is often invoked is that the fragmentation of the particle is instantaneous and complete. We return to this point in Section 7.4. For now, this means that the particle is often (but not always) considered to be spatially homogeneous, with a single effective diffusivity. The effective diffusivity in the macrograin, D_{eff}, is often estimated using the conventional expression for effective diffusivity in porous heterogeneous catalysts

$$D_{\text{eff}} = \frac{\varepsilon D_b}{\tau} \qquad (7.7)$$

where D_b is the monomer bulk diffusivity in the reaction medium, and ε and τ are the void fraction and tortuosity of the polymer particle, respectively. The fact that this representation is not necessarily valid for the entire range of pore sizes present in a particle, and that ε and τ are likely to vary as a function of the degree of fragmentation and expansion of the macroparticle is certainly one of the difficulties in getting a good estimate for D_{eff}. This means that D_{eff} is often treated as an adjustable parameter in the model. It is usually considered to be constant and independent of the position in the particle, but if we allow the porosity in Eq. (7.7) to vary as a function of time and/or position, then D_{eff} will change during the simulations.

Equation (7.1) also contains a reaction term that represents the volumetric consumption rate of species i at a given time and a given radial position. Since, in the MGM, the polymerization is assumed to take place only at the surface of the microparticles, this term couples the macrograin model to the micrograin model. In order to simplify the discussion here, we choose to calculate the reaction rate,

R_i (mol l^{-1} s^{-1}) using an expression similar to Eq. (5.1)

$$R_i = k_p [M_{i,p}]_{AS} [Y_0] \tag{5.1}$$

where $[Y_0]$ is the concentration of active sites and the $[M_{i,p}]_{AS}$ is the concentration of monomer i at the active sites situated at a given radial position in the macrograin. In the case of a copolymerization reaction, k_p will be the pseudo kinetic rate constant for propagation (Eqs. (5.21–5.23)). If the active site is not covered by polymer then $[M_{i,p}]_{AS} = [M_i]$. However, the active sites will already be surrounded by a polymer layer only fractions of a second after they are exposed to monomer in the reactor; therefore, to calculate the local polymerization rate, we need to know the concentration of monomer in the polymer layer just above the active sites in the micrograins. It is at this level that particle morphology really begins to come into play.

According to the MGM, the mass balance around a micrograin of spherical geometry, with respective initial and boundary conditions, are

$$\frac{\partial [M_{i,p}]}{\partial t} = -(\nabla \times N_i) = \frac{1}{r_s^2} \frac{\partial}{\partial r_s} \left(r_s^2 D_p \frac{\partial [M_{i,p}]}{\partial r} \right)_i \tag{7.8}$$

$$t = 0, [M_{i,p}] = [M_{i,p}^0] \tag{7.9}$$

at $r_S = R_c$, $N_{i,p}- = k_p [M_{i,p}]_{AS} [AS^*]$ \hfill (7.10)

at $r_S = R_S$, $[M_{i,p}] = [M_i^{eq}]$ \hfill (7.11)

where the subscript p refers to values in the primary particles and $[AS^*]$ is the concentration of active sites per unit surface area of the micrograin fragment. It is related to the term $[Y_0]$ as follows

$$[Y_0] = [AS^*] a_S \tag{7.12}$$

where a_S is the specific surface area of the support (m^2 of micrograin fragments per unit m^3 of particle). The monomer diffusivity in the primary particle, D_p, is a complex function of polymer chain crystallinity and immobilization of the polymer amorphous phase by the polymer crystallites, as well as of the temperature, pressure, and concentration of the different penetrants in the polymer layer. Experiments with suspension magnetic balances, for instance, can be used to estimate D_p values under well-controlled conditions for single penetrants on model materials. However, at least for the time being, we can at best hope to obtain an order of magnitude estimate for this parameter; D_p is generally used as an adjustable parameter in the model.

Notice that Eq. (7.8) does not contain a polymerization term. Because the MGM assumes that polymerization takes place at the surface of the catalyst fragment embedded within the primary particle, the reaction term appears as one of the two required boundary conditions. The inner boundary condition at the surface of the catalyst fragment, Eq. (7.10), states that the diffusional flux at the surface of the fragment ($r_S = R_c$) is equal to the polymerization rate. The boundary condition at the interface of the micrograin and the pore space ($r_S = R_S$), Eq. (7.11), states that the concentration of monomer in the polymer layer at the interface is in equilibrium with the monomer concentration in the pores of the macroparticle. It is possible that rapid changes in the pore space concentration would lead to an incorrect estimation of the concentration in the polymer layer due to nonequilibrium effects. However, one could probably assume that such changes would be rare and that Eq. (7.11) can be used with reasonable confidence. If a partition coefficient (K_{eq}) is used between the concentrations of monomer in the pores of the macroparticle and absorbed in the polymer layer surrounding the microparticle, we could write

$$[M_{i,p}]_{r_S=R_S} = [M_i^{eq}] = K_{eq}[M_i] \tag{7.13}$$

Of course, one can also use a more sophisticated equation of state to replace the equilibrium coefficient in Eq. (7.13). This will be particularly important in instances where we are dealing with multiple component vapor–liquid equilibrium (VLE), for example, in LLDPE production or condensed mode operation, since cosolubility effects might be important and require a more accurate representation of the equilibrium state.

The energy balance for the macrograins and micrograins can be written in an analogous manner. The energy balance equation and initial and boundary conditions for the macroparticle are

$$\overline{\rho C_p} \frac{\partial T}{\partial t} = \frac{1}{r_L^2} \frac{\partial}{\partial r_L}\left(r_L^2 k_{f,L} \frac{\partial T}{\partial r_L}\right) + (-\Delta \overline{H}_p) R_p \tag{7.14}$$

$$t = 0, T = T_0 \tag{7.15}$$

$$r_L = 0, \frac{\partial T}{\partial r} = 0 \tag{7.16}$$

$$r_L = R_L, \begin{cases} -k_f \frac{\partial T}{\partial r_L} = h\left(T_{r_L=R_L} - T_{\text{bulk}}\right) & (7.17) \\ \text{or} & \\ T_{\text{bulk}} = T_{r_L=R_L} & (7.18) \end{cases}$$

In these equations, T is the temperature at coordinates (r_L,t) in the macrograin, $\overline{\rho C_p}$ the average value of the heat capacity per unit volume of the macroparticle, $k_{f,L}$ the effective thermal diffusivity in the macrograin, $-\Delta \overline{H}_p$ the average enthalpy of polymerization, and h the average convective heat transfer coefficient between the particle and its surroundings. In a manner similar to the mass balance boundary conditions for the macrograin shown above, one can use the boundary condition expressed in Eq. (7.18) if it is possible to ignore any resistance to heat transfer in the external boundary layer. However, unless simulations or experiments show that we can neglect these resistances, it is recommended to use the more general boundary condition given in Eq. (7.17).

Certain hypotheses are needed to write the set of Eqs. (7.14–7.18). We assume that the energy is transported from the polymerizing macrograin via convection alone, as stated in Eq. (7.17), and we also assume that energy transfer through the particle is essentially via conduction through the solid phase alone. The first of these simplifying assumptions is not always valid, especially for small, highly active particles (which, as we see below, are the ones most prone to overheating) that come in contact with other particles, or reactor walls. In these cases, conductive heat transfer might, for a short timescale and for very short distances, make a nonnegligible contribution to energy transfer from the particle. In particular, the collision of small, hot particles with larger, cooler ones (an event much more likely in a continuous reactor than a small–small collision) will help in heat removal. Nevertheless, it is possible that the choice of a good empirical correlation to estimate the overall heat transfer coefficient might account for these additional mechanisms. The second hypothesis seems reasonable for most cases of interest.

For the micrograins, the model equations and boundary and initial conditions are

$$\rho C_p \frac{\partial T_S}{\partial t} = \frac{1}{r_S^2} \frac{\partial}{\partial r_S} \left(r_S^2 k_{f,p} \frac{\partial T_S}{\partial r_S} \right) \tag{7.19}$$

$$t = 0, T_S = T_{S,0} \tag{7.20}$$

$$r_S = R, T_S = T \tag{7.21}$$

$$r_S = R_c, k_{f,p} \frac{\partial T_S}{\partial r_S} = k_p \left[M_{i,p} \right]_{AS} [AS^*] \left(-\Delta H_p \right) \tag{7.22}$$

where the notation is as above, and $k_{f,p}$ is the thermal conductivity of the polymer layer around the particle fragments. In the boundary condition expressed in Eq. (7.21), we have implicitly assumed that the temperature at the surface of the micrograins situated at a given point in the macroparticle will be at a temperature T_S. This will be valid if the temperature rise inside a micrograin is negligible.

The set of Eqs. (7.2–7.6), (7.8–7.11), and (7.14–7.22) represents the well-known MGM for the prediction of concentration and temperature gradients inside growing polymer particles. Provided that we can find an appropriate set of polymerization kinetic constants, physical properties, and VLE data, and assuming that the hypotheses made in the MGM development are valid, this set of equations can be solved numerically to predict the impact of different polymerization conditions and particle sizes on polymerization rates and on certain polymer microstructural properties.

7.3.2
The Polymer Flow Model (PFM)

As mentioned above, assuming that the rates of change are small enough to be neglected in Eqs. (7.6), (7.8), (7.14), and (7.19) allows us to reduce the stiffness of the model and to obtain reasonable estimates for the evolution of concentration and temperature profiles as a function of radial position. However, given the structure of this model and the number of assumptions invoked to develop it, in particular about particle morphology, it is useful to consider a simpler way of setting up the problem. In fact, most authors no longer use the full MGM with the micrograin/macrograin morphology description since it is somewhat onerous to use and the accuracy of its predictions are based on parameter values, such as effective diffusivities or convective heat transfer coefficients, that are very difficult to know with any degree of confidence. In fact, it turns out that the qualitative predictions of the MGM are equivalent to those of a more simplified model, referred to as the *polymer flow model*.

The basic premise of the PFM is that resistances to heat and mass transfer at the level of the micrograins are negligible, and we can therefore treat the macrograin as a pseudo homogeneous medium where the concentration at the active sites at a radial position r_L is assumed to be in equilibrium with $[M_i]$ in the pore space. This is reasonable if, paradoxically, we accept the MGM description of the particle morphology and the relative structure of the particles remains unchanged throughout the polymerization, as illustrated in Figure 7.6b. To understand why the PFM is not an unreasonable simplification, let us consider the characteristic times for mass transfer in a fictitious particle with an MGM structure. If we ignore mass transfer by convection for the moment, and consider only mass transfer by diffusion, then the characteristic length scales for diffusion will govern the rate of mass transport in the particle. The characteristic diffusion time, τ_d, is related to the characteristic length scale, L, and the representative diffusivity by the expression

$$\tau_d \propto \frac{L^2}{D_{\text{eff}}} \tag{7.23}$$

For the macrograin, the representative length scale is R_L, and for the micrograin it is R_S (or R_C at $t = 0$). If we know the support specific surface area (A_S, typically reported in square meters per gram of support) and the density (ρ_c), then R_C can be found from

$$A_S = \frac{4\pi R_c^2}{\frac{4}{3}\pi R_c^3 \rho_c} \Rightarrow R_c = \frac{3}{A_S \rho_c} \tag{7.24}$$

The density of anhydrous $MgCl_2$ is in the order of 2.3 g cm^{-3}, and specific surface areas for commercial supports can run upwards of 300 m^2 g^{-1}. Under typical polymerization conditions, D_p will be on the order of $10^{-11} - 10^{-9}$ m^2 s^{-1} for the diffusion of an olefin monomer through a polyolefin, on the order of $10^{-10} - 10^{-8}$ m^2 s^{-1} for the monomers in the pores of the growing particle in a slurry polymerization, and $10^{-8} - 10^{-7}$ m^2 s^{-1} in the pores of a gas-phase system. This means that, for the same value of L^2, the characteristic time for diffusion in the growing particles will be thousands of times higher in the pores than in the polymer layer surrounding the micrograins. Figure 7.11 shows the orders of magnitude of the characteristic diffusion time for different systems (slurry versus gas) at the upper and lower limits of the diffusivities cited above and different specific surface areas (or values of R_C). We can see that the characteristic timescale for diffusion in the micrograins is typically very short with respect to diffusion in

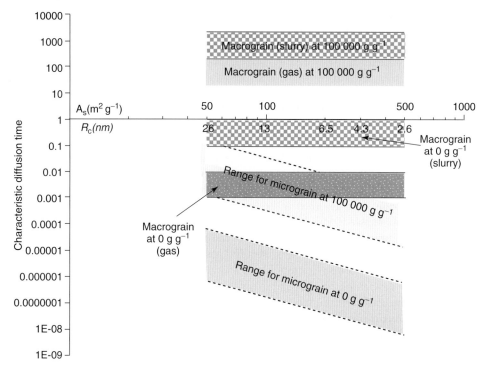

Figure 7.11 Characteristic diffusion times for micrograin and macrograins. The values of R_C are calculated with Eq. (7.24) by specifying a value of the specific surface area. The characteristic length scale for the micrograins, R_S, is shown at productivities of 0 and 100 000 g polymer per g support and is thus taken as $R_S = 44.7 \times R_C$. The macrograin domains are extended along the same range of values of A_S as the micrograins for the purposes of illustration, but are obviously independent of this value.

the macrograins for most systems. In other words, mass transfer resistance in the pores of the macrograin will almost always be more significant than that in the micrograins, and we can neglect (for the most part) mass transfer resistance at the micrograin scale *in the event that the original geometry of the catalyst support is replicated throughout the entire polymerization*. Some exceptions to this rule would be for very large micrograins (for instance, for low values of specific surface area), and highly crystalline polymers, such as HDPE, where there may be low values of the diffusion coefficient in the polymer layers around the micrograins. It is possible to do the same type of analysis for the characteristic heat transfer times; however, the results are the same: we can typically neglect resistances associated with heat transfer at the micrograin scale if we can make the same assumption about the replication of the particle geometry (once again, this may be an inadequate assumption for some of the polymer particles shown in Figure 7.9).

To summarize, the PFM can be viewed as a simplified version of the MGM, where the particle balances used are Eqs. (7.8–7.11) and (7.14–7.18). The model predicts the concentrations and temperature profiles in the pores of an imaginary pseudo homogeneous macroparticle, and it is assumed that the concentrations at the active sites at radial position r_L in the macroparticle are in equilibrium with the concentration in pores at the same location. The same approximation holds true for the temperature.

7.3.3
An Analysis of Particle Growth with the MGM/PFM Approach

Solving the macrograin balances (using either MGM or PFM) with diffusion as the only mechanism of mass transfer in the particles will produce profiles of concentration, polymerization rates, and macrograin temperatures such as the ones shown in Figures 7.12–7.14. Figure 7.12 shows general concentration profiles of homopolymerizations where the dimensionless monomer concentration $([M_i]_r/[M_i]_{bulk})$ decreases from the surface $(r_L/R_L = 1)$ to the center at several polymerization times. The steepness of the profile increases as the size of the initial catalyst particle increases, as the polymerization rate increases (in particular the initial polymerization rate), and as the effective diffusion coefficient decreases. For typical Ziegler–Natta catalysts that show a rapid activation and a decay-type rate curve, the steepest gradients will be seen at the beginning of the reaction (t_1), and the concentration profiles will slowly flatten out as time passes because the catalysts deactivate and the active sites get increasingly more diluted in the polymer phase. In extreme cases of very high initial rates and/or large fresh catalyst particles, one might even observe profiles such as the one indicated with the dashed line, Case (4) – in Figure 7.12, where the model predicts that the monomer concentration

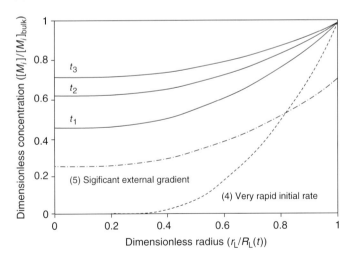

Figure 7.12 Dimensionless concentration profiles for a model catalyst with fast activation rate. The different times are such that $t_1 < t_2 < t_3$. The solid lines show moderate polymerization rates, dashed line indicates an example of an extremely rapid initial polymerization rate, and dashed–dotted line indicates a case with significant boundary layer resistance.

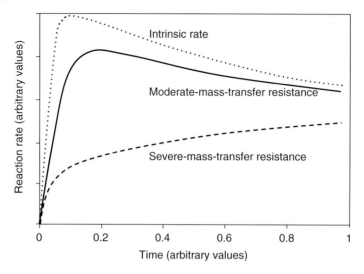

Figure 7.13 Polymerization rates (arbitrary values) for different levels of mass transfer resistance in a simulated semibatch experiment. The dotted line represents the intrinsic rate (obtained with no mass or heat transfer resistance), the solid line corresponds to moderate resistances, and the dashed line corresponds to the rate that would be observed with very severe mass transfer resistance (solid and dashed lines represent the same cases as in Figure 7.12).

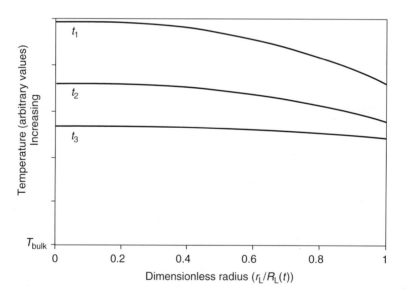

Figure 7.14 Dimensionless profiles of temperature for a simulated, rapidly activating catalyst. The different times are such that $t_1 < t_2 < t_3$.

drops to zero somewhere in the center of the particle. The end result of this phenomenon is that the outer shell of the particle would grow rapidly, but the inner core would not, leading to a hollow particle. Finally, Case (5) in Figure 7.12 illustrates the case where there are significant external gradients in the particle boundary layer. Note that this figure shows the concentration gradients for just one monomer. In the case of a copolymerization (e.g., LLDPE or ethylene–propylene rubber (EPR)), we would expect to see similar trends for both monomers. In other words, we would have two sets of profiles, with the concentration of each monomer decreasing toward the center of the particles. In cases where one monomer is significantly more reactive than the other (e.g., C_2H_4 with respect to C_4H_8 in the case of LLDPE), the profiles of the least reactive monomer will be much flatter.

As one would expect, internal mass transfer resistances will be greater as the intrinsic polymerization rate and diameter of the catalyst particle increase because the polymerization rate is proportional to the number of active sites in the catalyst particle, and the rate of mass transfer into the particle is proportional to its surface area. If we compare two catalysts prepared in the same way, ideally one would expect the number of active sites in a particle to be proportional to its volume, in other words, to the cube of its radius. Since the rate of monomer entry in the particle is proportional to the surface area (the square of the radius), the ratio of the monomer entry rate divided by the rate of monomer consumption varies as one over the particle radius. All other things being equal, it is harder to supply a large fresh catalyst particle with the monomer it needs to polymerize at the intrinsic reaction rate than it is to supply the needs of a smaller particle.

However, one needs to be careful here and not confuse fresh catalyst particles with growing polymer particles. As the particles grow, the resistance to mass transfer will decrease (if the intrinsic activity does not continue to increase once the reaction begins). Once again, the rate of monomer consumption in the particle is proportional to the number of active sites in the original catalyst particle (assuming that most sites are activated quickly). If the number of active sites does not change during the polymerization then the amount of monomer required by the particle to ensure a rate equal to the intrinsic rate will not change either. Thus, as the particle grows, the surface area grows, and the mass flux of the particle increases along with it. This is the reason that the profiles in Figure 7.12 flatten out over time.

As a quick indication of what will influence the external concentration gradient, let us consider the quasi steady-state mass balance across the boundary layer

$$\frac{R_p}{mw} \times \frac{4}{3}\pi R_c^3 \times \rho_c = k_s 4\pi R_L^3 \left([M_i]_{bulk} - [M_i]_{r=R_L}\right) = k_s 4\pi R_L^2 \Delta M_i \quad (7.25)$$

where mw is the molecular weight of the polymerizing species. Equation (7.28) can be rearranged to

$$\Delta M_i = \frac{R_p \times R_c^3 \times \rho_c}{3k_s R_L^2 mw} \quad (7.26)$$

Equation (7.26) tells us that the external gradients will be significant for high-polymerization rates, large initial catalyst particles, small R_L values (low-growth rates early in the polymerization), and low values of the boundary layer mass transfer coefficient. The mass transfer coefficient k_s can be estimated from Sherwood number (Sh) correlations such as the one given below

$$\text{Sh} = \frac{2R_L k_s}{D_{\text{bulk}}} = 2 + 0.6 \text{Sc}^{\frac{1}{3}} \text{Re}^{\frac{1}{2}} \tag{7.27}$$

Equation (7.27) is the well-known Ranz–Marshall correlation [2] that relates Sh to two other common dimensionless groups, Sc, the Schmidt number and Re, the Reynolds number

$$\text{Sc} = \frac{\mu}{\rho_{\text{bulk}} D_{\text{bulk}}} \tag{7.28}$$

$$\text{Re} = \frac{\rho_{\text{bulk}} u (2R_L)}{\mu_{\text{bulk}}} \tag{7.29}$$

Many other correlations for Sh (or k_s directly) are available from the chemical engineering literature, and an in-depth analysis is beyond the scope of the discussion here. It suffices to say that while different correlations will take different forms, the functionality will remain similar. What we see from this, and other correlations, is that k_s decreases as

- the diffusivity in the bulk phase decreases. We expect more external resistance in liquid phase reactions than in gas phase;
- the viscosity (μ_{bulk}) increases;
- the relative fluid particle velocity (u) decreases, such as in poorly mixed zones of a reactor.

Also, as the particles get larger, k_s will also increase.

As with the internal gradients, external gradients will be more important for rapid polymerizations during the early stages of the polymerization and for large initial catalyst particle diameters, and we also expect that mass transfer resistance will become less important as the polymerization proceeds.

The effect of mass transfer resistance is shown in Figure 7.13, where we see the polymerization rate as a function of time in a semibatch experiment with constant monomer pressure and temperature. The dotted line in this figure represents the intrinsic rate of polymerization: the theoretical polymerization rate that would be seen in the absence of any mass transfer resistance. In other words, this is the polymerization rate that would be observed if the monomer concentration in the pores of the particle were equal to the bulk monomer concentration and if the monomer concentration at the active sites were in equilibrium with that in the pores. In the case of moderate mass transfer resistance (e.g., corresponding to the case of the solid lines in Figure 7.12), there is a slight reduction in the observed rate with respect to the intrinsic rate during the early stages of the reaction because the concentration of monomer near the center of the particle was lower than that near the outer edges. This means that the active sites in the center polymerize more slowly and that the overall rate of polymerization in the particles is less than it potentially could be. If the resistance is only moderate, the difference between the observed and intrinsic rates becomes less significant as time goes on (i.e., the surface area for mass transfer increases as the particles grow). In the case of severe mass transfer resistance, we might never see a peak in the activity, and there will be significant differences between the observed and intrinsic rates that persist for the entire reaction.

Note that if experimental data show polymerization rate curves such as the last case of severe mass transfer resistance, there might be explanations other than mass transfer resistance for the shape of the rate curve. It is entirely possible that the catalyst formulation might be such that one obtains a slowly increasing rate, such as is characteristic of chromium catalysts.

Typical temperature profiles are shown in Figure 7.14 (here the units are arbitrary since the upper value depends on the simulation). As with the concentration profiles, the highest gradients are to be found at the beginning of the polymerization for a typical rapidly activating catalyst, and the gradients can eventually flatten out as the particles grow larger. Most of the time, one will observe the largest gradients of temperature in the particle boundary layer (i.e., between the surface and the bulk). This is especially true of gas-phase reactors where one can rarely neglect the possibility of significant external temperature gradients, in particular at short times. The model usually predicts that the highest temperature in the system will be at the center of the particles, but that the internal temperature gradient can often be neglected.

Once again, we can invoke the quasi steady-state hypothesis, and rearrange the energy balance in Eq. (7.14) to obtain an estimate of the temperature gradient across the boundary layer (ΔT_{BL})

$$\Delta T_{BL} = \frac{R_p \left(-\Delta H_p\right) \times R_c^3 \times \rho_c}{3h R_L^3 \text{mw}} \quad (7.30)$$

As with the external concentration gradient, Eq. (7.30) tells us that the temperature gradient in the boundary layer will be more significant:

- at high-polymerization rates;
- for large initial catalyst particles;
- during the early stages of the polymerization (smaller values of R_L);
- low values of the heat transfer coefficient h.

As noted above, one way to avoid particle overheating is to use a prepolymerization step. Since R_p is set at low values during this step, one can grow the particles until R_L is large enough for the particles to be injected into the main reactor(s) with minimal risk of overheating. If it is possible to use a slowly activating catalyst, one does not need a prepolymerization step. As shown in Figure 7.15, a slowly activating catalyst will avoid the critical zone for heat and mass transfer since a catalyst that takes a certain time to reach full activity undergoes a sort of prepolymerization *in situ*.

The coefficient h is typically estimated using correlations for the Nusselt number (Nu). Several modeling efforts have used the Ranz–Marshall correlation for

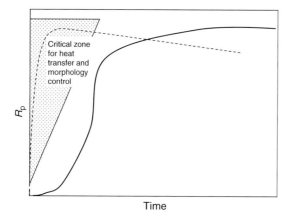

Figure 7.15 Slowly activating catalysts (solid curve) allow one to avoid the tendency for small, rapidly activating catalysts to overheat and lose control of the particle morphology.

this purpose

$$\text{Nu} = \frac{(2R_\text{L})\,h}{k_f} = 2 + 0.6\text{Pr}^{\frac{1}{3}}\,\text{Re}^{0.5} \tag{7.31}$$

where k_f is the thermal conductivity of the continuous phase, and Pr and Re are the Prandtl and Reynolds numbers. While it is widely accepted that this correlation is quite accurate for isolated particles, it is not a good idea to use it for the prediction of heat transfer coefficients of systems where particle–particle and particle–wall interactions cannot be neglected. Indeed, interactions between the polymerizing particles and their immediate environment might have a strong influence on the heat transfer coefficient. Another difficulty with correlations of this type is that users are tempted to use an average particle size to calculate a single value of h for all the particles in a reactor. In fact, one of the pitfalls of using SPMs, such as the ones developed here, is the temptation to use a "one size fits all" approach to predicting the behavior of an average particle in the reactor – such an approach falls short of describing the complex particle size distribution of the polymerizing particles. It is, therefore, preferable to use a correlation that takes into account as many bed-related parameters as possible, for instance, the particle density, energy dissipation, solids loading, and the particle size distribution.

Nevertheless, the Ranz–Marshall correlation shows that heat transfer in the boundary layer is improved for fluids with high values of k_f and at high-Reynolds numbers. Values of k_f and c_p are higher for liquids than for gases, so we can expect the boundary layer heat transfer coefficient to be higher for slurry-phase reactions than for gas-phase reactions. Experience also shows that one can expect the temperature gradient across the boundary layer to be an order of magnitude higher in gas-phase polymerization than in slurry reactions. A summary of very generalized conclusions on the significance of external and internal gradients that can be expected in different phases according to the MGM/PFM modeling of single particle growth is given in Table 7.1. Note that these are approximate guidelines,

Table 7.1 Summary of mass and heat transfer resistances.

Phase	Resistance to	Particle boundary layer	Macroparticle
Liquid	Mass transfer	Might be important for large, active particles at the beginning of polymerization	Can be important, especially at beginning of reaction or for high activities
	Heat transfer	Negligible	Negligible
Gas	Mass transfer	Negligible	Generally negligible except for large, active particles at beginning of reaction
	Heat transfer	Cannot be neglected; will be especially important for large, active particles at the beginning of polymerization	Generally negligible except for large, active particles at the beginning of polymerization

and it is recommended to those attempting to develop an accurate particle model to test these simplifying hypothesis carefully.

It should be evident from the model development presented in this section that mass transfer and heat transfer resistances will be intimately linked. In other words, polymerization conditions that lead to significant mass transfer resistances will probably also cause measurable temperature gradients in the particle and/or boundary layer around the particles.

7.3.4
Convection in the Particles – High Mass Transfer Rates at Short Times

The MGM/PFM analysis discussed above allows us to understand a great deal about olefin polymerization. By correctly choosing the appropriate model parameters and carefully identifying the proper boundary and initial conditions, one can obtain reasonable predictions of polymerization rates, molecular weight and composition distributions, and use the models for reactor control. Nevertheless, the conventional form of the mass balance shown in Eq. (7.6) may be enhanced by including certain additional phenomena.

For instance, the MGM/PFM approach does not allow for self-induced convection inside porous particles. This might be a factor that needs to be considered when one reactant is much more reactive than another (or if there are inert compounds present in large quantities). In general, mixed transport modes, such as convection and diffusion, provide higher mass transfer rates than diffusion alone and can be used to better illustrate what occurs during the initial moments of the polymerization when polymerization rates are high and mass transfer limitations are potentially at their greatest.

It has been widely discussed in the literature that one of the shortcomings of the PFM approach to single particle modeling is its inability to fit experiments with extremely high initial polymerization rates. In order to understand why, let us return to first principles and look at the complete expression of the molar flux inside the growing particles that includes both diffusive and convective terms. Using the notation of Bird et al. [3] the flux term of Eq. (7.6) is

$$N_i = c_i v_i = c_i (v_i - v^*) + c_i v^* = J_i^* c_i v^* \tag{7.32}$$

where c_i is the concentration of transported species i, v_i is the average velocity of component i, and v^* is the molar averaged velocity defined as

$$v^* = \frac{1}{c} \sum_{i=1}^{n} c_i v_i = \frac{1}{c} \sum_{i=1}^{n} N_i \tag{7.33}$$

where c is the total molar concentration in the fluid and n is the number of species in the system.

The diffusive flux (J_i^*) is the flux of species i with respect to the molar average velocity, expressed in a multicomponent mixture as

$$J_i^* = c_i (v_i - v^*) = -c \sum_{j=1}^{n-1} D_{ij} \nabla x_j \tag{7.34}$$

where D_{ij} are the Fickian diffusion coefficients of the i–j pair in the multicomponent mixture, and are concentration dependent. Combining Eqs. (7.32–7.34) and using the Stefan–Maxwell approach, one arrives at the net molar flux of species i

$$N_i = \frac{-c \sum_{j=1}^{n-1} D_{ij} \nabla x_j + x_i \sum_{\substack{j=1 \\ j \neq 1}}^{n} N_j}{1 - x_i} \tag{7.35}$$

Equation (7.35) is valid for any system of interest. Looking at this equation, it is apparent that the flux can only be reduced to the pure diffusive expression used in Eq. (7.6) for small values of x_i (i.e., $x_i \to 0$). Equation (7.6) is, therefore, a reasonable approximation of mass transfer flux inside the macropores of the macroparticle in the case of slurry polymerizations in hydrocarbon diluents (for instance, ethylene in isobutene or hexane) where the monomer concentration is relatively small, or in cases where the monomer phase is highly concentrated (such as in liquid propylene slurries). For larger values x_i, Eq. (7.35) is to be preferred.

As an example, let us consider the simulations shown in Figure 7.16. These results show the rate of polymerization of a semibatch gas-phase polymerization at 24 bar of ethylene and 6 bar of butene. We have neglected mass transfer resistance in the boundary layer and chosen a reasonable value for the diffusion coefficient in the pores of the macroparticle (to simplify the simulation, we have set the diffusion coefficient equal for both components). It can be seen in Figure 7.16a that if we use a mixed-mode mass transfer mechanism, the predicted rate increases more quickly and mass transfer limitations are not nearly as strong as in the case of diffusion-only mass transfer. Figure 7.16b shows that including convection in the model eliminates the large ethylene gradient predicted using the PFM.

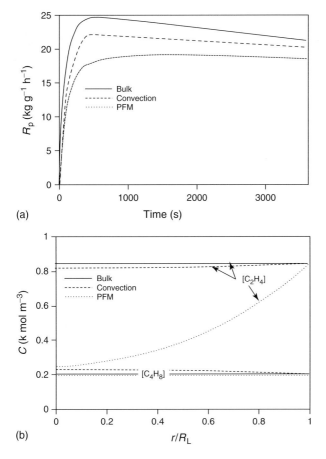

Figure 7.16 (a) Simulations of "observed" rates of polymerization with flux described by Eq. (7.35) (with convection), and with Fickian diffusion only (PFM). (b) Monomer concentration profiles after 100 s (maximum gradient). Gas-phase polymerization of ethylene with $D_{eff} = 4.1 \times 10^{-7}$ m^2 s^{-1}, 24 bar C$_2$H$_4$, 6 bar C$_4$H$_8$, $R_{L,0} = 20$ μm. $k_p = k_{p,max}\left[1 - \exp\left(-3t/t_p\right)\right]\exp\left(-k_d t\right)$ where $t_p (= 300$ s$)$ is a parameter controlling the time it takes to reach peak activity, and $k_d (= 5 \times 10^{-5}$ s$^{-1})$ is the deactivation constant. $k_{p,max}$ was adjusted to give a maximum activity of 25 000 g PE g^{-1} catalyst h^{-1} assuming no mass transfer resistance. $T = 70\,^\circ$C.

Of course, this is a highly simplified simulation where we have made a number of assumptions, including an isothermal particle (which is highly unlikely). As we have discussed above, gas-phase reactions with fast initial rates might lead to overheating of the growing particles. If the particle does not melt, limited overheating will lead to an acceleration of the polymerization rate. However, it turns out that the rates of mixed-mode transport are much more sensitive to particle size than to the polymerization rate itself. In other words, if we use the model that includes a convective flux, increasing the reaction rate from 25 to 100 kg g^{-1} catalyst h^{-1} will do very little (on paper!) to the predicted concentration profiles. On the other hand, changing the radius of the catalyst particle does have an impact as one might expect based on our discussion of the PFM above; in short, the larger the particle, the greater the gradients.

However, there is one distinct difference in the case where we allow convection to occur, and it is that the concentration of the inert, or less reactive components, do not decrease toward the center of the particle; instead, they increase. Close inspection of Figure 7.16b reveals that while the PFM predicts that concentration profile of butene (here assumed to have a k_p for butene 50 times less than that of ethylene) decreases toward the center of the particle, the full mixed model predicts the opposite, because ethylene is consumed extremely rapidly, and the flux created by its movement into the particles entrains the butene. Since there is more butene in the center of the particle, the polymer in the center will actually be richer in butene than that at the exterior of the particle. This same reasoning can be extended to systems where the reactor contains inert species as well. They could be entrained along with the principal-polymerizing monomer, and if there is enough accumulation, this could have the effect of diluting the monomer inside the particles, causing the polymerization rate to drop with respect to a polymerization with the same partial pressure of monomer, but no inert components.

To summarize, it might be important to consider convective and diffusional mass transfer of monomers inside polymerizing particles in certain gas-phase reactions. Using the full flux expression of Eq. (7.35) allows one to better model experimentally observed rapid reaction rates using realistic diffusion coefficients.

7.4
Limitations of the PFM/MGM Approach: Particle Morphology

Support fragmentation and subsequent (or concurrent) particle expansion are caused by the buildup of hydraulic forces due to the local accumulation of polymer and the dissipation of this energy by the deformation of the particle through breakup of the continuous particle matrix (fragmentation), particle expansion (growth due to the accumulation of polymer), or changes in local structure (e.g., compression of the polymer substructures). The rate at which these changes happen, and the resulting particle morphology, will depend on the rate at which forces build up, the rate at which energy is dissipated, and on the physical properties of the polymer during the reaction. The physical properties of the polymer will depend on local temperatures, densities, and degree of swelling by monomer, among other factors. As we saw in Figure 7.9, this can lead to very different particle structures, some of which are quite distinct from the idealized morphology model shown in Figure 7.10.

To understand the implications of this deviation, let us examine a highly simplified simulation. We will use the standard PFM that neglects mass transfer resistance in the micrograins to simulate the slurry polymerization of ethylene in heptane on an $MgCl_2$-supported Ziegler–Natta catalyst for the highly idealized cases shown in Figure 7.17. It is assumed that the particle remains isothermal since the

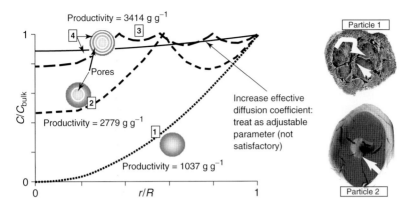

Figure 7.17 Simplified schema of the role of particle morphology in mass transfer and observed productivity for different particle configurations. Calculations were made using the PFM for the slurry polymerization of ethylene in heptane at 6 bar of monomer, with initial particle diameter of 10 μm and an intrinsic activity of 25 kg polyethylene g^{-1} catalyst h^{-1}. Profiles 1–3 were calculated with an effective monomer diffusivity in the macroparticle of $D_{eff} = 2 \times 10^{-10}$ m^2 s^{-1}. Profile 4 was calculated with $D_{eff} = 10^{-8}$ m^2 s^{-1}. Profiles 1–4 and the productivities indicated are dimensionless monomer concentrations inside the growing particle after 300 s of polymerization.

polymerization is taking place in a slurry reactor (this is not a perfect assumption, but the simplification changes nothing in the conclusions of this example).

In the first case, we will use a pseudo homogeneous particle typical of the standard PFM model. The other cases show model predictions for a more open morphology, with one or two "pores" of diameter 1 μm that goes around the center particle in a spherical shell. These pores are assumed to be connected to the bulk phase, and therefore, the concentration of monomer in the pores will be equal to that of the bulk reactor phase. Of course, this is admittedly an overly simplistic way of describing the real morphology, but it should give us an idea of how large influxes of monomer could alter the mass transfer scenario. First of all, we see once again that significant concentration gradients are predicted for the standard PFM at short times for this reasonably active catalyst. However, if we have one or more large "pores" in the particle, the significance of the mass transfer resistance is attenuated. This is reflected in the values of the productivity, defined as

$$P = \iint R_p dV dt \tag{7.36}$$

The higher productivities predicted for the pores with an open structure clearly show that a morphology that creates shorter length scale for diffusion leads to higher polymerization rates and less mass transfer resistance. Note that the only factor that was changed between profiles 1, 2, and 3 was the structure of the starting particle. Even though this is an extremely simplified example, it shows that polyethylene powders with different pore structures can have distinct mass transfer rates. Finally, if we use the diffusion coefficient as an adjustable parameter and increase its value by two orders of magnitude (profile 4), it is possible to use the conventional morphological representation of the original PFM (no macropores or voids) to predict similar productivities to those simulated with the morphology corresponding to profile 1.

Thus, it is clear that an appropriate adjustment of the effective diffusion coefficient in the PFM can help to fit experimental data, and therefore, simplified models can serve as a basis for process design and control *under well-defined conditions*. What we mean by "well-defined conditions" is that this type of model fitting will provide reasonably reliable estimates for a given catalytic system in a specific medium (gas, slurry, supercritical) under given polymerization conditions (temperature, pressure, and reactant concentrations). If these conditions are changed, then the model will need to be refitted if the structure of the growing particles changes with the polymerization conditions even if the "intrinsic" polymerization kinetics are known for the new process conditions.

One of the real limitations in single particle modeling, therefore, arises from the fact that we have no real tools (i.e., morphology models) capable of predicting what the morphology of the particles will be at a given moment during the reaction. In other words, in the example presented, it is not possible to know a priori whether we should use an open structure with short diffusion paths through a pseudo homogeneous macrograin or a more isotropic structure in the definition of our boundary conditions.

As an extension of this rationale, the question arises of whether the two-tier macrograin–micrograin structure on which the MGM is based, and on which the PFM simplifications rely, is actually a valid representation of particle morphology. Is it reasonable to say that, in all circumstances, the micrograins present in the original catalyst particle really persist throughout the polymerization as assumed in the MGM/PFM particle model? Visual evidence from Figure 7.9 would suggest not, and other studies of mass transfer rates in nonreactive powders also suggest that this is not the case. If that is so, then the hypotheses that the length scales are short enough that gradients of temperature and concentration at the micrograin scale can be neglected are not valid.

Where does all this leave polyolefin reaction engineer practitioners? All is not lost since the PFM/MGM approach to single particle modeling as presented above can be implemented to help predict reaction rates and product quality and to help design and control polymerization reactors. Typically, in order to do so, we obtain rate and property data from the system we want to model, and adjust the model parameters, including kinetic rate constants, and diffusivities so that the model predictions match the data. In this sense, a well-designed PFM particle model is a useful tool for a number of tasks. As we saw in the example of Figure 7.17, by adjusting the diffusivity it is possible to account for changes in particle morphology (changing length scales). The cost of this approach is that, in this particular case, the diffusivity is no longer a constant that depends on physical properties alone, but becomes an adjustable model parameter. This means that if an MGM/PFM approach plus a given set of parameters can be used to predict how a polymerization would unfold under a particular conditions, it might be difficult to predict how the same catalyst would behave for a wider range of reaction conditions. *This is in large part because particle morphology has a particularly strong influence on the rate of mass transfer inside the growing particles.*

References

1. Severn, J.R. and Chadwick, J.C. (eds) (2008) *Tailor-Made Polymers: Via Immobilization of Alpha-Olefin Polymerization Catalysts*, Wiley-VCH Verlag GmbH.
2. Ranz, W.E. and Marshall, W.R. (1952) *Chem. Eng. Prog.*, **48**, 141.
3. Bird, R.B., Stewart, W.E., and Lightfoot, E.N. (2001) *Transport Phenomena*, 2nd edn, John Wiley and Sons, Inc., New York.

Further Reading

The earliest published work on the modeling of olefin polymerization on supported catalysts dates from 1970. Yermakov, Y.I., Mikhaichenko, V.G., Beskov, V.S., Grabovskii, Y.P., and Emirova, I.V. (1970) *Plast. Massy*, **9**, 7 (original in Russian) were actually the first researchers to propose a model that included two levels of morphological organization. However, since it was published in Russian, it was larger unknown to the scientific community outside the Soviet Union for some time. Crabtree, J.R., Grimsby, F.N., Nummelin, A.J., and Sketchley, J.M. (1973) *J. Appl. Polym. Sci.*, **17**, 959 looked at the influence of particle structure on the diffusion process during ethylene polymerization, and Nagel, E.J., Kirilov, V.A., and Ray, W.H. (1980) *Ind. Eng. Chem. Prod. Res. Dev.*, **19**, 372 used an MGM-type model to describe the molecular weight distribution during ethylenepolymerization. Schmeal, W.R. and Street, J.R. (1971) *AIChE J.*, **17**, 1189 and Singh, D. and Merrill, R.P. (1971) Molecular weight distribution of polyethylene produced by Ziegler–Natta catalysts. *Macromolecules*, **4**, 599–604 described particle growth using what amounts to the PFM in the early 1970s (but without the clear justifications proposed by Floyd, S., Choi, K.Y., Taylor, T.W.,k and Ray, W.H. (1986) *J. Appl. Polym. Sci.*, **32**, 2935.

Chiovetta, M.G. and Laurence, R.L. (1983) Heat and mass transfer during olefin polymerization from the gas phase, in *Polymer Reaction Engineering: Influence of Reaction Engineering on Polymer Properties* (eds K.H. Reichert and W. Geisler), Hanser Publishers, Munich, and then Ferrero, M.A. and Chiovetta, M.G. (1987) *Polym. Eng. Sci.*, **27**, 1436; (b) Ferrero, M.A. and Chiovetta, M.G. (1987) *Polym. Eng. Sci.*, **27**, 1448; (c) Ferrero,M.A. and Chiovetta, M.G. (1991) *Polym. Eng. Sci.*, **31**, 886–903; (d) Ferrero,M.A. and Chiovetta, M.G. (1991) *Polym. Eng. Sci.*, **31**, 904 were the first to propose an MGM that includes particle fragmentation, and thus some inkling of the influence of morphological changes on the rates of reaction, heat-generation rate, and mass transfer during the early stages of polymerization. They modeled the macrograin as an agglomeration of concentric layers of microparticles, and the full set of mass and energy balances was solved for both levels of particle morphology. They reasoned that since diffusion occurs from the external edge of the particle, that polymer would accumulate there first, and fragmentation would therefore take place on a layer-by-layer basis of micrograins from the outside to the center of the particle. They calculated the growth factor $\phi = R_s/R_c$ of each layer, and when ϕ reached a critical value the layer was considered fragmented. They assigned different values for the macroparticle diffusivities in the nonfragmented and fragmented layers and assumed that the nonfragmented core had lower porosity than the fragmented layers. Simulations showed that the development of temperature and concentration profiles depended on fragmentation speed, and that rapid fragmentation led to high-temperature excursions in the macroparticle because this enhanced the mass transfer rate, and thus the monomer concentration in the particles.

During the same period, the group of W. Harmon Ray began to publish extensively on single particle modeling. They performed an in-depth analysis of the potential significance of internal Floyd, S., Choi, K.Y., Taylor, T.W., and Ray, W.H. (1986) *J. Appl. Polym. Sci.*, **32**, 2935 and external Floyd, S., Choi, K.Y., Taylor, T.W., and Ray, W.H. (1986) *J. Appl. Polym. Sci.*, **31**, 2231 gradients that formed our basis for the formalization of the multigrain versus PFMs that we discussed above, as well as systematically exploring the impact of these gradients on polymer properties.

Floyd, S., Heiskanen, T., Taylor, T.W., Mann, G.E., and Ray, W.H. (1987) *J. Appl. Polym. Sci.*, **33**, 1021–1065.

Further publications in the series looked at the possibility of multiple steady states in growing particles, and also conditions for overheating.

Hutchinson, R.A. and Ray, W.H. (1987) *J. Appl. Polym. Sci.*, **34**, 657.

They also combined modeling with experiments to demonstrate that mass transfer rates are linked to polymer morphology.

Hutchinson, R.A. and Ray, W.H. (1991) *J. Appl. Polym. Sci.*, **43**, 1271–1285.

Kittilsen, P., Svendsen, H., and McKenna, T.F.L. (2001) *Chem. Eng. Sci.*, **56**, 3997 developed a PFM that included convective effects. This approach was extended by Parasu Veera, U., McKenna, T.F.L., and Weickert, G. (2007) *J. Sci. Ind. Res.*, **66**, 345 using the dusty gas model. These papers present a more in-depth treatment of the development of the expressions in Section 7.3.4.

Soares, J.B.P., and Hamielec, A.E. (1995) *Polym. React. Eng.*, **3**, 261.

Soares and Hamielec used a version of the PFM that combined mass and heat transfer with the use of Stockmayer distribution to predict the MWD and CCD of polyolefins as a function of particle radial position and polymerization time, combining the modelling approaches discussed in Chapter 6 with those seen in the present chapter.

As we mentioned in Section 7.3.1, it is commonly assumed that the heat generated in the particles is removed entirely by convection. However, McKenna, T.F.L., Cokljat, D., and Spitz, R. (1999) *AIChE J*, **45**, 2392 showed that this is not always the case, and that particle–particle interactions can help to remove some the heat generated by the reaction. Eriksson, E.J.G. and McKenna, T.F.L. (2004) *Ind. Eng. Chem. Res.*, **43**, 7251 used the same approach to further explore the importance of the relative size and spacing of the particles on the heat removal, as well as the impact of the state of the reactor walls on particle temperatures (particles near fouled walls heated faster than particles near clean walls).

Eriksson, E.J.G., Weickert, G., and McKenna, T.F.L. (2010) *Macromol. React. Eng.*, **4**, 95.

A few years later, the group of Kiparissides proposed what is probably the most complete version of a SPM at the time this book is written.

Kanellopoulos, V., Dompazis, G., Gustafsson, B., and Kiparissides, C. (2004) *Ind. Eng. Chem. Res.*, **43**, 5166.

Their version of the PFM is called the random pore polymer flow model (RPPFM), and was developed for gas-phase polymerization (although it can be used for any type of polymerization with the appropriate parameter choices). The RPPFM includes convective effects as described by Kittilsen, P., Svendsen, H., and McKenna, T.F.L. (2001) *Chem. Eng. Sci.*, **56**, 3997 in the definition of the diffusivity in the macrograin mass balance, and an adjustment of the effective diffusivity as a function of changing fractions of polymer space and pore space. They also provide a correlation to calculate h in Eq. (7.14) as a function of the energy dissipation rate in a bed, and the particle concentration in the reaction.

Some of the major points with regard to different types of models, including "morphology models" (models that help predict the shape of the growing particles as a function of catalyst type, polymer properties, reaction medium, etc.) were discussed by McKenna, T.F.L. and Soares, J.B.P. (2001) *Chem. Eng. Sci.*, **56**, 3931. As the discussion above shows, increasing the length scale for diffusion in the polymer phase of the particle can lead to significant mass transfer resistance. One of the first papers to look at the impact of changing morphology on mass transfer rates was that presented by Debling, J.A. and Ray, W.H. (1995) *Ind. Eng. Chem. Res.*, **34**, 3466, who proposed a modification for the MGM where, instead of remaining rigid and incompressible, the microparticles of impact polypropylene (reactor blends of semicrystalline isotactic polypropylene and amorphous EPR made in at least two reactors in series) were allowed to deform. The EPR was considered to be practically immiscible with the isotactic polypropylene and to "ooze out" of the microparticles, thus progressively filling the pores of the macroparticle. In this way, the particle could change from multigrain to polymer flow morphology, depending on the amount of compression observed in the polymer particles (according to a compressibility factor that allowed the microparticles to deform and to fill any voids in the neighborhood). Hoel, E.L., Cozewith, C., and Byrne, G.D. (2006) *AIChE J*, **40**, 1669 used a more sophisticated version of the PFM model to describe particle growth during ethylene/propylene copolymerization to form amorphous polymer. They found that the diffusivities of the different monomers changed as a function of time in the particles, an observation they attributed to an increase in the effective length scale for diffusion in the polymer phase as the porosity decreased.

Given that the development of morphology models is in its infancy, we did not discuss this topic in any detail in this chapter. Some recent articles, such as those by Kittilsen, P., McKenna, T.F.L., Svendsen, H., Jakobsen, H.A., and Fredriksen, S.B. (2001) *Chem. Eng. Sci.*, **56**, 4015; Kittilsen, P., Svendsen, H.F., and McKenna, T.F. (2003) Viscoelastic model for particle fragmentation in olefin polymerisation. *AIChE J.*, **49** (6), 1495–1507, or Di Martino, A., Weickert, G., Sidoroff, F., and McKenna, T.F.L. (2007) *Macromol. React. Eng.*, **1**, 338 show that there is a well-defined relationship between the particle morphology and the parameters mentioned above. The group of Grof, Z., Kosek, J., Marek, M., and Adler, P.M. (2003) *AIChE J.*, **49**, 1002 have developed the most in depth and extensive set of models for the study of particle morphology in the area of olefin polymerization on supported catalysts. They have looked at the modeling of particle morphology from different angles: examining ways to develop a three-dimensional (3D) reconstruction of particle in order to overcome some of the limitations mentioned above in terms of loss of detail in the evolution of particle morphology. They have also looked at advanced morphological representations of the particles to use a force balance to predict particle morphogenesis.

Grof, Z., Kosek, J., and Marek, M. (2005) *Ind. Eng. Chem. Res.*, **44**, 2389. Grof, Z., Kosek, J., and Marek, M. (2005) *AIChE J.*, **51**, 2048.

Depending on the initialization conditions, the authors simulated the formation of various morphologies in the course of particle growth, such as compact particles of low porosity, particles with oriented macrocavities, hollow particles, fines generation, onionlike structure, attrition of microelements from the surface, and cobweb structures. Despite the great interest and potential usefulness of these approaches, much work remains to be done before they can be combined with heat and mass balances to provide us with a full predictive model that links transport phenomena on the one hand to the evolution of particle morphology on the other.

8
Developing Models for Industrial Reactors

> A model's just an imitation of the real thing.
> *Mae West (American actress, 1892–1980)*

> Engineering problems are under-defined, there are many solutions, good, bad and indifferent. The art is to arrive at a good solution. This is a creative activity, involving imagination, intuition and deliberate choice.
> *Sir Ove Nyquist Arup (1895–1988)*

8.1
Introduction

Throughout this book, we have discussed how phenomena taking place at different length scales influence polyolefin microstructure, particle growth, and olefin polymerization rates. We have learned that the catalyst type ultimately determines the polymer microstructure for a given set of polymerization conditions such as temperature and concentrations of monomer, comonomer, and chain transfer agent, and also that the polymerization conditions themselves are a consequence of the type of support and reactor used to produce the polyolefin. We have also seen that different reactor technologies offer a distinct means of contacting the catalysts and reactor fluids, and that the different residence time distributions (RTDs) and macroscopic flow fields will all have an effect on the properties of polyolefins made at the active sites.

Let us reconsider Figure 7.1, reproduced for convenience as Figure 8.1, but this time in the context of developing models for industrial reactors. In the previous chapter, we used this figure to illustrate the concept that a supported catalyst particle is at the heart of a polymerization process, acting as a filter between the macroscopic events such as mixing, bulk heat transfer, liquid injections and the impact of particle interactions on the one side, and the reaction, crystallization, and other molecular level events that control the architecture of the polyolefin molecules on the other. However, in light of the remarks we made above, we can also consider this figure in the context of developing models for industrial reactors. It also shows the interconnection between the different characteristic length scales

Polyolefin Reaction Engineering, First Edition. João B. P. Soares and Timothy F. L. McKenna.
© 2012 Wiley-VCH Verlag GmbH & Co. KGaA. Published 2012 by Wiley-VCH Verlag GmbH & Co. KGaA.

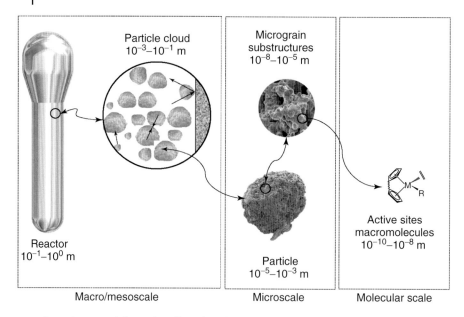

Figure 8.1 Rough hierarchy of length scales in an olefin polymerization process, placing the particle at the heart of events as a filter between macroscopic events on the left-hand side and molecular events on the right-hand side.

of the phenomena occurring inside the polymerization reactor. For example, in the case of a fluidized bed reactor (FBR), the injection of gas (and eventually liquids) will fluidize the particles in the bed. This will cause a complex flow field to develop, leading to the mixing (and perhaps segregation) of the powder. Particles can therefore be exposed to different velocity fields, and thus heat transfer conditions. The heat transfer conditions will have an impact on the particle temperature and, therefore, on intraparticle rate of polymerization and on the polymer molecular weight. Depending on temperature, molecular weight, and α-olefin content, the polymer may soften if overheated. If the mixing conditions in the particle cloud favor particle–particle contact, then the softened particles can eventually agglomerate, leading to lump or chunk formation. If this becomes serious, then bed operation can become compromised, and one may be faced with a "single particle" with a volume of several cubic meters such as the one shown in Figure 4.4.

A complete phenomenological mathematical model for olefin polymerization in industrial reactors should, therefore, ideally include a description of phenomena taking place from the molecular scale to macroscale, where the models can be grouped according to the length scale that they describe, as suggested in Figure 8.1. A summary of the different phenomena that might be included at each length scale is given in Table 8.1. It may come as a disappointment, but perhaps not as a surprise once we remember that "a model is just an imitation of the real thing", to learn that many mathematical models for industrial reactors ignore several details. For instance, many models assume that the conditions in the

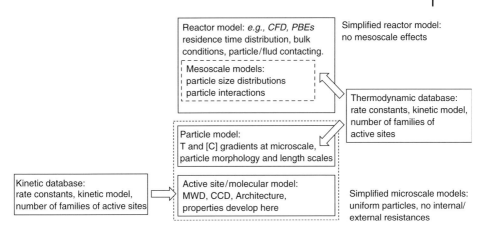

Figure 8.2 Hierarchy of the modeling of industrial polyolefin reactors. Dashed lines indicate common simplifications.

polymerization reactor are uniform (top box in Figure 8.2) and that the equations derived in Chapters 5 and 6 can be directly applied to bulk values of temperature and concentrations of monomer, comonomer, and hydrogen (bottom box in Figure 8.2). As we mentioned earlier, this last simplification might be a good approximation for solution polymerization reactors under uniform, well-mixed conditions but will depart from reality (but, not necessarily, by much) for supported catalysts in slurry and gas-phase processes.

If this hypothesis is an oversimplification, why is it so often made? Mostly, because the cost associated with the additional effort required to integrate microscale, mesoscale, and macroscale phenomena in a single model, and to acquire accurate physical and kinetic data, does not necessarily lead to better quantitative predictions when it comes to industrial reactors. Uncertainties on model parameter values can be too high to try to decouple "true" polymerization kinetic parameters from mass and heat transfer effects, particularly for multiple-site-type complexes such as Ziegler–Natta and Phillips catalysts. As much as adjustments to the diffusivity allow us to use the simplified polymer flow model (PFM) to predict polymerization rates inside a growing polymer particle, as well as providing useful theoretical insight into physical phenomena taking place during particle growth, it is possible that apparent kinetic parameters will do an equally good job from an engineering perspective.

The objective of this chapter is to bring together the different elements that we have discussed in the seven previous chapters into a coherent whole. We will not attempt to develop models of full-scale industrial reactors, as the nature of the model strongly depends on the objectives of the model builder and the information available. Rather, we use some simplified examples to demonstrate the concepts involved in model building, as well as the importance (and cost) of the choices that can be made in terms of model complexity. To begin with, let us first consider a simplified example of a reactor model, and then discuss how this simplified model could be altered to provide more detail.

Table 8.1 Description of phenomena to be modeled at different length scales of olefin polymerization reactors.

Length scale	Macro/mesoscale (reactor scale)	Microscale (particle scale)	Particle (molecular scale)
Dimensions	$10^{-3} - 10$ m	$10^{-8} - 10^{-3}$ m	10^{-10} m
Phenomena	Macromixing: reactor scale concentration and temperature gradients	Particle morphology: porosity, relative length scale of polymer and pore phases, particle shape, and distribution of phases in high impact polypropylene	Polymerization kinetics: propagation, transfer, LCB formation, site activation, and deactivation
	Segregation of particles Residence time distribution and particle size distribution	Interphase heat and mass transfer phenomena Internal heat and mass transfer phenomena	Microstructure formation: MWD, CCD, and LCB Chain crystallization
	Reactor temperature profiles, hot spots	Observed kinetics, rate-limiting steps	–
	Process safety and reactor stability	Phase equilibrium, monomer sorption and desorption in polymer phase, diffusion	–
	Fouling and sheeting, static electricity	Particle agglomeration	–
	Particle interactions, aggregation	Micromixing	–
	Hydrodynamics and local velocity fields	–	–
	Overall heat and mass balances	–	–

How can we combine the MGM and effects of reactor RTD to obtain a more complete description of polymerization in industrial reactors? Some of these aspects have been discussed in the literature but would require a lengthier treatment than space allows here. We, alternatively, show a rather simple, but elegant, treatment that combines some of the aspects we covered in our discussion on microscale, mesoscale, and macroscale phenomena to model polyolefin reactors under steady-state operation.

Let us consider the case of a continuous olefin polymerization process taking place in a single well-stirred autoclave reactor, operating with a heterogeneous

Ziegler–Natta or supported metallocene catalyst. Initial assumptions for a straightforward reactor model include the following:

- The reactor is well mixed and its conditions are uniform so that the RTD is that of an ideal CSTR.
- The RTD of the polymer/catalyst particles is independent of their size.
- All catalyst particles have the same size.
- The reactor is operating at steady state.

The general expression for the monomer molar balance under steady-state conditions is

$$0 = -\overline{R_p}V + F_{M,in} - sM \tag{8.1}$$

where V is the reactor volume, $\overline{R_p}$ the average polymerization rate per unit volume of the reactor, $F_{M,in}$ the molar flow rate of the monomer feed, M the monomer concentration in the continuous particle phase, and s the reciprocal of the average residence time (V/q where q is the volumetric flow rate at the outlet).

$\overline{R_p}$ can be written as

$$\overline{R_p} = \overline{R'_p} N_s \tag{8.2}$$

where N_s is the number of macroparticles per unit volume of the reactor and $\overline{R'_p}$ is the average polymerization rate per particle, given by the expression:

$$\overline{R'_p} = \int_0^\infty E(t)\, R_p(t)\, dt \tag{8.3}$$

In Eq. (8.3), $R_p(t)$ is the polymerization rate in a polymer particle (secondary particle, according to the MGM terminology) with a residence time t in the reactor, which can be calculated by integrating the single particle material balances presented in Chapter 7, and $E(t)$ is the RTD of the reactor. We can recall from our classic reactor engineering studies that the RTD, $E(t)$, of a CSTR is given by

$$E(t) = \frac{1}{\bar{t}} \exp\left(\frac{-t}{\bar{t}}\right) \tag{8.4}$$

where $\bar{t}$ is the average residence time of the reactor. For practical purposes, a value of 5–6 average residence times is enough for the upper limit of the integral in Eq. (8.3).

Equations (8.1) and (8.3) can be solved by assuming a value for M, calculating $R_p(t)$ with a single particle model such as the MGM, $\overline{R'_p}$ with Eq. (8.3) and $\overline{R_p}$ with Eq. (8.2), and finally recalculating the value of M with Eq. (8.1) until convergence. At this point, an additional simplification occasionally introduced is to ignore heat and mass transfer limitations in the particles and assume the particles are at the same temperature as the continuous phase and the monomer concentration at the active sites is in equilibrium with the concentration in the continuous phase. Similar considerations apply to calculate the concentrations of comonomers, hydrogen, and any other reactant in the system.

If we consider a stirred autoclave to be an ideal CSTR, the steady-state energy balance to model nonisothermal reaction operation is

$$\sum_i C_{pi} w_{mi} N_i \frac{dT}{dt} = 0 = UA(T_w - T) + R_p(-\Delta H_r)V \\ - \sum_i F_{in,i} w_{mi} C_{pi}(T - T_{io}) - \dot{m}_{vap}(-\Delta H_{vap}) \quad (8.5)$$

where the subscript i indicates monomer type, hydrogen, nitrogen, diluent, or impurities; N is the number of moles of a given component in the reactor; C_p the average heat capacity; w_{mi} the molecular weight of the component; $(-\Delta H_r)$ the heat of the reaction; R_p the polymerization rate; U the global heat transfer coefficient; T_w the coolant temperature; T_{i0} the temperature of feed stream to reactor; A the total heat transfer area of the reactor; $\dot{m}_{vap}$ the vaporization rate of any volatile component; and $(-\Delta H_{vap})$ the heat of vaporization. If our well-mixed CSTR is operating at steady state, it will of course be isothermal. This balance will need to be solved in order to design the heat-exchange system of the reactor, and, as with the dynamic version of Eq. (8.1), in nonsteady state conditions such as grade changes.

For heterophase systems, the monomer concentration in the continuous phase of the reactor must be converted to concentration in the polymer phase surrounding the active sites with a thermodynamic relationship. The simplest methods for accomplishing this rely on constant partition coefficients,

$$K = \frac{[M_p]}{[M_{fluid}]} \quad (8.6)$$

where $[M_p]$ is the concentration in the polymer and $[M_{fluid}]$ the concentration in the fluid (bulk or pore space). Similarly, for diluent slurry reactors where the monomer is introduced in the gas phase, a partition coefficient such as Henry's law constant must also be used to calculate the concentration of monomer in the diluent, which, in turn, is used to estimate the concentration of monomer in the polymer phase surrounding the active sites. The advantage of this approach clearly lies in its simplicity and ease of integrating into a full reactor model. However, this approach makes it difficult to account for changes in polymer properties and reaction conditions or cosolubility effects in multicomponent systems.

Evidently, more sophisticated thermodynamic relationships relating to the concentration of the monomer in the gas phase, diluent, and polymer can be used, but, from a practical point of view, are only justified when we know the polymerization kinetic constants very well. When this is the case, models such as the Sanchez-Lacombe equation of state [1] or a perturbed-chain statistical associating fluid theory (PC-SAFT) approach [2] are to be preferred. In the event that cosolubility effects are judged important, the latter approach appears to be the only one currently applicable. Evidently, the consequence of such a (semi) mechanistic approach is the need for sound parameter estimates and a more complex model solution.

Once the concentration of the monomer in the reactor is known, any additive property of the polymer, $X(t)$ (moments of living and dead chains, chain length averages, and average comonomer compositions), produced in particles with a given residence time t can be calculated combining the balances in the particles and the equations in Chapter 6. If we treat the polymer particles as a macrofluid, the average value of any property $\overline{X}$ is given by

$$\overline{X} = \int_0^\infty E(t)X(t)dt \tag{8.7}$$

For the nonadditive properties, such as the chain length distribution, $w(r)$, the average value can be obtained by the equation:

$$\overline{w(r)} = \frac{\int_0^\infty E(t)v_1(t)w(r,t)dt}{\int_0^\infty E(t)v_1(t)dt} \tag{8.8}$$

Among the assumptions we made above, the weakest was to consider that all catalyst particles had the same size, since we know that industrial catalysts have a distribution of particle sizes. Defining the catalyst particle size distribution (PSD) (in volume) as $n_c(v)$ such as,

$$\int_0^\infty n_c(v)dv = 1 \tag{8.9}$$

we can easily extend the treatment proposed above for catalysts that have a distribution of particle sizes,

$$\overline{X} = \int_0^\infty n(v) \left(\int_0^\infty E(t)X(t)dt \right) dv \tag{8.10}$$

$$\overline{R'_p} = \int_0^\infty n(v) \left(\int_0^\infty E(t)R_p(t)dt \right) dv \tag{8.11}$$

$$\overline{w(r)} = \frac{\int_0^\infty n(v) \left(\int_0^\infty E(t)v_1(t)w(r,t)dt \right) dv}{\int_0^\infty n(v) \left(\int_0^\infty E(t)v_1(t)dt \right) dv} \tag{8.12}$$

Equations (8.3) and (8.7–8.12) are valid for any RTD and any PSD provided that expressions for $E(t)$ and $n(v)$ are available. As discussed in Chapter 4, it is common practice to approximate the RTDs of a number of the continuous reactors with that of an ideal CSTR. Depending on the quality of the information one has about the process, physical and kinetic parameters, and the requirements of the model, this might be an acceptable simplification. In the case of overall mass and energy balances and the dimensioning of a reactor, this is probably the case. However, if a more detailed picture of the reactor is required, for instance, if one wishes to obtain an accurate picture of how the reactor behavior influences the final molecular properties of the polymer, a better picture of the RTD might be required.

As an example, let us consider an FBR. As we mentioned in Chapter 4, fluidized beds for olefin polymerization typically operate in the bubbling regime (i.e., at three to eight times the minimum fluidization velocity, u_{mf}). This means that the bed will contain three phases (ignoring any condensed material for now): an emulsion phase where the gas flow rate is approximately that required for minimum fluidization, a bubble phase (only gas), and a wake phase that trails behind the bubble (well-mixed region of particles and gas at lower solids than the emulsion phase). The simplest approximation to the RTD of the bed is that of a CSTR where the emulsion phase (including the solids in the wake) and the bubble phases are well mixed, as shown to the left of the schema in Figure 8.3. Here, the emulsion phase is said to be well mixed, and the bubble size of the gas phase is uniform throughout the reactor. This last point is important since the heat and mass transfer coefficients that determine the flow rates indicated in the diagram do in fact depend on the bubble size. This approach allows us to calculate the overall molecular weight and composition distributions, provided we have the necessary thermodynamic and kinetic data. In addition, with the appropriate correlations of bubble size as a function of the particle size and gas flow rates, it is also possible to get a rough understanding of the impact of changing fluidization conditions on the average bed temperature, and thus the observed rates and properties.

Further refinements are brought by considering that the bubble phase is in fact divided into N well-mixed compartments (the middle schema in Figure 8.3), each of which exchanges heat and mass with the emulsion phase. Each gas-phase compartment is considered well mixed, and the bubbles in this cell will have the same size, but this size can vary as we move up through the bed from the distributor plate to the freeboard zone. In this model, allowing the bubble size to change means that it is possible to calculate bubble-size-dependent heat and mass transfer coefficients as a function of bed height, and therefore to develop a gas-phase temperature profile, and a better estimate of the impact of fluidization conditions and average particle sizes on the reactor behavior. Note that both the constant bubble size model and the bubble growth (with well-mixed powder) approaches consider that all of the particles in the bed are of the same size (or at least that any variations in particle size do not affect the fluidization conditions or transfer coefficients).

The most complete approach is that shown in the third schema of Figure 8.3. Here, the emulsion and bubble phases are divided into J and N different compartments, respectively. Note that most of the time all of the bubble compartments are of equal volume, as are those of the emulsion phase (although emulsion and bubble compartments do not have to be of equal volume). This representation of the RTD of the fluidized bed allows one to account for many different phenomena, including bubble growth (as mentioned above); a PSD; particle growth, agglomeration, and rupture; entrainment of monomer up (or down) the bed in the solid phase; and bed segregation. Of course, in order to include all of these phenomena in the model, it is necessary to use models relying on population balance equations (PBEs) as well (note that population balances can also be used with the other representations, but with the necessary limitations that this will not influence the interaction between the emulsion and the bubble phases). A detailed discussion of population balance modeling is outside the scope of this chapter, but the concept can be explained by considering Figure 8.4. Here, each of the J compartments of the emulsion phase will contain powder with a given PSD. This PSD is divided into a certain number of sections according to particle size (or volume), and the population balance is a mathematical expression that keeps track of the quantity of particles in each of the given size ranges. Different mechanisms for particles to change their size and

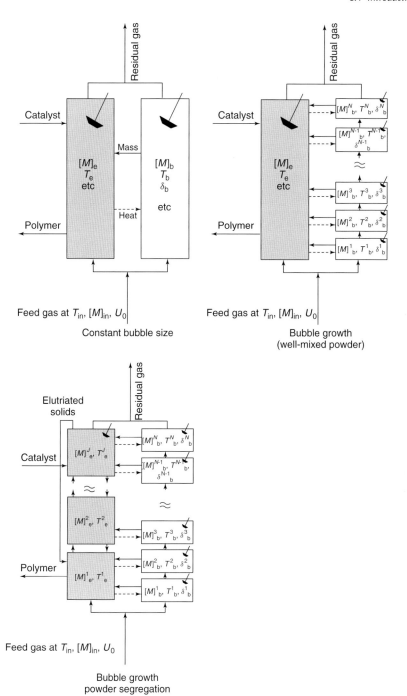

Figure 8.3 Different representations of the RTD of the phases in an FBR allow one to understand the impact of different process parameters.

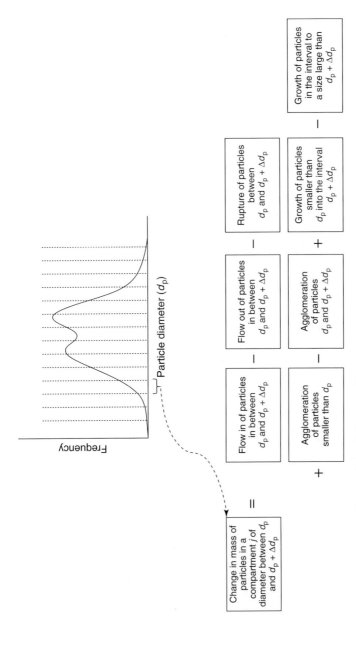

Figure 8.4 Concept of a population balance model.

Figure 8.5 Schematic representation of the possible residence time distribution of a slurry loop reactor.

either add to or subtract from the quantity in a given size range are suggested in Figure 8.4. Of course, each of these phenomena need specific data and parameters associated with them (for instance, agglomeration kernels that link changes in size to physical causes such as softening and changes in velocity), adding exponentially to the complexity and cost of the model.

Note that the concept of PBE modeling is of course totally independent of the RTD; it is quite reasonable to use a PBE to calculate the evolution of particle size, and so on, even with a simple representation of the residence time, such as the constant bubble diameter model in Figure 8.3 (or equivalently, the CSTR discussed above).

Certain reactor types, such as the horizontal stirred bed used in the Innovene and Horizon PP processes will require the imposition of more complex RTDs, $E(t)$, than the CSTR model given by Eq. (8.5). The HSBR is said to have an RTD roughly equivalent to four to six stirred tanks in series (or plug flow with dispersion). Other reactors, such as the slurry loops can be approximated by that of a CSTR. However, if a more accurate representation of particle evolution and kinetics is required, then the reactor might be better represented by two CSTRs (corresponding to the recirculation pump and the settling legs) interconnected by plug flow reactors (eventually with dispersion) of different lengths (Figure 8.5).

The mathematical treatment we described above combines the MGM (or any other single particle model, such as the PFM), the polymerization kinetic models described in Chapter 5, and the different reactor configurations discussed in Chapter 7 (through their respective RTDs, $E(t)$). There is no "best" level of complexity for "standard" modeling challenges. A cost–benefit analysis must be done by the individual(s) building the model. Unless a quantitative understanding of polymerization kinetics steps, particle growth, and interparticle and intraparticle mass and heat transfer resistances is available, the integration of all these phenomena in a single model that details the RTD and includes particle agglomeration, rupture, and growth in a PBE-based approach will simply be a stimulating academic exercise, but will hardly lead to mathematical models with improved predictive capabilities. A well-defined, simple model using an approximate residence, a single particle model that uses diffusivity as an adjustable parameter, and a reasonable kinetic model based on some solid laboratory data might actually give a more robust picture of what is going on inside the reactor.

References

1. Novak, A., Bobak, M., Kosek, J., Banaszak, B.J., Lo, D., Widya, T., Ray, W.H., and de Pablo, J.J. (2006) Ethylene and 1-Hexene sorption in LLDPE under typical gas-phase reactor conditions:

experiments. *J. Appl. Polym. Sci.*, **100**, 1124–1136.

2. Banaszak, B.J., Lo, D., Widya, T., Ray, W.H., and de Pablo, J.J. (2004) Ethylene and 1-Hexene sorption in LLDPE under typical gas phase reactor conditions: a priori simulation and modeling for prediction of experimental observations. *Macromolecules*, **37**, 9139–9150.

Further Reading

The following papers by Jorge Zacca *et al.* from Professor Harmon Ray's group at the University of Wisconsin (Madison) are two of the most widely cited papers we have run across in terms of the development of models of the effect of RTDs on olefin polymerization. While these papers use a PBE-based approach, it was assumed that the RTD is independent of the particle size. Despite this limitation, these two papers give an excellent overview of the issue involved in modeling the RTD, and its potential impact on polymer properties and reactor behavior.

Zacca, J.J., Debling, J.A., and Ray, W.H. (1996) Reactor residence time distribution effects on the multistage polymerization of olefins – 1. Polymer properties: bimodal polypropylene and linear low density polyethylene. *Chem. Eng. Sci.*, **51**, 4859–4886.

Zacca, J.J., Debling, J.A., and Ray, W.H. (1997) Reactor residence time distribution effects on the multistage polymerization of olefins – 2. Basic principles and illustrative examples, polypropylene. *Chem. Eng. Sci.*, **52**, 1941–1967.

And a final article looks at nonidealities in the mixing of loop reactors:

Reginato, S., Zacca, J.J., and Secchi, A.R. (2003) Modeling and simulation of propylene polymerization in nonideal loop reactors. *AIChE J.*, **49**, 2643–2654.

Soares, J.B.P., and Hamielec, A.E. (1995) *Macromol. Theory Simul.*, **4**, 1085.

Soares and Hamielec also developed a model to account for RTD effect on the PSD of polyolefin particles. Their model was generic and applicable to any RTD and PSD shape.

Another article from the group of Professor Kiparissides presents an interesting treatment of a cascade of slurry loop reactors for ethylene polymerization that looks at the dynamics of running two loops (treated like CSTRs with semicontinuous product removal) in series.

Touloupides, V., Kanellopoulos, V., Pladis, P., Kiparissides, C., Mignon, D., and Van-Grambezen, P. (2010) Modeling and simulation of an industrial slurry-phase catalytic olefin polymerization reactor series. *Chem. Eng. Sci.*, **65**, 3208–3222.

Two separate articles provide an interesting treatment of the RTD of the HSBR reactors used in the Innovene and Horizon processes. The article from Caracotsios is the first to have appeared in the literature, whereas the second by Dittrich and Mutsers raises interesting questions about what can influence the RTD in this type of reactor.

Caracotsios, M. (1992) Theoretical modelling of Amoco's gas phase horizontal stirred bed reactor for the manufacturing of polvolefine resins. *Chem. Eng. Sci.*, **47**, 2591–2596.

Dittrich, C.J. and Mutsers, S.M.P. (2007) On the residence time distribution in reactors with non-uniform velocity profiles: the horizontal stirred bed reactor for polypropylene production. *Chem. Eng. Sci.*, **62**, 5777–5793.

The three articles below provide an interesting overview of the progress of fluidized bed models with increasing degrees of complexity much along the lines of the three images shown in Figure 8.3. The first is the definitive constant bubble diameter model of an FBR, and the second article from K.Y. Choi's group is an extension of the first article by Choi and Ray to allow for bubble growth. The last article from the group of Professor Kiparissides is certainly one of the most complete treatments of FBR modeling in the open literature at the time the book was written (short of including computational fluid dynamic simulations to model the RTD), and it provides an excellent treatment of how population balance models can be applied to this type of problem.

Choi, K.Y. and Ray, W.H. (1985) The dynamic behaviour of fluidized bed reactors for solid catalysed olefin polymerization. *Chem. Eng. Sci.*, **40**, 2261–2279.

Kim, J.Y. and Choi, K.Y. (2001) Modeling of bed segregation phenomena in a gas phase fluidized bed olefin polymerization reactor. *Chem Eng. Sci.*, **56**, 4069–4083.

Dompazis, G., Kanellopoulos, V., Touloupides, V., and Kiparissides, C. (2008) Development of a multi-scale, multi-phase, multi-zone dynamic model for the prediction of particle segregation in catalytic olefin polymerization FBRs. *Chem. Eng. Sci.*, **63**, 4735–4753.

The two articles by Khare *et al.* provide an interesting overview of a simplified approach that can be used to model an entire process for an HDPE slurry process, and a gas-phase PP process using a commercial simulation software package. They demonstrate the type of information required and the fact that a certain amount of process improvements can be obtained using well-defined, but manageable, reactor and unit operations models.

Khare, N.P., Seavey, K.C., Liu, Y.A., Ramanathan, S., Lingard, S., and Chen, C.C. (2002) Steady-state and dynamic modeling of commercial slurry high-density polyethylene (HDPE) processes. *Ind. Eng. Chem. Res.*, **41**, 5601–5618.

Khare, N.P., Lucas, B., Seavey, K.C., Liu, Y.A., Sirohi, A., Ramanathan, S., Lingard, S., Song, Y., and Chen, C.C. (2004) Steady-State and Dynamic Modeling of Gas-Phase Polypropylene Processes Using Stirred-Bed Reactors. *Ind. Eng. Chem. Res.*, **43**, 884–900.

Index

a

agglomeration   286
Alfrey–Goldfinger equation   255
Arrhenius law   152, 153
autoclaves   105–106

b

Bernoullian model   145, 146, 147, 212, 232, *233*
bimetallic mechanisms   83
bimodal polyethylenes   8
Borstar process   118, *127*
buildup rate   117
bulk density   277

c

chain length distribution (CLD)   188, 190, 191–194, 199
– non-steady-state reactor operation effect on   195–197
chain transfer   79, *80*
– agent   311
– steps, for terminal model   213
chain walking   67–68, *68*
– Monte Carlo simulation of polyethylene SCB distribution by   259–262
chemical composition distribution (CCD)   6–9, 12, 54, 55, 256
– crystallizability-based techniques   29–40
– high-performance liquid chromatography   40–43
– multiple-site catalysts   222–232
– single-site catalysts   212–222
cocatalyst   57, 66, 77–78, *78*, *80*, *81*, 85, 100, 131, 132, , 135, 144, 155, 162, 165, 263
comonomer effect   148
– on polymerization rate   173–179

comonomer sequence length distribution (CSLD)   232–237
condensed mode operation   96–97
constrained geometry catalyst (CGC). *See* half sandwich metallocene
convection   292, 293
– in particles   301–304
coordination catalysts   76–86
copolymerization   5, 6, 10, 63, 67, 70, 83, 122, 126, 132, 145–149, 170, 173, *174*, *175*, 177–178, 179, 212, 214, 237, 251, 290, 297
– models   213
– terminal model   80, *81*
Cossee's mechanism   81, *82*, 83
cross-fractionation techniques   43–46
crystallizability-based techniques   29–40
crystallization analysis fractionation (CRYSTAF)   33–36, 218–220, *219*, 224
crystallization elution fractionation (CEF)   36, *37*, 39
cumulative distribution   34

d

deashing   112, 122
differential distribution   34
differential scanning calorimetry (DSC)   15, 38–40
diffusion   18, 28, 179, 181, 276, 288, 289, 293–294, 295, 301, 302, 304, 305–306
donors   59
dormant β-agostic interaction site models
– comparison with slow metal hydride   167–169
– hydrogen effect and α-olefin concentration on ethylene copolymerization rate using   177–178
Dowlex process   120
drop-in technology   66

Polyolefin Reaction Engineering, First Edition. João B. P. Soares and Timothy F. L. McKenna.
© 2012 Wiley-VCH Verlag GmbH & Co. KGaA. Published 2012 by Wiley-VCH Verlag GmbH & Co. KGaA.

e

effective diffusivity   289
elution volume   17
evaporative light scattering detector (ELSD)   41
external donors   59

f

Fick's law   28
field flow fractionation   27–28
fines generation   277, 286
Flory distribution   188, 190, 191, 192, 193, 194, 198–201, 207, 216
– MWD deconvolution of polyethylene sample with heterogeneous Ziegler–Natta catalyst using   208–212
Flory–Schulz most probable distribution. *See* Flory distribution
fluidized bed reactor (FBR)   89, 91–97, 115, 116, *118*, 312, 317, *319*

g

gas-phase reactors   90–91, 132
– fluidized bed   91–97
– horizontal stirred   99–102
– multizone circulating   102–104
– vertical stirred bed   97–99
gel permeation chromatography (GPC)   17, 22–23, 24, 26–28, *45*, 229
good morphology   277

h

half sandwich metallocene   63
heat transfer   272, 273, 288, 291, 292, 293, 294, *296*, 300, 301
high-density polyethylene (HDPE)   4–5, 9
high-performance liquid chromatography (HPLC) techniques   15, 40–43
homopolymerization   134–145
– trigger mechanism for   *157*
Horizone process   124
horizontal stirred gas-phase reactor   99–102
Hostalen process   112, *113*
hydrogen effect   83–85
– on polymerization rate   161–173

i

industrial reactors, developing models for   311–321
Innovene process   116, 124
internal donors   59
isotactic polypropylene   80

l

late transition metal catalysts   67–70
linear low-density polyethylene (LLDPE)   5–7, *8*, *9*, 228
long chain approximation (LCA)   147, 148
long-chain branching   2, 5, 9, 10, 46–51, 237–250
– microstructure investigation, from average polymer properties   245–248
loop reactors   *105*, 106–107, 114, *321*
low-density polyethylene (LDPE)   2, 3, 4, 5, 10

m

macromonomers   63, 77, 78
mass transfer   173, 179, 272, 273, 278, 286, 287, 288, 289, 293, 294, 295, *296*, 297, 298–299, 301–304, 305
metallocenes   2, 5, 9, 11, 12, 31, *35*, 36, *40*, 42, 44, 54, 55, 62–67, 70, 73–75, *76*, 78, 81, 104, 114, 117, 120, 137, 139, 154, 156, 162, 169, 218, *281*, 283
methylaluminoxane (MAO)   62, 65–66
micrograins   *274*, 275, 292
microstructural characterization   15–17
– chemical composition distribution
– – crystallizability-based techniques   29–40
– – high-performance liquid chromatography   40–43
– cross-fractionation techniques   43–46
– long-chain branching   46–51
– molecular weight distribution   17
– – field flow fractionation   27–28
– – size exclusion chromatography   17–27
microstructural modeling   187
– instantaneous distributions   188
– – chemical composition distribution   212–232
– – comonomer sequence length distribution   232–237
– – long-chain branching distribution   237–250
– – molecular weight distribution   188–212
– – polypropylene   250–251
– Monte Carlo simulation   251–252
– – dynamic models   262–268
– – steady-state models   252–262
Mitsui Hypol process   126
molar flow rate   315
molecular weight distributions (MWDs)   54, 55, 61, 63, 95, 101, 103, 106, 108, 110, 111, 114, 120, 125, *230*
– deconvolution   200, 201–206, 208–211, 225–229

- experimental profiles prediction measured by GPC 248–249
- field flow fractionation 27–28
- multiple-site catalysts 199–212
- single-site catalysts 188–199
- size exclusion chromatography 17–27
monomer diffusivity 289, 290
Monte Carlo simulation 251–252
- dynamic models 262–268
- steady-state models 252–262
multigrain model (MGM) 287–292
- limitations 304–307
multiple-site catalysts 149–152
- chemical composition distribution 222–232
- molecular weight distributions 199–212
- triad distributions for copolymers with model 234–237
multizone circulating reactor 102–104

n

negative polymerization orders, with late transition metal catalysts 179–181
Novolen process 123

o

olefin polymerization *See individual entries*
α-olefins 1–2, 5–8, 19, 26, 29, 31, 38, 41, 47, 48, 50, 61, 63, 76, 77, 81, 83, 89, 107, 120, 173–174, 177–178, 179, 187
orthodichlorobenzene (ODCB) 17, 30

p

particle size distribution 317, 318
pen-penultimate model 212
penultimate model 146, 212
Phillips catalyst 2, 5, 53, 54, 55, 56, 61, 66, 76, 78, 83, 132, 149, 151, 200, 201, 205, 313
Phillips process 114
plug flow reactor 98
polydispersity 192, *197*
polyethylene manufacturing processes
- gas-phase processes 115–118
- mixed-phase processes 118–119
- slurry (inert diluent) processes) 112–115
- solution processes 119–121
polyethylene resins 4–10
polymer density 277
polymer flow model (PFM) 287, 292–295, 313, 321
- limitations 304–307
polymerization catalysis and mechanism 53–56
- catalyst types

- – late transition metal catalysts 67–70
- – metallocenes 62–67
- – Phillips catalysts 61
- – Ziegler–Natta catalysts 56–61
- with coordination catalysts 76–86
- supporting single-site catalysts 70–76
polymerization kinetics 131–133
- fundamental model for 134
- – kinetic constants temperature dependence 152–154
- – multiple-site catalysts 149–152
- – number of moles of active sites 154–156
- – single-site catalysts 134–149, 157–161
- nonstandard models 156
- – comonomer effect on polymerization rate 173–179
- – hydrogen effect on polymerization rate 161–173
- – negative polymerization orders with late transition metal catalysts 179–181
- – polymerization orders greater than one 156–161
- for second-order deactivation and noninstantaneous activation 144–145
- vapor-liquid-solid equilibrium considerations 181, 183–184
polyolefins 1–4. *See also individual entries*
- polyethylene resins 4–10
- polypropylene resins 10–12
polypropylene 250–251
- manufacturing processes 121–122
- – gas-phase reactors 122–125
- – mixed-phase processes 125–128
- – slurry (inert diluent) processes 122
- resins 10–12
poor morphology 277
population balance model 318, *320*, 321
prepolymerization 278
pseudo constants method 80
pseudokinetic constants 175
pseudo-propagation rate constant 146, 147, 148

r

Ranz–Marshall correlation 298, 300
reactors and processes 87
- configurations and design 89–90, *109*
- – gas-phase reactors 90–104
- – slurry-phase reactors 104–107
- – solution reactors 107–108
- olefin polymerization processes 109–112
- – polyethylene manufacturing processes 112–121

reactors and processes (*contd.*)
– – polypropylene manufacturing processes 121–128
relaxation 28
replication phenomenon 276, 284
residence time distributions (RTDs) 89, 91, 93, 94, 98, 99, 101, 103, 107, 108, 114, 315, 311, 317, *319*, *321*
retention time 17

s

sandwich compounds 62
short chain branches (SCB) 2, 5, 6, 67, 68, 69, 230
– formation, by chain walking 259–262
single particle modeling 271–275, 286–287
– convection in particles 301–304
– multigrain model (MGM) 287–292
– – limitations 304–307
– particle fragmentation step 275–284
– particle growth 284–286
– – analysis with MGM/PFM approach 295–301
– polymer flow model (PFM) 292–295
– – limitations 304–307
single-site catalysts 157–161
– chemical composition distribution 212–222
– copolymerization 145–149
– homopolymerization 134–145
– molecular weight distributions 188–199
– polymerization catalysis 70–76
– polymerization kinetics 134–149
size exclusion chromatography 17–27
slurry-phase reactors 104–105, 132, *162*, 179, 183
– autoclaves 105–106
– loop reactors 106–107, *321*
solution reactors 107–108
Spherilene line of processes 116
Spheripol process 125
Spherizone process 126
Stockmayer distribution 215, 216, 217, 218, 220, 221, 223, 227, 250, 256

stopped-flow reactor 263
– simulation, with dynamic Monte Carlo model 266

t

temperature gradient interaction chromatography (TGIC) 41–42
temperature rising elution fractionation (TREF) 30–31, 33, 36, *37*, *38*, 221, 222, 224, 225, 226
terminal model 145–146, 147, 148, 212, *233*
– chain transfer steps for *213*
– Monte Carlo simulation flowchart of terpolymerization with *257*
trichlorobenzene (TCB) 17, 30, 41
true fragmentation mechanism 281
turnover frequency (TOF) 179

u

Unipol processes 115, *116*, 122
universal calibration curve 20, *21*

v

vapor-liquid-solid equilibrium considerations 181, 183–184
vertical stirred bed reactor 97–99
viscosity branching index 23

z

Ziegler–Natta catalysts 6, *7*, *8*, 11–12, 30, *31*, 34, *35*, *43*, 53, 54, 55, 56–61, 63, 66, 70, 76, 78, 81, 83, *84*, *85*, *86*, 101, 104, 112, 116, 117, 120, 132, 149, 150, *151*, *162*, 169, 173, 200, 217, *225*, 225–228, *229*, 234, *274*, *277*, 278, 283, 284, *285*, 295, 313
– MWD deconvolution of polyethylene sample with heterogeneous 201–212
– simultaneous MWD-FTIR deconvolution of ethylene/1-butene copolymer made with 231–232
Zimm–Stockmayer correction factors, for radius of branched molecules gyration 248–249

Epilogue

"There is no pleasure in having nothing to do; the fun is in having lots to do and not doing it."

Mary Wilson Little (1880-?)

"There are no facts, only interpretations."

Friedrich Nietzsche (1844–1900)

After completing the book, the authors took a well-needed "rest".